FUTURE TRENDS
IN MICROELECTRONICS

FUTURE TRENDS IN MICROELECTRONICS

The Nano Millennium

Edited by

SERGE LURYI
JIMMY XU
ALEX ZASLAVSKY

The Institute of Electrical and Electronics Engineers, Inc., New York

A JOHN WILEY & SONS, INC., PUBLICATION

Library of Congress Cataloging-in-Publication Data is available.

ISBN 0-471-21247-4

Printed in the United States of America

10 9 8 7 6 5 4 3 2 1

Contents

2   THE FUTURE BEYOND SILICON: SEMICONDUCTORS, SUPERCONDUCTORS, PHASE TRANSITIONS, DNA

3   THE FUTURE ALONGSIDE SILICON: OPTICAL

4   THE FUTURE WAY BEYOND SILICON: OTHER PARADIGMS

Preface

S. Luryi
Dept. of Electrical and Computer Engineering
State University of New York at Stony Brook, Stony Brook, NY 11794-2350, U.S.A.

J. M. Xu
Div. of Engineering, Brown University, Providence, RI 02912, U.S.A.

A. Zaslavsky
Div. of Engineering, Brown University, Providence, RI 02912, U.S.A.

This book is the brainchild of the third meeting in the *Future Trends in Microelectronics* (FTM) workshop series, which took place in the summer of 2001 at Ile de Bendor, a beautiful little French island in the Mediterranean off the coast between Marseille and Nice. The main purpose of the FTM meetings is to provide a forum for a free-spirited exchange of views among leading professionals in industry, academia, and government. It is a common view among the leading professionals in microelectronics that its current explosive development will likely lead to profound paradigm shifts in the near future. Identifying the plausible scenarios for the future evolution of microelectronics presents a tremendous opportunity for constructive action today — especially since, for better or worse, our economy and indeed our civilization is destined to be based on electronics and semiconductors for the foreseeable future.

Ever since the invention of the transistor, and especially after the advent of integrated circuits, semiconductor devices have expanded their role in our life. Electronic circuits entertain us and keep track of our money, they fight our wars and decipher the secret codes of life, and one day, perhaps, they will relieve us from the burden of thinking and making responsible decisions. That day has not yet arrived and dark clouds looming on the microelectronics horizon. The dramatic downturn of the industry world-wide is compelling us all to fend for ourselves. The key to success is to have a clear vision of where we are heading in these turbulent times.

The celebrated Si technology has known a virtually one-dimensional path of development: reduction of the minimum size of lithographic features. This dramatic evolution has both led us to the threshold of nanotechnology and at the same time brought about doubts regarding future development. Our crystal ball is cloudy. This is going to be the nano millennium ... but how about femto-electronics? There are clearly physical limits, but can we be reasonably sure that electronics will not dip below 1 nm in the next 1000 years?

New electronic materials, the most powerful enablers of new technologies, are naturally a central theme in the FTM program. The evolution of semiconductor electronics has always been intimately connected with advances in material science and technology. Differentiating itself from the usual materials meetings, our workshop considers materials prospects and fundamentals only in the context of future technologies. The first revolution in electronics, which replaced vacuum tubes with transistors, was based upon doped semiconductors and relied on newly discovered methods of growing pure crystals. The early semiconductors could not be properly termed "doped" — they were just dirty. Today, semiconductors routinely used in devices are cleaner (in terms of the concentration of undesirable foreign particles) than the vacuum of vacuum tubes.

The subsequent evolution of transistor electronics has been associated with progress in two areas: the miniaturization of device design rules, brought about by advances in lithographic resolution and doping by ion implantation; and the development of techniques for layered-crystal growth and selective doping, culminating in epitaxial technologies that are capable of monolayer control over doping and chemical composition. Of these two areas, the first has definitely had a greater impact in the commercial arena, whereas the second has been mainly setting the stage for the exploration of device physics.

These roles may well be reversed in the future. Development of new and exotic lithographic techniques with nanometer resolution will be setting the stage for the exploration of various physical effects in mesoscopic devices, while epitaxially grown devices (especially heterojunction transistors integrated with optoelectronic elements) will be gaining commercial ground. When (and whether) this role reversal will take place, will be determined perhaps as much by economic as by technical factors. It is anticipated that the progress in lateral miniaturization may face diminishing returns once the speeds of integrated circuits and the device packing densities become limited primarily by the delays, the complexity, and the power dissipation in the interconnecting metal rather than the individual transistors. Further progress may then require circuit operation at cryogenic temperatures or heavy reliance on optical interconnections. Implementation of the latter within the context of silicon VLSI may usher in hybrid-material systems with heteroepitaxial islands of foreign crystals grown on Si substrates.

All these anticipated developments are likely to be heavily dependent on the progress of material science and techniques, such as epitaxy, and on new architectures. However cloudy our crystal ball may be regarding the future trends in microelectronics, one trend appears clearly: the device designer of tomorrow will be thinking in terms of multilayer structures defined on an atomic scale and new information processing paradigms.

The present volume contains a number of original papers, some of which were presented at FTM-3 in oral sessions, others as posters. From the point of view of the program committee, there was no difference between these types of contributions in weight or importance. There was, however, a difference in style and focus — and that difference was imposed intentionally by the organizers. All oral presenters were asked to focus on their views and projections of future directions, assessments or critiques of important new ideas or approaches, and *not* on their own achievements. This latter point is perhaps the most innovative and distinguishing feature of the FTM workshops. Indeed, we are asking scientists not to speak of their own work! This format has proven to be very successful, however, in eliciting powerful and frank exchanges. The presenters were asked to be provocative and/or inspiring. The latest advances made and results obtained by the participants could be presented in the form of posters and group discussions.

Most working days of FTM-3 were concluded by evening panel sessions that attempted to further the debates on selected controversial issues connected to the theme of the day. Each such session was chaired by one forceful character who invited 2–3 attendees of his or her choice to lead with a position statement, all other attendees being the panelists. The debate was moderated forcefully and irrelevant digressions were cut off without mercy. Moderators were also assigned the hopeless task of forging a consensus on critical issues.

To accommodate these principles, the FTM takes a format that is less rigid than normal workshops to allow and encourage uninhibited exchanges and sometimes confrontation of different views. A central theme is designed together with the speakers for each day. Another traditional feature of FTM workshops is a highly informal vote by the participants on the relative importance of various fashionable current topics in modern electronics research. Results of these votes are perhaps too bold and irreverent for general publication, but Horst Stormer carefully maintains them and makes them available to every new generation of FTM participants.

To produce a coherent collective treatise, the interactions among FTM participants begin well before their gathering at the workshop. All the proposed presentations are posted on the web in advance and can be subject to change up to the last minute to take into account peer criticism and suggestions. After the workshop is over, these materials (not all of which have made it into this book) remain on the web indefinitely, and the reader can peruse them starting at the FTM website (*www.ece.sunysb.edu/~serge/FTM.html*).

To start the debate at FTM-3 and help focus the discussion, the following list of questions was raised as "topics for discussion," with the knowledge that this list should be neither exclusive nor exhaustive:

- What is the technical limit to shrinking devices? Is there an economic sense in pursuing this limit? In the distributed and local memory markets? In the microprocessor market?

- What lies beyond nanoelectronics? Is there such a thing as femtoelectronics? Ditto: *molecular electronics? Spintronics?* Are we pursuing the remaining degrees of freedom or false hopes for "life after the end of miniaturization?"

- Are we having an evolution or revolution in computing architectures? Should we expect a departure from the binary serial paradigm — the root of the twin crises of power dissipation and wiring? Is *quantum computing* a realistic way out? Or is it a replay of linear optics in electronic signal processing?

- Optical interconnects — so far a marginal player on the digital electronics platform. Would it give impetus to a computer architecture revolution?

- What if such a revolution occurs under our noses undetected? Perhaps we are witnessing the emergence of a computing system the size of our planet and including optical networked distributed storage?

- Where are the big-stake market pulls and pushes for new semiconductor technologies? The Internet, mobile electronics, 3D displays? Human-machine interfaces? Bio-informatics? Genomics and proteomics?

- We can make *quantum-effect devices*, but can we make them into useful circuits and systems? Can we solve problems inherent in large-scale integration of quantum devices, like critical biasing, wiring, stochastic fluctuations? Ditto for *resonant tunneling* and *single electronics*?

- What is happening with counter-culture technologies like nonlithographic nanofabrication, nonsemiconductor electronics, macro-electronics, non-von Neumann computing? Shall we see localized photons again? Are there green pastures beyond semiconductor technologies?

- Telecom acronyms: are PDH, SDH, SDL, and ATM fundamentally flawed and limited, but long-live ATM and everything else? Is *dynamic photonic networking* the obvious solution? Ditto, *zero switching network*? Bandwidth efficiency versus access efficiency?

- Should low-earth-orbit satellite networks, satellite-on-wafer, and free-space optics cause us to rethink the fiber-optics system?

- Can we expect bioelectronics with self-produced designer cells? Any prospect for DNA computing? DNA-assisted self-assembly and packaging? Chip-in-brain, bio-battery running on body fluids — perhaps do-able but is it acceptable?

- Carbon nanotubes — the black gold — what are the prospects for applications? Is it a one-size-fits-all new material in the making?

Needless to say, not all of these topics received the same scrutiny, but some proved controversial indeed — particularly the combative quantum computing debate that followed Michel Dyakonov's skeptical take on the prospects of this currently fashionable direction, elaborated in his contribution to this book. Scientific controversies are a hallmark of FTM workshops, with the 1998 FTM-2 meeting eliciting a historic bet between Horst Stormer and Nikolai Ledentsov on the replacement of quantum well lasers by quantum dot devices in mainstream technology within five years[1] — a bet that will come due at the next FTM-4 meeting, tentatively scheduled for the summer of 2003. There is no doubt that other fruitful controversies will arise at FTM-4 as well and since the time is short, the reader and potential attendee is invited to begin thinking at once!

The first of the FTM meetings, "FTM-1: Reflections on the Road to Nanotechnology,"[2] gathered in 1995, also at Ile de Bendor. The second workshop,

"FTM-2: The Road Ahead,"[1] took place in 1998 on a larger island in the same area, Ile des Embiez. Instead of going to a still larger island, the third workshop, "FTM-3: The Nano Millennium," went back to its origins on Ile de Bendor in 2001. To compensate, the next FTM is planned to take place on the biggest island of them all, Corsica. Normally the FTM workshops gather every three years; however, the next FTM-4 workshop is tentatively scheduled one year ahead of the usual schedule, in the summer of 2003.

Like all FTM meetings to date, FTM-4 will be fairly small (less than 100 people) by invitation only. If you, the reader, wish to be invited, please consider following a few simple steps outlined on the conference website. This website (*www.ece.sunysb.edu/~serge/FTM.html*) contains links to all past and planned workshops in the series, their programs, publications, and sponsors. The attendee lists are also included, and our attendees have been an illustrious lot. Suffice it to say that among past FTM participants we find four Nobel laureates (Zhores Alferov, Herbert Kroemer, Horst Stormer, and Klaus von Klitzing, of which the first three received their Nobels *after* attending FTM) and countless others poised for a similar distinction. To be sure, being invited to FTM does not require high distinction, but only the ability and desire to bring and discuss fresh ideas. The result the last FTM-3 meeting is this collective treatise, authored not only by the individual contributors but all the workshop participants.

Acknowledgments

The 2001 FTM-3 workshop at Ile de Bendor and therefore this book were made possible by the generous support of the following agencies, companies, and institutions:

- U.S. National Science Foundation: DMR, ECS, INT programs

- U.S. Department of Defense: ARO, ARL-ERO, DARPA, ONR, ONRIFO

- NASA: JPL

- French government: CNRS, DRET

- Industry: Motorola, Nortel Networks, Philips

- Academia: State University of New York at Stony Brook

On behalf of all the Workshop attendees, sincere gratitude is expressed to the above organizations for their generous support and especially to the following individuals whose initiative was indispensable:

- Leon Alkalai

- François and Alix Arnaud d'Avitaya

- Herbert Goronkin

- Gail Habicht

- Letitia Harrison
- LaVerne D. Hess
- Rajinder Khosla
- Yoon-Soo Park
- Claudine Simson
- Michael A. Stroscio
- Robert Trew
- Claude Weisbuch
- Colin Wood
- John Zavada

Finally, the organizers want to thank all of the contributors to this volume and all the attendees for making the workshop a rousing success.

References

1. S. Luryi, J. M. Xu, and A. Zaslavsky, eds., *Future Trends in Microelectronics: The Road Ahead*, New York: Wiley, 1999. The historic bet is recorded in Ref. 46 of N. N. Ledentsov's contribution, see p. 236.
2. S. Luryi, J. M. Xu, and A. Zaslavsky, eds., *Future Trends in Microelectronics: Reflections on the Road to Nanotechnology*, NATO ASI Series E Vol. 323, Dordrecht: Kluwer Academic, 1996.

1 THE FUTURE OF SILICON

Contributors

1.1 S. M. Sze

1.2 J. Benschop

1.3 P. M. Solomon

1.4 T. P. Smith III, H. A. Maudar, J. W. Liu, and S. J. Farrell

1.5 H. Iwai, S. Ohmi, A. Akama, *et al.*

1.6 D. Esseni, C. Fiegna, and M. Mastrapasqua

1.7 F. Allibert, J. Pretet, A. Zaslavsky, and S. Cristoloveanu

1.8 T. Grasser and S. Selberherr

1.9 P. Palestri, L. Selmi, A. Dalla Serra *et al.*

Microelectronics Technology: Challenges in the 21st Century

S. M. Sze
National Nano Device Laboratory
National Chiao Tung University, Hsinchu, Taiwan, R.O.C.

1. Introduction

The earliest semiconductor device was the metal–semiconductor contact studied in 1874 to reveal that the resistance of a contact between a metal and a semiconductor depends on the magnitude and polarity of the applied voltage.[1,2] The earliest technology related to semiconductor processing was the lithography technique invented in 1798. In this process, the pattern, or image, was transferred from a stone plate (litho).[3,4]

In the past two centuries, the most important milestones in semiconductor development were the inventions of the bipolar transistor[5] in 1947 and the integrated circuit[6] in 1959. The bipolar transistor ushered in the modern electronics era and the integrated circuit laid the foundation for the rapid growth of the microelectronics industry. In the past four decades, there has been phenomenal progress in microelectronics technology and enormous expansion of the global semiconductor market. The electronics industry has grown to be the largest in the world with global sales of over one trillion dollars. The foundation of the electronics industry is microelectronics technology.

Figure 1 shows the sales volume of the semiconductor-device-based electronics industry in the past 20 years and projects sales to the year 2010. Also shown are the gross world product (GWP) and the sales of the automobile, steel, and semiconductor industries.[7] Electronics industry sales surpassed automobile industry sales in 1998. If the current trends continue, in year 2010 the sales volume of the electronics industry will constitute about 4.5% of the GWP. The semiconductor industry will grow even faster, to surpass the steel industry in the early 21st century and to constitute 30% of the electronics industry in 2010.

We shall review the key achievements in the past four decades and consider some major challenges we face in the 21st century. These challenges include:

- the growth of super-large wafers,

- the sub-100-nm resolution of lithography systems,

- the ultra-small dimensions of logic and memory devices,

- the parasitic *RC* delay of multi-level interconnect, and

- the huge capital investment of the microelectronics industry.

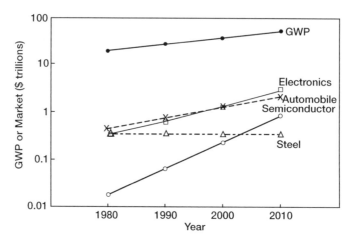

Figure 1. Gross world product (GWP) and sales volumes of the electronics, automobile, semiconductor, and steel industries[7] from 1980 to 2000 and projected to 2010.

We must develop innovative technologies and new products to meet these challenges and to move successfully toward the nanoelectronic era. If we succeed, it is predicted that by the year 2030, the semiconductor industry and the electronics industry may reach 3 trillion dollars and 10 trillion dollars in sales, respectively.

2. Key achievements

Since 1959, we have had many achievements in the microelectronics industry. Of particular significance were:

1. the developments of DRAM (1967),[8] the nonvolatile semiconductor memory (1967),[9] and the microprocessor (1971);[10]

2. phenomenal progress in design and process technology with a 190 times reduction in minimum feature length and four million times increase in DRAM density; and

3. the enormous expansion of the global market with a 400 times increase in sales volume and a 50 billion times increase in annual shipments of transistors.

This progress is summarized in Table 1. Also shown in the table are the increase in Si wafer diameter by a factor of twelve, the 2×10^4 times increase in microprocessor clock rate, the 5×10^5 times increase in nonvolatile memory density, and over 10^7 times decrease in average transistor price. If the automobile industry could achieve the same kind of progress as the microelectronics industry, a Rolls Royce would cost only 25 cents, get over 3,000,000 miles per gallon of gas, and could deliver enough power to drive the Queen Elizabeth II ocean liner.[11] No other industries in human history have had such an enormous impact on the global economy and society so quickly.

Figure 2 shows the density of DRAM chips in the past 20 years and projects the density to the year 2010. Figure 3 shows the microprocessor computational power (in million instructions per second or MIPS) over the same period. We note

Year	1959	1970–71	2001	Ratio
Design rule (µm)	25	8	0.13	190 ↓
V_{DD} (V)	5	5	1.25	4 ↓
Wafer diameter (mm)	25	30	300	12 ↑
Devices per chip	6	2×10^3	2×10^9	3×10^8 ↑
DRAM density (bit)	—	1K	4G	4×10^6 ↑
Nonvolatile memory density (bit)	—	2K	1G	5×10^5 ↑
Microprocessor clock rate (Hz)	—	108 k	2G	2×10^4 ↑
Transistors shipped/year	10^7	10^{10}	5×10^{17}	5×10^{10} ↑
Average transistor price ($)	10	0.3	5×10^{-7}	5×10^{-8} ↓

Table 1. Progress in microelectronics.

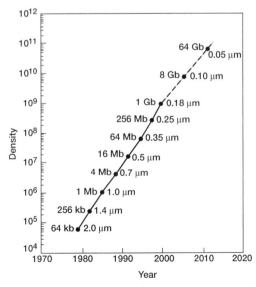

Figure 2. Exponential increase of DRAM density *vs.* year based on the SIA Roadmap.[13]

that both DRAM density and microprocessor power increase exponentially with time, doubling every 18 months, or a factor of four every 3 years — the well-known Moore's Law.[12]

Table 2 shows the International Technology Roadmap for Semiconductors[13] based on Moore's Law. For example, in year 2010 the DRAM chip area will be increased to 14 cm^2, the minimum feature length will be scaled to 50 nm, and the cost per transistor will be reduced to 10 microcent.

The main concern we have is how long can Moore's Law remain valid — that

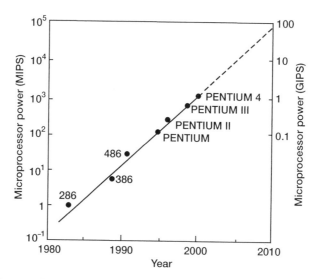

Figure 3. Exponential increase of microprocessor computational power *vs.* year.

Parameter	Projected algorithm for every 3 years	Year 2010
Chip area	1.5×	14-cm^2 DRAM
Min. feature length	30% reduction	50 nm
Components/chip	4×	64-Gb DRAM
On-chip circuit clock	1.5×	50-GHz μP
Cost/transistor	> 50% reduction	10^{-7} $
Fab cost	2×	> $24 billion

Table 2. International Technology Roadmap for Semiconductors.[13]

is, how long can the microelectronics industry maintain its historical rate of performance and cost improvement? There are many challenges we face in the 21st century. We shall consider some major challenges in the next section.

3.   Major challenges

We shall consider five major challenges: wafer, lithography, device, interconnect, and economy.

The *first major challenge* is super-large-diameter wafers. Figure 4 shows silicon wafer diameter and ingot weight in the past 50 years and projects to the year 2010. We note that since the mid-1950's, the wafer diameter has increased exponentially, doubling every 12 years.[14] The most advanced process lines have already adopted 300-mm (12-inch) wafers. Figure 5 shows a 300-mm diameter ingot (left) and a 400-mm diameter ingot. It is very difficult to grow such large-diameter ingots, because the weight of the ingot is over 200 kg for 300 mm wafers and 350 kg for 400 mm wafers. In addition, special arrangements (such as movable magnetic coils) are needed to damp thermal convection for the large volume of melt in the Czochralski crystal growth. Also we have to eliminate the crystal-originated pits (COPs), which are voids with diameters of 10–100 nm. For 200-mm wafers the density of COPs is typically 10^5 cm^{-3}. However, for 300-mm wafers, complete elimination of COPs is necessary for acceptable device yield.[15]

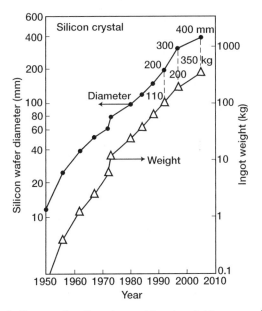

Figure 4. Increase of silicon wafer diameter and ingot weight *vs.* year.[14]

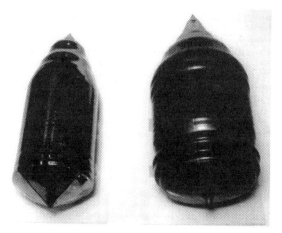

Figure 5. Photographs of 300-mm (12 inch) and 400-mm (16 inch) Czochralski-grown silicon ingots (courtesy of Shin-Etsu Handotai Co., Tokyo).

Parameter		Al/SiO$_2$ 0.18 μm	Cu/low-κ 0.13 μm
200 mm	cost/wafer ($)	1700	2122
	cost/cm^2 ($)	4.58	6.76
300 mm	cost/wafer ($)	2663	3328
	cost/cm^2 ($)	3.77	4.71
From 200 mm to 300 mm	cost/wafer (%)	+57	+57
	cost/cm^2 (%)	−30	−30

Table 3. Comparison of production costs for 200-mm and 300-mm wafers.[16]

Table 3 compares production costs for 200-mm and 300-mm wafers. Consider the case of a wafer with copper metalization, a low-κ interlayer dielectric, and a 0.13-μm design rule. There is a 57% increase in cost per wafer when we go from 200- to 300-mm wafers, however, the cost per unit area for the 300 mm wafer is 30% lower.[16] Therefore, 300-mm and even larger diameter wafers will be used as long as the production cost per unit area can be reduced. If an epitaxial layer of 2–3 μm is required, there will be additional cost for the epi process. For 300-mm wafers, we require a thickness tolerance of ± 4% and a density of structural defects of less than 0.002 defects per cm^2; even tighter controls are needed for larger wafers.

The *second major challenge* is lithography systems with sub-100-nm resolution. It is expected that the 193-nm ArF laser projection lithography system will support the 100-nm technology node, using resolution enhancement techniques such as phase-shifting masks (PSMs) and optical proximity correction (OPC). For even smaller design rules, there are five possible lithography tools, as shown in Fig. 6. [17]

The 157-nm F_2 laser projection lithography system is a strong candidate for the 70-nm technology node. Electron-beam projection systems, such as SCALPEL (scattering with angular limitation projection electron-beam lithography), are limited by electron proximity effects. X-ray lithography systems are limited by mask fabrication complexities. Ion-beam systems are limited by stochastic space-charge effects. The most likely candidate for sub-70-nm technology nodes is the extreme ultraviolet (EUV) system with a wavelength of 13 nm. The EUV system potentially is capable of providing resolutions down to 20 nm with high wafer throughput. To produce such EUV systems before the year 2005, we need close international collaboration and a tight development schedule.

The *third major challenge* is ultra-small devices for logic and memory applications. Currently, the dominant logic device is the MOSFET. Figure 7 shows the evolution of MOSFET architecture. The conventional MOSFET structure can be scaled down to 70 nm.[18] For even smaller devices, an SOI (silicon-on-insulator) substrate or pulse doping may be needed. Eventually a dual-gate structure may be needed. We also can envision a lithography-independent process with a vertical structure by turning the device 90°. Recently, an experimental MOSFET with a 20-nm gate length has been demonstrated,[19] illustrated in Fig. 8. The gate oxide thickness is only 0.8 nm. The transconductance is very high, over 1200 mS/mm for *n*MOS and 700 mS/mm for *p*MOS. The gate delay is very short, less than 0.8 ps for *n*MOS and 1.7 ps for

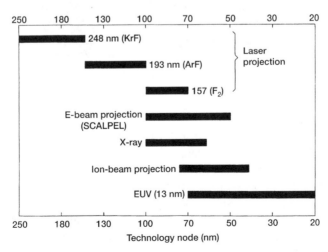

Figure 6. Introduction of new lithography tools for future technology nodes.[17]

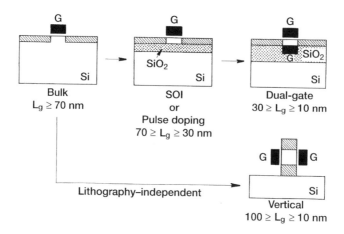

Figure 7. Evolution of MOSFET architecture.[18]

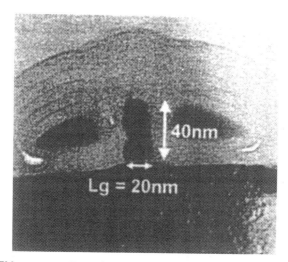

Figure 8. TEM cross section of an experimental *n*MOS transistor with a 20-nm gate length.[19]

*p*MOS. These results indicate that the MOSFET will remain the key device for logic circuits for the foreseeable future.[26]

At present, the dominant memory devices are the DRAM (mainly for office equipment) and the nonvolatile semiconductor memory (NVSM such as flash memory, mainly for portable systems). The highest density for DRAMs is probably around 64 Gbit because it becomes very difficult to control the amount of charge in the storage capacitor at this density. The most likely candidate for high-

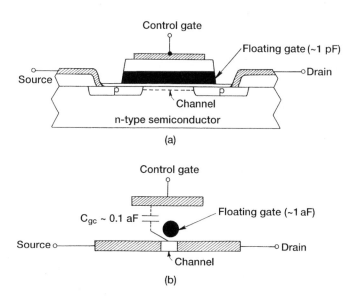

Figure 9. Conventional floating-gate nonvolatile semiconductor memory (NVSM) (a).[9] Single-electron memory cell (b)[20] — a limiting case of NVSM.

density memory beyond 64 Gbit is the single-electron memory cell (SEMC).[20] The SEMC is a limiting case of NVSM, as shown in Fig. 9(b). When we reduce the floating gate of a conventional NVSM, Fig. 9(a), to an ultra-small floating dot (~10 nm), the capacitance of the floating dot is very small, on the order of 0.1–1 aF (10^{-19}–10^{-18} F) range. When an electron tunnels into the quantum well of the floating dot, the potential in the well will increase to block the entrance of another electron — the Coulomb blockade. Figure 10 shows the density of SEMC *vs.* the minimum feature length.[21] Room-temperature operation of a 265-Tbit (250×10^{12} bits) SEMC is projected for a minimum feature length of 1 nm.

To fabricate ULSI circuits with ultra-small devices, we need extensive material innovations, such as silicon-on-insulator substrates (*e.g.* the smart-cut method using hydrogen implantation[22]), high-dielectric materials (*e.g.* barium strontium titanate and Ta_2O_5) for DRAM, low-dielectric materials for interlayer isolation (*e.g.* black diamond and fluorinated amorphous carbon), and highly electromigration-resistant materials for metalization (*e.g.* copper).

The *fourth major challenge* is the parasitic *RC* delay due to interconnects. As we continue to reduce the minimum feature length and increase the circuit complexity, both the parasitic resistance and capacitance increase. At the sub-100-nm technology node, the interconnect parasitic *RC* delay becomes orders of magnitude larger than the intrinsic gate delay.[23]

To minimize the *RC* delay, multilevel interconnect schemes have been developed with copper replacing aluminum to improve the electromigration

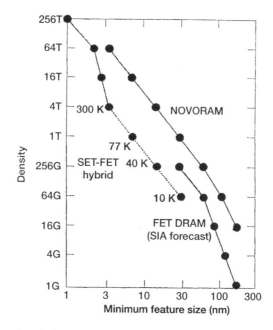

Figure 10. Density of single-electron memory cell *vs.* minimum feature length. Also shown is the density of DRAM.[21]

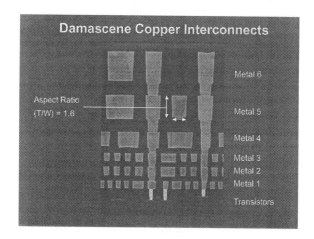

Figure 11. Six-level damascene copper interconnect for a 180-nm design rule.[24]

performance. A six-level damascene copper interconnect for a 180-nm design rule is shown in Fig. 11.[24] For even smaller design rules, 8 or more levels are required. Another approach to minimum *RC* delays is to develop system-on-chip ICs.

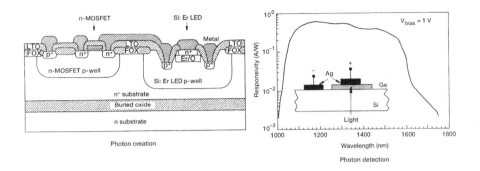

Figure 12. Components for silicon microphotonics:[25] photonic generator — silicon-erbium LED (left), and photonic detector — SiGe alloy detector (right).

An elegant method to solve the interconnect problem is silicon microphotonics.[25] For such an approach, we need three photonic components to take care of photon creation, photon propagation, and photon detection. On the left side of Fig. 12 we show a possible means for photon creation by employing the silicon-erbium LED. The photon propagation can be accomplished easily by using on-chip Si/SiO_2 optical fiber; since the refractive index ratio is quite large (3.5/1.5), the LED output will be confined in the silicon core region. For photon detection, we can use a SiGe alloy as shown on the right side of Fig. 12. The responsivity is quite high over a wide range of wavelengths (*e.g.* higher than 0.3 A/W from 1080 nm to 1520 nm). If the silicon microphotonic approach can be successfully implemented, we can substantially reduce *RC* delays, provide precise clock distribution and system synchronization, reduce power dissipation, improve voltage isolation and impedance matching, and minimize cross-talk and pin inductance.

The *fifth major challenge* is economy. We need low-cost manufacturing and new applications to broaden the electronics market. The cost per fab line has doubled every 3 years. In 1995, the cost of a fab line was about $750 million; now it is $3 billion; in ten years, it will cost $24 billion. We have to find ways to reduce the fab-line cost and to improve its productivity and yield.

Based on Fig. 1, we have extrapolated the GWP, electronics sales, and semiconductor sales to the year 2040, as shown in Fig. 13. If we assume that the GWP will maintain its current annual growth rate of 3% in the next 40 years, it will reach $100 trillion in 2030, and $134 trillion in 2040. If the electronics industry can maintain its current 7% annual growth rate, it will reach $10 trillion in 2030, *i.e.*, 10% of the GWP. It is unlikely that any one industry can constitute more than 10% of the GWP. Therefore, after 2030 the electronics sales will level off, and increase at the same ~ 3% rate as the GWP.

The global semiconductor industry will most likely maintain its current high

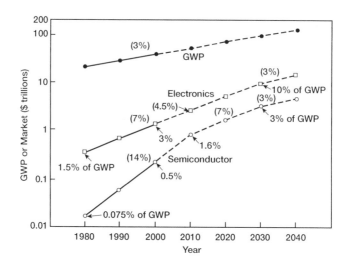

Figure 13. Gross world product (GWP) and sales volumes of the electronics and semiconductor industries from 1980 to 2000 and projected to 2040.

growth rate of 14% to the year 2010. By then, semiconductor sales will reach $800 billion to constitute about 30% of electronics sales. This ratio is probably an upper limit for semiconductor sales because there are many other components in an electronic system such as the software, display, and packaging.

Therefore, after 2010, the growth rate of semiconductor sales will be reduced to 7%, and after 2030 it will be further reduced to 3%. However, in the event that electronics sales reach beyond 10% of the GWP and that semiconductor sales go beyond 30% of electronics sales, the semiconductor industry would enjoy a higher than 7% growth rate after 2010, and a higher than 3% growth rate after 2030.

4. Conclusion

In the past 40 years, microelectronics has been responsible for the rapid growth of the global electronics industry, which is now the largest industry in the world (> $1 trillion). There are many major challenges in microelectronics: large wafers, sub-100-nm lithography, ultra-small devices, interconnect, and economic challenges.

We believe that super-large-diameter wafers will be adopted as long as the production cost per unit wafer area can be reduced. For lithography, the 157-nm F_2 system will most likely be used for the 70-nm technology node, and EUV for the 50-nm and even smaller technology nodes. The dominant logic device will continue to be the MOSFET, and the dominant memory device will be the single-electron memory cell — the ultimate floating-gate nonvolatile semiconductor memory.

Integrated-circuit performance will be limited by interconnect. An elegant solution is silicon microphotonics, since all the required photonic components are readily available. We need low-cost manufacturing processes and broadened electronics markets to maintain the growth of the microelectronics industry. We expect that the global electronics industry will reach $10 trillion in the year 2030. However, we must develop innovative microelectronic and nanoelectronic technologies to meet the aforementioned challenges.

References

1. F. Braun, "Über die Stromleitung durch Schwefelmetalles," *Ann. Phys. Chem.* **153**, 556 (1874).

2. Most of the classic device papers are collected in: S. M. Sze, ed., *Semiconductor Devices: Pioneering Papers,* Singapore: World Scientific, 1991.

3. M. Hepher, "The photoresist story," *J. Photo. Sci.* **12**, 181 (1964).

4. For a historical review of semiconductor device and technology developments, see, for example, S. M. Sze, *Semiconductor Devices: Physics and Technology,* 2nd ed., Chapter 1, New York: Wiley, 2002.

5. J. Bardeen and W. H. Brattain, "The transistor, a semiconductor triode," *Phys. Rev.* **71**, 230 (1948); W. Shockley, "The theory of *p-n* junction in semiconductors and *p-n* junction transistors," *Bell Syst. Tech. J.* **28**, 435 (1949).

6. R. N. Noyce, "Semiconductor device-and-lead structure," U.S. Patent 2,981,877 (filed 1959, granted 1961); J. S. Kilby, "Invention of the integrated circuit," *IEEE Trans. Electron Dev.* **ED-23**, 648 (1976), U.S. Patent 3,138,743 (filed 1959, granted 1964).

7. *2000 Annual Report on Semiconductor Industry,* Industrial Technology Research Institute, Hsinchu, 2000.

8. R. M. Dennard, "Field effect transistor memory," U.S. Patent No. 3,387,286 (filed 1967, granted 1968).

9. D. Kahng and S. M. Sze, "A floating gate and its application to memory devices," *Bell Syst. Tech. J.* **46**, 1283 (1967).

10. The inventors of the microprocessor are M. E. Hoff, F. Faggin, S. Mazor, and M. Shima. For a profile of M. E. Hoff, see R. Slater, *Portraits in Silicon,* Cambridge: MIT Press, 1987, p. 175.

11. J. S. England and R. W. England, "The reliability challenge: New materials in the new millennium, Moore's Law drives a discontinuity," *Proc. IEEE Int. Reliability Physics Symp.,* Reno, 1998.

12. G. Moore, "VLSI, what does the future hold," *Electron. Aust.* **42**, 14 (1980).

13. *The International Technology Roadmap for Semiconductors,* Semiconductor Industry Association (SIA), San Jose, 1999.

14. W. Zulehner, "Historical overview of silicon crystal pulling development," *Materials Sci. Eng.* **1373**, 7 (2000).

15. H. Dietrich *et al.*, "Three hundred-mm wafers, a technological and an economic challenge," *Microelectronic Eng.* **45**, 183 (1999).

16. SEMATECH, 2000.

17. R. A. Lawes "Future trends in high-resolution lithography", *Applied Surf. Sci.* **154**, 519 (2000); see also contribution by J. Benschop in this volume.

18. S. Cristoloveanu, "State-of-the-art and future of silicon on insulator technologies, materials, and devices," *Microelectronics Reliability* **40**, 771 (2000).

19. R. Chau, "30 nm and 20 nm physical gate length NMOS transistors," *Digest 2001 Silicon Nanotechnol. Workshop*, Kyoto, Japan (2001), p. 2.

20. S. M. Sze "Evolution of nonvolatile semiconductor memory: From floating-gate concept to single-electron memory cell", in: S. Luryi, J. M. Xu, and A. Zaslavsky, eds., *Future Trends in Microelectronics: The Road Ahead*, New York: Wiley, 1999, pp. 291-304.

21. K. K. Likharev, "Single-electron devices and their applications," *Proc. IEEE* **87**, 606 (1999).

22. M. Bruel, "The history, physics, and applications of the Smart-cut process," *MRS Bulletin*, p. 35 (1998).

23. R. Liu *et al.*, "Interconnect technology trend for microelectronics," *Solid State Electronics* **43**, 1003 (1999).

24. Private communication, Intel Corporation, 2001.

25. L. C. Kimerling, "Silicon microphotonics," *Appl. Surf. Science* **159**, 8 (2000).

26. An even smaller device has been reported recently: B. Yu *et al.*, "15 nm gate length planar CMOS transistor," *Tech. Digest IEDM* (2001), p. 937.

Trends in Microlithography

J. Benschop
ASML, P.O. Box 324, 5500 AH Veldhoven, The Netherlands

1. Introduction

Over the last 30 years the IC industry has continued to shrink the minimum feature size on an IC, effectively doubling IC capacity every 18–24 months. This trend is commonly referred to as "Moore's Law" after Gordon Moore,[1] the man who predicted this phenomenon in 1965.

Shrinking the minimum feature size has lead to a boom in the IC industry, as shrinking size has resulted in a simultaneous reduction in cost (the same functionality occupies less wafer area and hence costs are reduced) and an increase in performance (smaller features increase speed and reduce power consumption). This shrinkage of the minimum feature size is predicted to continue for the next decade, as can be seen in Fig. 1.

Figure 1 illustrates another interesting phenomenon: the National Technology Roadmap for Semiconductors (NTRS) in 1994 and 1997, as well as the International Technology Roadmap for Semiconductors (ITRS) in 1999 predict a

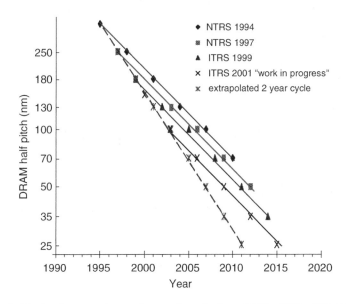

Figure 1. Technology roadmaps for semiconductors, 1994–2001 editions.

three-year cycle (30% shrinkage of features every 3 years). However, the roadmap continues to be pulled in by one year every time a revision is published. The net effect is that the industry has been on a two-year cycle for the last 8 years. Extrapolating this two-year cycle into the future means that the 50-nm node (which corresponds to 50-nm lines in a 100-nm pitch or 35-nm isolated lines) will be taken into production in the year 2007. This schedule is well ahead of the latest official publication of the 1999 ITRS roadmap, which predicts that the 50-nm node will be taken into production by the year 2011.

The vast majority of the lateral features on an IC are defined by imaging a mask on the wafer using optical projection lithography. In Section 2, I will describe how optical lithography has been able to define 30% smaller features every 2–3 years. A discussion about where the limits of optical lithography might be is given in Section 2 also. Section 3 introduces alternatives to optical lithography, the so-called "next generation lithography" (NGL), and the associated main challenges are discussed. The impact of tool cost, tool throughput, mask cost, and mask usage on Cost of Ownership (CoO) are explained in Section 4.

2.   Limits of optical lithography

The resolution R of optical lithography is given by

$$R = k_1 \lambda / NA ,   \tag{1}$$

where λ is the wavelength of the light and NA is the numerical aperture of the lens (given by the refractive index of the surrounding medium multiplied by the sine of the angular semi-aperture of the lens).[2] Resolution has been improved by the combination of:

- wavelength reduction: 436 nm (g-line mercury lamp), 365 nm (i-line mercury lamp), 248 nm (KrF excimer laser), 193 nm (ArF excimer laser), 157 nm (F_2 excimer laser)

- NA increase from 0.4 to 0.8

- k_1 reduction

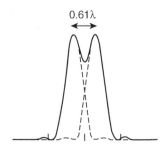

0.61λ

Figure 2. Resolution according to Rayleigh's criterion: $k_1 = 0.61$.

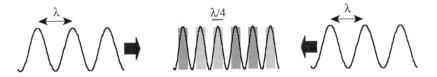

Figure 3. Physical limit of resolution for equal lines and spaces: $k_1 = 0.25$.

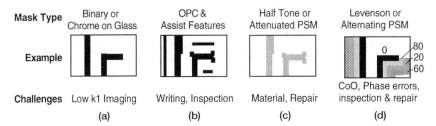

Mask Type	Binary or Chrome on Glass	OPC & Assist Features	Half Tone or Attenuated PSM	Levenson or Alternating PSM
Example				
Challenges	Low k1 Imaging	Writing, Inspection	Material, Repair	CoO, Phase errors, inspection & repair
	(a)	(b)	(c)	(d)

Figure 4. Evolution of mask technology allows lower k_1 imaging but adds complexity and cost. The simplest mask is an exact copy (apart from the magnification factor) of the feature that has to be printed in resist (in this case a vertical line and an elbow structure are depicted) (a); enhancement of imaging by applying assist features (like scatter bars) and optical proximity correction (OPC) features (b). Further improvement can be obtained by applying attenuated (c) or alternating (d) phase shift masks (PSMs).

If Rayleigh's criterion[3] is applied, whereby two point-spread functions (PSFs) of homogeneously filled pupils are summed incoherently and the limit of resolution is defined where the first minimum of one PSF coincides with the peak of the other (see Fig. 2), the limit of k_1 equals 0.61. The fact that imaging in a lithographic system is partially coherent has some effect, but in any case,[4] $0.57 < k_1 < 0.83$.

The criterion that resolution is defined by the peak of one PSF on top of the minimum of the other is somewhat arbitrary. A more fundamental limit on the period of an image is depicted in Fig. 3. The minimum period that can be obtained corresponds to a situation in which two waves coming from opposite directions interfere with each other. In this case a standing wave is obtained with a period of half of the wavelength. If resolution is defined as the line width in an equal-lines-and-spaces pattern, the minimum feature corresponds to a quarter of the wavelength or $k_1 = 0.25$. Over the years, optical lithography has been used in production with ever decreasing k_1 values, made possible by applying various resolution enhancement techniques to the tool (annular or quadruple illumination), the mask (see Fig. 4), and the resist (thin-layer imaging).

Optical lithography systems					
Wavelength NA		365 nm	248 nm	193 nm	157 nm
Source		i-line	KrF	ArF	F$_2$
NA		0.65	0.63 → 0.80	0.63 → 0.85	0.70 → 0.85
Method:	Process factor K$_1$	Application range or limit (nm)			
Conventional Stepper	0.60	335	235 → 185		
+ Off axis illumination	0.50	280	195 → 155	155 → 115	110 → 90
PSM and / or strong OAI + OPC	0.40		175 → 125	125 → 90	90 → 75
All the tricks thin layer imaging	0.30		120 → 90	90 → 70	70 → 55
Physical limit equal lines and space	0.25	140	100 → 75	75 → 55	55 → 45

Figure 5. Application range and limit of optical lithography.

Figure 5 summarizes the most likely application range for optical projection lithography. ArF lithography (193-nm wavelength) will be applied down to and below the 100-nm node. F$_2$ lithography (157 nm) will be applied down to and below the 70-nm node, with the 70-nm node corresponding to 70-nm lines in 140-nm pitch. This node is very close to the physical limit, as can been seen from Fig. 5, and hence all possible image enhancement technologies (for tool, mask, and resist) will have to be applied.

3. Next generation lithography

Beyond the 70-nm node various technologies, commonly referred to as "next generation lithography" (NGL), are competing. The main contenders are electron beam projection lithography (EPL), ion beam projection lithography (IPL), hard X-ray (1 nm wavelength) lithography and extreme ultraviolet projection lithography (EUVL).

- *Electron beam projection lithography*
  It is well known that electrons can be used for sub–70-nm imaging. To improve the throughput of an e-beam system, EPL has been pursued since the early 1970's.[5] Originally, absorbing stencil masks were used but they suffer from excessive mask heating. To overcome this problem, Berger and Gibson[6] proposed the use of scattering masks and back focal plane filtering as depicted in Fig. 6. Typically in EPL, a mask area of 1×1 mm^2 is illuminated and 4× reduction of the image onto the wafer occurs.

  The major challenge in EPL is to maintain an acceptable throughput for reduced feature sizes. Two physical phenomena lead to a tendency for throughput to decrease with decreased feature sizes: electron-electron (Coulomb) interaction and shot noise in the resist.

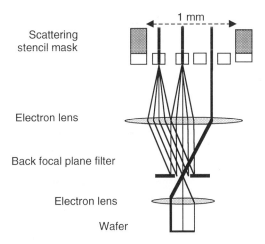

Figure 6. Schematic illustration of electron-beam projection lithography (EPL).

As far as electron-electron interaction is concerned, stochastic as well as global electron-electron interaction in the column gives rise to blur in the image.[7] Because the amount of blur that can be tolerated is proportional to the resolution, any given EPL tool will have to reduce the current in the column and hence decrease the throughput when being used to image smaller features.[8]

As for shot noise, a 35×35-nm^2 pixel covered with a 4-μC/cm^2 resist will be exposed by only 307 electrons. Because of Poisson shot noise, the dose fluctuation per pixel will therefore be equal to $(307)^{-1/2} \sim 6\%$. This number exceeds the total dose fluctuation budget for lithographic systems and would result in an unacceptably large line-width variation. The only way to overcome this fundamental limit is to increase the number of electrons, hence decrease the resist sensitivity. However, decreased sensitivity leads to reduced throughput.

Electron-beam projection lithography is currently being pursued by IBM/Nikon.[9]

- *Ion beam projection lithography*

In IPL, charged particles are again used for the 4$\times$ reduction of a mask image onto a wafer. However, there are a number of fundamental differences between EPL and IPL. The ions have a much shorter wavelength (wavelength of 100 keV He ions $\sim 5\times10^{-5}$ nm) than electrons (wavelength of 100 keV electrons $\sim 4\times10^{-3}$ nm). Thus the optics in IPL have a much smaller *NA* than for EPL (μrads for IPL *vs.* mrads for EPL). This smaller *NA* allows a much larger field size to be exposed than for EPL. Typically, 12.5×12.5 mm^2 fields can be exposed.

Because ions are absorbed quickly by most materials, only stencil masks (consisting of openings in a thin membrane) can be used. Contrast is achieved by absorption of the ions in the mask. Clearly closed structures (like a donut shape) cannot be made with a single stencil mask. In these cases, two complementary masks are used to obtain the required image on the wafer.

A more extensive overview of the technology can by found elsewhere.[10] Here it suffices to note the main challenges for IPL. First, stochastic Coulomb interactions and shot noise in the resist decrease throughput for smaller feature sizes. Second, as explained above, some features require the stitching of images on two complementary masks. To control the feature size over the stitching boundary, the placement of both images with respect to one another is very critical. This problem results in an image placement requirement of a few nanometers.[11]

IPL is being pursued by a consortium in Europe,[12] but the worldwide support for this technology remains critically low.

- *X-ray lithography*

In x-ray lithography, a 1× mask is exposed by 1-nm photons normally generated by a synchrotron. An image is obtained by proximity printing of a thin (few microns) membrane mask brought in close proximity (tens of microns) to the wafer. An extensive overview of the technology can be found elsewhere.[13]

Since the original proposal,[14] x-ray lithography has been pursued by a number of large programs in the USA[15] and Japan.[16] The major challenge of x-ray lithography involves the 1× mask. The requirements for the mask writer and mask generation process are much more stringent than for a 4× reduction mask (used in optical lithography and all the other NGL technologies). The mask has proven a major stumbling block to the progress of x-ray lithography, and the worldwide activity in x-ray lithography has seen a sharp decline over the last few years.

- *Extreme ultraviolet lithography*

Like optical lithography, EUVL uses scanning masks and wafer stages to achieve the reduction of the mask pattern onto the wafer. An extremely short wavelength of 13 nm is used for exposing. A plasma source is used to generate 13-nm photons, and all-mirror optics is used to illuminate the mask and achieve the reduction of the mask pattern onto the wafer.

The major challenges of EUV are: a defect-free mask, a sufficiently high-power source, and atomically flat multilayer coated mirrors. A typical EUV mask requires 80 alternating layers of molybdenum (Mo) and silicon (Si) with a period of 6.7 nm on a substrate without introducing a defect larger than 55 nm in diameter. In 1999 the median of added defects was equal to $0.5/cm^2$ (particles > 130 nm). In one year this figure was improved significantly to a median of added defects equal to $0.05/cm^2$ (particles > 90 nm).[17] Nevertheless, significant further improvement is required to get down to an acceptable defect level of $0.003/cm^2$ (particles > 55 nm), to obtain a mask blank yield of 70%.

Furthermore, for acceptable operating costs, EUVL requires a high (> 80 300-mm wfr/hr) throughput. Such a high throughput will require that at least 60 W of source power be collected in a bandwidth of 0.24 nm.[18] The power of sources

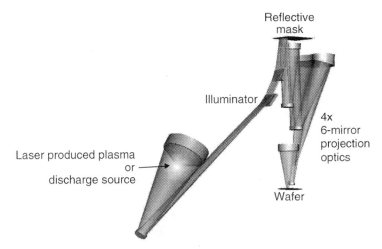

Figure 7. Principle of extreme ultraviolet lithography.

used today is limited to approximately 1 W. Several suppliers have roadmaps to achieve more than 60 W by the time EUVL is commercialized (the timeframe is the year 2005).

Finally, atomically flat EUVL mirror substrates have figure (that is, low spatial frequency) and roughness (that is, high spatial frequency) requirements in the 0.1–0.2 nm RMS range. Results obtained by Carl Zeiss are close to this final specification.[19]

Despite these significant challenges, EUVL currently has by far the most support worldwide of all the NGL technologies. All the major lithographic equipment makers (ASML, Nikon and Canon) are pursuing it with active programs in Europe, Japan, and the USA.[20–22]

4. Cost of ownership

The CoO for lithography can be described as follows:

$$\text{CoO} \left[\$/\text{wfr} \right] = C_1 \frac{\text{tool cost} \left[\text{M}\$ \right]}{\text{throughput} \left[\text{wfr/hr} \right]} \times \frac{\text{mask cost} \left[\$ \right]}{\text{mask usage} \left[\text{wfr/mask} \right]} + C_2 \, . \quad (2)$$

Equation (2) indicates that CoO is dominated by two factors: tool cost divided by tool throughput and mask cost divided by mask usage.

Over the last 20 years, the tool cost has increased from a few $100,000 to more than $10,000,000.[23] However, this trend has been compensated mostly by the fact that the throughput has increased by roughly a factor of two over the last 15 years together with a simultaneous increase of wafer size from 150 mm (6 inches) to 300 mm (12 inches). This increase has resulted in an increase of the area exposed per hour by a factor of 8.

The mask cost, like the tool cost, has increased dramatically over the last decade. The cost of a binary mask has risen from $1500 (at the 800-nm node) to approximately $20,000 (at the 180-nm node). As explained in Fig. 5, low k_1 imaging requires OPC and/or PSMs. The cost of a PSM for the 70-nm node is expected[23] to be at least $30,000.

The number of wafers exposed per mask, the so-called mask usage, is highly dependent upon the product type and area of application. A survey[24] has shown that ASIC manufacturers expose, on average, only 500 wafers per mask, whereas DRAM manufacturers expose, on average, 5000 wafers per mask.

The resulting CoO as a function of the IC generation is shown in Fig. 8 for a situation of 2800 exposures per mask (which equals the industry average). Figure 8(a) shows that the cost per wafer has gone up significantly, largely due to the relative increase in mask cost. In fact, for a large segment of the industry, the cost of the mask is beginning to dominate — see Fig. 8(a). Fortunately, the cost per pixel continues to fall, as shown in Fig. 8(b), due to larger wafer size and smaller resolution.

By correlating the CoO model in Eq. (2) to a more extensive model by Sematech,[24] which covers a wide range of technologies (157 nm, EUVL, and EPL), throughput (10–80 wfr/hr), and mask usage (500–5000 wfr/mask), one finds

$$C_1 = 90 \ [10^{-6}/\text{hr}] \text{ and } C_2 = 8 \ [\$] \ . \tag{3}$$

The good correlation ($R^2 = 0.98$) between both models justifies the use of the simple model given in Eq. (2).

By inserting the values for C_1 and C_2 given in Eq. (3) into Eq. (2), it is possible to calculate what throughput (as a function of the mask usage) has to be achieved for a $30,000,000 mask-less tool to be cost competitive with a $30,000,000 mask-projection tool having a throughput of 100 wfr/hr and using a $80,000 mask. The result is shown in Fig. 9.

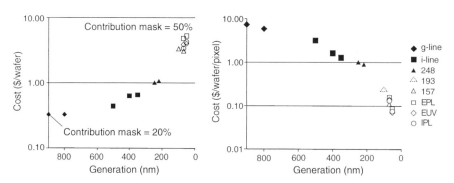

Figure 8. Cost-of-ownership of lithography as a function of IC generation. Both graphs are normalized to 1 at the 250-nm generation.

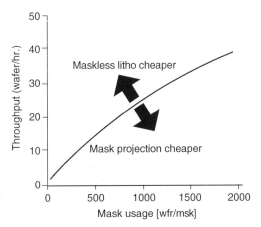

Figure 9. The mask-less lithography tool ($30M tool) is cheaper than the mask-projection lithography tool ($30M tool, 100 wfr/hr, $80K mask) for throughput values that lie above the plotted curve.

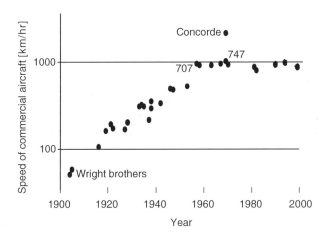

Figure 10. The historical trend in aircraft speed.

Various methods for mask-less lithography have been proposed.[25–28] The required throughputs and related high data rates (>> 1 Tbit/s) are yet to be demonstrated, but a move to mask-less lithography in the future cannot be ruled out.

Figure 10 shows that CoO can have a major impact on industry. In the aircraft industry, the speed of aircraft followed a semi-logarithmic trend from the year 1903 to the year 1957. After 1957 the trend in the speed of aircraft leveled out at approximately 1000 km/hr. It has remained at this level for the last 44 years. In 1969, with the development of the Concorde aircraft, it was proven that it was possible to continue on the semi-logarithmic trend in aircraft speed. However, the Concorde aircraft has never been an economic success.

5. Conclusions

It is expected that the minimum feature size on ICs will continue to shrink by 30% every 2 to 3 years for the foreseeable future. Optical lithography using F_2 (157-nm wavelength) excimer lasers, high-NA (>0.80) lenses and advanced mask and resist schemes will be able to print dense lines and spaces of 70 nm and below.

It is expected that the NGL will be taken into production at the 50-nm node. This change could occur as early as the year 2007. EUVL is the leading candidate but major challenges, such as defect-free multilayer coated masks, high power sources, and atomically flat mirrors, still need to be addressed.

References

1. G. Moore, "Cramming more components onto integrated circuits," *Electronics* **38**, 114 (1965).
2. M. Born and E. Wolf, *Principles of Optics*, Oxford: Pergamon Press, 1980, p. 187.
3. Lord Rayleigh, "Investigations in optics with special reference to the spectroscope," *Phil. Mag.* **8**, 261 (1879).
4. H. H. Hopkins and P. M. Barham, "The influence of the condenser on microscopic resolution," *Proc. Phys. Soc.* **63**, 737 (1950).
5. M. B. Heritage, "Electron-projection microfabrication system," *J. Vac. Sci. Technol.* **12**, 1135 (1975)
6. S. D. Berger and J. M. Gibson, "New approach to projection-electron lithography with demonstrated 0.1 μm linewidth," *Appl. Phys. Lett.* **57**, 153 (1990).
7. M. M. Mkrtchyan, J. A. Liddle, S.D. Berger, and L. R. Harriott, "Stochastic scattering in charged particle projection systems: a nearest neighbor approach," *J. Appl. Phys.* **78**, 6888 (1995).
8. X. Zhu, E. Munro, J. A. Rouse, and W. K. Waskiewicz, "Comprehensive analysis of electron optical design of SCALPEL-HT/Alpha," *SPIE Symp. on Emerging Lithographic Technologies IV*, Vol. 3997, 170 (2000).
9. H. C. Pfeiffer, "PREVAIL: IBM's e-beam technology for next generation lithography," *SPIE Symposium on Emerging Lithographic Technologies IV*, Vol. 3997, 206 (2000).
10. J. Melngailis *et al.*, "A review of ion projection lithography," *J. Vac. Sci. Technol. B* **16**, 927 (1998).
11. A. Ehrmann, T. Struck, E. Haugeneder, H. Loeschner, J. Butschke, *et al.*, "IPL stencil mask distortions: Experimental and theoretical analysis," *SPIE Symposium on Emerging Lithographic Technologies IV*, Vol. 3997, 385 (2000)

12. G. Gross *et al.*, "Ion projection lithography: Status of the MEDEA project and United States/European cooperation," *J. Vac. Sci. Technol. B* **16**, 3150 (1998).
13. F. Cerrina, "X-ray Lithography," in: P. Ray-Choudhury, *Handbook of Microlithography, Micromachining, and Microfabrication, Vol.1*, Bellingham, WA: SPIE Press, 1997, pp. 251-320.
14. D. L. Spear and H. I. Smith, "High-resolution pattern recognition using soft x-rays," *Electronics Lett.* **8**, 102 (1972).
15. C. Wasik, G. P. Murphy, A. C. Chen, A. A. Krasnoperova, A. L. Flamholz, *et al.*, "X-ray fills the gap," *Proc. SPIE* **3331**, 150 (1998).
16. Y. Tanaka *et al.*, "130 nm and 150 nm line-and-space critical-dimension control evaluation using XS-1 x-ray stepper," *J. Vac. Sci. Technol. B* **16**, 3509 (1998).
17. C. C. Walton, P. A. Kearney, P. B. Mirkarimi, J. M. Bowers, C. J. Cerjan, *et al.*, "Extreme ultraviolet lithography — reflective mask technology," *SPIE Symp. Emerging Lithographic Technol. IV*, Vol. 3997, 496 (2000).
18. V. Banine and J. Moors, "Extreme ultraviolet sources for lithography applications," to be published in *SPIE Symp. Emerging Lithographic Technol. V*, Vol. 4343, 203 (2001).
19. U. Dinger, F. Eisert, M. Lasser, M. Mayer, A. Seifert, *et al.*, "Mirror substrates for EUV-lithography: Progress in metrology and optical fabrication technology," *Proc. SPIE* **4146**, 35 (2000).
20. J. P. H. Benschop, U. Dinger, and D.C. Ockwell, "EUCLIDES, first phase completed!" *SPIE Symp. Emerging Lithographic Technol. IV*, Vol. 3997, 34 (2000).
21. H. Kinoshita, T. Watanabe, Y. Li, A. Miyafuji, T. Oshino, *et al.*, "Recent advances of three-aspherical-mirror system for EUVL," *SPIE Symp. Emerging Lithographic Technologies IV*, Vol. 3997, 70 (2000).
22. C. W. Gwyn *et al.*, "Extreme ultraviolet lithography," *J. Vac. Sci. Technol. B* **16**, 3142 (1998).
23. See http://www.sematech.org/public/resources/coo/ist2h99.pdf
24. See http://www.sematech.org/public/resources/coo/coo113099.pdf
25. D. J. D. Carter, "Maskless x-ray lithography," *J. Vac. Sci. Technol. B* **16**, 3426 (1998)
26. L. P. Murray, "Advances in arrayed microcolumn lithography," *J. Vac. Sci. Technol. B* **18**, 3099 (2000).
27. E. Yin, "Electron optical column for a multicolumn, multibeam direct-write electron beam lithography system," *J. Vac. Sci. Technol. B* **18**, 3126 (2000).
28. T. Sandstrom and U. Ljungblad, "High-Performance laser pattern generation using spatial light modulators (SLM) and deep-UV radiation," preprint (2001).

Strategies at the End of CMOS Scaling

P. M. Solomon

IBM T. J. Watson Research Center, Yorktown Heights, NY 10598, U.S.A.

1. Historical perspective

Throughout the latter third of the 20th century Moore's Law has ruled the growth of the semiconductor industry. This astounding exponential growth at a compound rate of 40% per year over 40 years has been the product of higher packing densities and larger chip sizes, where the higher packing densities have been achieved both by finer lithography patterning as well as by innovative self-aligned device structures. The non-lithographic contribution has been very substantial, since while feature sizes have been reduced by a factor of ~10^2 during this time, circuit count has increased by a factor of ~10^8. For the past decade or so, Moore's law has been codified as the SIA (Semiconductor Industry Association) or, more recently, ITRS (International Technology Roadmap for Semiconductors) roadmaps. In these roadmaps, detailed goals are set forth, from device structural parameters to system performance, for both memory and logic. For all of this time the silicon MOSFET has been the dominant device used in integrated logic and memory chips, and for the past twenty years, complementary MOS (CMOS) has been the dominant circuit type.

Everywhere the trend has been toward replacing other solutions with integrated CMOS, rather than the other way around. For instance, in the early 1990's CMOS replaced the Si bipolar transistor for most high-end applications, even though the bipolar transistor remains the faster switching device to this day. Gallium arsenide technology was developed rapidly in the 1980's to challenge the dominance of silicon in high-end switching applications. While III-V technology (including GaAs) did secure a niche for itself in the high-frequency analog domain, and even in some specialized digital applications such as high-speed test equipment, it never mounted a serious challenge against CMOS. Indeed, the existence of the III-V competition spurred CMOS toward more aggressive scaling where the combination of high performance, high integration level, and relatively low power made it difficult to beat, especially since the electron transport advantages of III-V devices diminish with scaling.[1] Recently CMOS itself has been encroaching into the lower end of the GaAs RF market. Witness the recent progress in silicon microwave circuits[2] and the continuing trend to displace analog circuits requiring specialized technologies with digital CMOS circuits.

Apart from superior technical attributes, the sheer momentum of entrenched CMOS — with its costly infrastructure of fabrication facilities, manufacture of fabrication equipment and, not to be overlooked, investment in circuit and system design — poses a huge barrier for any new technology to overcome.

2.   The end of CMOS scaling?

Now silicon CMOS is approaching the end of its scaling path, although there is some uncertainty as to exactly where the endpoint is. In some sense this question is not well posed. Transistors become leakier as they get smaller, due to leakier gate oxides, tunneling leakage in the silicon itself, and thermal leakage over reduced barriers due to reduced gate voltage swings. The latter effect is the major leakage component today, the familiar sub-threshold leakage, but the others become proportionately more important as scaling progresses. As noted by Frank et al.,[3] different applications have varying sensitivity to leakage current, leading to various scaling scenarios and endpoints depending on the application. In some applications that require very low leakage, the most notable being DRAM[3] and flash EEPROM,[4] device channel-length scaling has already effectively stopped, but density continues to increase because of improvements in cell layout, back-end ground rules (wiring and vias), *etc*. On the other end of the applications spectrum are the high-end servers (descendents of high-end mainframe computers). Here high performance is the primary concern and power is secondary, although even in this environment, where Si bipolar transistors used to dominate, CMOS wins because of its higher density, hence lower cost, and lower power dissipation. It is in this environment that scaling can be carried the furthest. A subset of Frank's table is reproduced in Table 1.

   For bulk silicon, scaling can be carried down to ~15 nm for the high-performance designs but only to ~25 nm for the low-power designs. As seen from the ITRS roadmap of Table 2, the predicament will be reached before 2014, by which point we will be unable to implement low-power designs in the most aggressive scaled technology. As will be discussed below, there are design

Application	Power (W/cm²)	T (°C)	V_{DD} (V)	I_{off}^{max} (nA/μm)	t_{ox}^{min} (t_{eq}) (nm)		L_{nom} (nm)	
					bulk	DG	bulk	DG
High performance	30–10³	85	0.8–1.2	100–10³	1.0–1.3	1.0–1.3	15–19	13*
		–40	0.7–1.0	"	1.0–1.2	1.0–1.3	14–18	13*
		–170	0.5	"	0.9–1.1	1.0–1.2	14*	14*
Med-high perf.	5–30	85	0.8–1.2	20–120	1.2–1.5	1.3–1.6	17-24	13*–17
Moderate perf.	0.5–5	85	0.6–1.0	2–25	1.3–1.6	1.3–1.7	17-27	13*–16
Low power	10⁻³–0.5	65	0.7–0.9	0.01–1	1.7–2.0	1.7–2.1	22-29	14–20
Ultra-low power	<10⁻³	40	0.7–1	<0.005	2.1–2.6	2.2–2.6	27-38	25–33

*Limited by source-drain tunneling

Table 1. Application-dependent limits (after Ref. 3) for various performance levels and bulk Si or double-gate (DG) device geometries.

Year		1999	2003	2005	2008	2011	2014
General Lithography	(nm)	180	130	100	70	50	35
DRAM							
Memory size	(GB)	1	4	8	24	68	194
Chip size	(mm^2)	130	160	170	199	229	262
LOGIC							
Gate length	(nm)	140	80	65	45	30	20
Equivalent oxide t_{eq}	(nm)	2.5	1.9	1.5	1.2	0.8	0.6
Gate delay metric	(ps)	11	6.9	5.7	3.7	2.6	2.4
Transistors/chip	(10^6)	24	95	190	530	1500	4300
V_{DD}: High performance	(V)	1.8	1.5	1.2	0.9	0.6	0.6
Low power		1.5	1.2	0.9	0.6	0.5	0.3
Power: High performance	(W)	90	140	160	170	170	180
Low power		1.4	2.1	2.4	2.0	2.2	2.4
Number of wiring levels		6	7	8	9	9	9

Table 2. Subset of the 1999 ITRS roadmap.

strategies that can mitigate the effect of increased leakage, permitting channel-length scaling to continue somewhat further.

The inability of a single device design to meet the range of applications is already felt in today's chip designs where some input/output devices are designed with thicker gate oxides and longer channel lengths in order to be able to accommodate higher voltages. In the future this trend is bound to become more pervasive, where only the highest performance devices, constituting a relatively small number of devices, will be aggressively scaled, while the bulk of the devices will have somewhat thicker oxide, in several flavors perhaps, and longer channel lengths. As DRAM has taught us, this limitation does not necessarily mean the end of all progress, since other paths can be explored to continue the performance and density improvements.

3.   Why CMOS still?

Given that CMOS scaling is nearing the end of a long and distinguished journey, why is CMOS still almost universally looked upon in the industry as the technology of the future as well as of the present? Instead of ceding applications to other technologies, it is taking over territory. The answer to this question lies in the sheer power of numbers and the ability of "brute force" scaling to improve performance. Figure 1 shows the numbers of transistors available for different types of applications. It is clear that there is substantial processing power still to be wrung out of CMOS before the "end." With billions of transistors potentially available the question revolves more around how to use them to good effect, given power constraints, than how to find a way to increase their numbers even more.

The power of CMOS lies in its versatility. The CMOS transistor is close to the ideal switch, drawing negligible gate current and able to be used in series or parallel at will. With complementary p-FET and n-FET switching types available, conducting paths can be cut off in both logic states. The versatility stems from the simplicity of CMOS device physics and the almost unique occurrence of the silicon/silicon-dioxide passivated interface. The wide band gap of silicon dioxide enables extremely thin oxide layers to be realized with small tunneling currents, and tunneling is suppressed in the silicon itself by silicon's indirect band gap and relatively large tunneling masses. While it is true that CMOS devices become leaky as the scaling limit approaches, less aggressively scaled devices may have orders of magnitude less leakage, so that a trade-off of speed $vs.$ leakage power can be built into the same chip.

There is an almost complete dearth of competitors to CMOS. Some III-V semiconductor devices are faster, especially those based in (In,Ga)As,[5] but a true III-V CMOS implementation is still far off.[6] It is this lack of versatility, the inability to accommodate both high-speed and high-density/low-power applications on a single chip, that makes III-V devices uncompetitive.

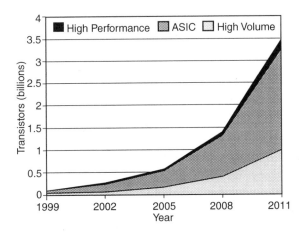

Figure 1. Number of transistors in each type of chip (400 mm^2) as projected from the SIA roadmap (reproduced with permission from an article by R. Nair to be published in *IBM J. Res. Develop.*, 2002).

There is a whole area of research today devoted to "beyond CMOS" technologies, invoking candidates such as carbon nanotubes,[7] molecular electronics,[8] and single-electron and superconducting logic.[9]   It should be recognized that all of the above candidates, except for the last, are electrostatically controlled devices.   Hence they are qualitatively subject to the same scaling restrictions as silicon CMOS, in that the gate insulator has to be smaller than the source-drain separation.   This restriction, coupled with the need to interconnect the circuits with the same wiring technology, means that radically different circuit densities are not to be expected for logic applications.   The same might not be true for memory, where a hybrid memory technology might be incorporated into a CMOS logic chip.

Today, system performance is more and more achieved by running multiple CPUs in parallel.   In the future this trend is expected to continue onto the chip. With this trend in mind, it is necessary to compare technologies not just on the basis of single CPU performance, but also on the ability to support multiple CPUs.   This ability is where the strength of CMOS lies and where the competitors fall short.

4.   Technological advantage is frozen-in

While CMOS is following Moore's Law, the advantages achieved by down-scaling dwarf all others, even though the down-scaling has required many device and materials innovations, *e.g.* self-aligned silicides, halo implants, and chemical-mechanical polishing for the back-end of line.   As the end of scaling approaches, more radical innovations can be considered, since they may be the only other way to obtain improved performance, or to extend scaling a little further.   Economic forces will also be at work.   Research into new device options is very expensive, and once the large increase in productivity embodied in Moore's Law no longer drives the semiconductor industry, the incentive to do further device research diminishes, so that any improvements made in this period will tend to become frozen-in.   In a way, the period we are in now is the golden age of semiconductor device research, when radical new device designs, all in CMOS, are being investigated in order to exploit the full potential of the technology.

5.   Scaling limits

Scaling limits are conveniently discussed in terms of a scale length λ.   This length is a measure of the basic ability of the gate to shield the channel against changes in drain and source potentials, and is closely related to the ability of the gate to turn the FET off.   As the channel length decreases, the off-current is influenced more and more by the drain voltage (drain-induced barrier lowering, DIBL) and likewise the threshold voltage is reduced with decreasing channel length (V_T roll-off).   These deleterious effects are kept within tolerable limits for channel lengths greater than $\sim1.5\lambda$.[10]   While the screening length describes the main exponential decay term, pre-exponential terms related to the source and drain boundary conditions can significantly affect the results.[10] First-order equations for the scale-length dependence of the FET geometry (for an idealized case) for both bulk and double-gated FETs were derived by Frank *et al.*[3] (the derivation of the scale length is simplified by limiting it to the subthreshold region, where the space charge of carriers in the channel can be neglected, and only Laplace's equation need be solved). The solutions are shown in Fig. 2.

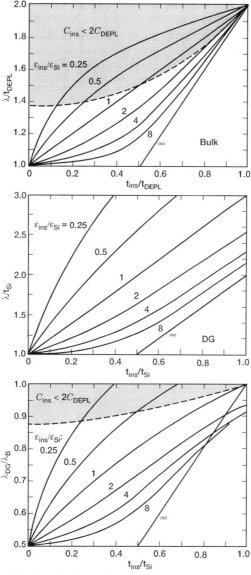

Figure 2. Normalized scale factor λ *vs.* normalized insulator thickness t_{ins} for MOS transistors in the subthreshold regime:[3] bulk FETs, normalized to depletion layer thickness (a); symmetric double-gate (DG) FETs, normalized to body thickness t_{Si} (b); comparison of bulk and DG-FETs, with $t_{ins} = 2t_{Si}$ (c).

In Fig. 2(a) the bulk case is presented, where both the scaling length and the insulator thickness are normalized to the depletion layer depth. Arbitrary ratios of insulator to silicon dielectric constants are considered in the figure with the technologically important SiO_2/Si ratio being ~1/3. The capacitance across the depletion layer cannot be too large, otherwise deterioration of the subthreshold slope results. This forbidden region is indicated by the shaded area in Fig. 2(a) and restricts the range of t_{ox}/t_{depl}, especially for low-κ insulators. Furthermore, band-to-band tunneling from drain to doped well in the bulk case further limits t_{depl} to values above ~5 nm, corresponding to an electric field in the depletion layer of ~2 MV/cm.[11] Assuming then silicon dioxide with $t_{ox} = 1$ nm, a minimum channel length (1.5λ) of 11 nm is derived, although this number should not be taken too literally because of the approximations made and the dependencies on boundary conditions and application criteria.

6.   Device design strategies

While the roadmap indicates 10–15 years before hard scaling limits are encountered at ~20-nm gate lengths, device research is pushing up against these limits today, and new device designs and materials are being introduced to best cope with these limits.

Three generic approaches are being considered:

- Maintain relatively long channel lengths, but improve performance by using higher mobility, yet CMOS-compatible, channel materials.

- Introduce new insulator and gate materials in order to increase the gate control and thereby extend scaling.

- Progress from FETs made on bulk silicon wafers to FETs having ultra-thin silicon bodies, either single- or double-gated, to extend scaling and increase flexibility of use.

We will discuss each of these approaches below.

- *Higher mobility channels*

There is considerable work today on improving the mobility of silicon by applying strain.[12,13] Silicon-germanium alloys had been used successfully to improve the performance of bipolar transistors. MOSFETs were expected to gain higher mobility from SiGe channels. While the SiGe channel itself was difficult to implement in CMOS because of the difficulty of growing a high-quality oxide, it was found experimentally, and supported theoretically,[14] that Si grown on a SiGe substrate exhibits higher mobility due to the strain caused by accommodating the overlying Si to the larger lattice constant of the SiGe. Experimentally, transconductance improvements of up to 50% have been reported in state-of-the art *n*-FETs.[13]

Other possibilities might be to introduce III-V materials selectively onto a Si chip, for example by selective epitaxy, or placement of III-V chiplets on a Si chip.

These approaches provide a modest one-time boost to performance, which might be economically justified according to our previous arguments. The compatibility of these options with ultimately-scaled silicon has to be evaluated, since some properties such as smaller bandgaps, lower densities of states, and more complex structures may run counter to scaling requirements. It is also essential to maintain the general-purpose nature of the rest of the chip, since only a small portion of such a chip is likely to be devoted to the highest performance devices.

- *New gate and gate-insulator materials*

The most important impediment to scaling is the exponentially increasing tunneling leakage currents in thin oxides. By replacing a thin SiO_2 layer by an insulator that is thicker but has a higher dielectric constant, it is hoped to be able to extend scaling further. There is much work today on a wide variety of alternate dielectrics, some of which indeed offer the promise of lower leakage currents. However, the benefits of high dielectric constant, κ, cannot simply be encapsulated by an equivalent oxide thickness $t_{eq} = (\kappa_{ins}/\kappa_{ox})t_{ox}$, because of two-dimensional effects.

The discussion is facilitated by returning to Fig. 2(a). At small insulator thicknesses the relationship $\lambda \cong \kappa_{Si}t_{ins}/\kappa_{ins} + t_{depl}$ holds, supporting the equivalent oxide thickness concept. But this relationship breaks down for large κ_{ins} or t_{ins}, where for large κ_{ins} the scaling length becomes less dependent on κ_{ins}. At the same time, other material properties, such as bandgap or mobility,[15] may degrade with κ_{ins}, so that the optimal κ_{ins} lies in the mid-range ($\kappa_{ins} \cong \kappa_{Si}$).

There is renewed interest in metal gates, including completely converting the polysilicon gate to silicide.[16] The metal gate eliminates depletion and, given a suitable work function, may improve channel mobilities by reducing channel doping. Another possible benefit is circumventing the mobility reduction that is predicted to occur in very thin oxides do to remote interactions between electrons in the channel and plasmons in the polysilicon gate.[17] Since many metals cannot withstand the high temperatures of CMOS processing, replacement-gate methods are being explored.[18]

- *Away from bulk silicon*

Today, some companies are moving away from bulk silicon to silicon-on-insulator (SOI)[19,20] partly to improve performance and partly to improve scaling. Improvements in scaling are predicted to occur only for extremely thin SOI, of the order of 10 nm thick or less.[21] Experiments have shown reasonable mobilities being maintained in SOI films down to ~5 nm thickness.[22] The thickness limit for thin SOI is probably ~3 nm (see Fig. 3) where quantum effects cause the bandgap (and hence the threshold voltage, as well as the channel potential) to become very sensitive to the thickness, so that slight fluctuations in thickness will cause large fluctuations in the potential.

The scaling potential of single-gated FETs (SG-FETs) on thin SOI is not particularly good. A simple scaling length cannot be defined for single-gated SOI as the screening of the drain potential by the gate is non-exponential.[21,23]

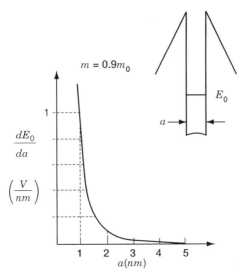

Figure 3. Sensitivity of lowest subband energy E_0 in a <100> silicon quantum well to the silicon thickness a.

The SOI has to be close to the thickness limit to achieve competitive gate lengths, because of the easy penetration of the electric field underneath the FET. Still, this configuration could be attractive due to its simplicity and has yielded excellent device results at 50-nm gate length.[20] Scaling could be enhanced if the buried oxide could be replaced with a low-κ dielectric or simply a void.[21]

The best scaling is obtained with double-gate FETs (DG-FETs) where there is a gate on both sides of a thin silicon body. The scaling length of a symmetric DG-FET is shown in Fig. 2(b). At first glance the scaling length appears to be longer for the DG-FET than for a bulk FET, especially for small κ (e.g. SiO₂, $\kappa_{ins}/\kappa_{Si} \cong 1/3$) because, for the DG-FET there is an insulator on both sides of the silicon body — compare Fig. 2(b) to Fig. 2(c). This difference is illusory, since the depletion layer plays the role of an insulator. Due to considerations of tunneling and capacitance, as discussed above, the minimum bulk depletion layer thickness is considerably larger than the minimum DG-FET body thickness. Bulk and DG FETs are compared in Fig. 2(c), where we have assumed a ratio of two for these thicknesses. As seen from the figure, while the improvement in scalability for the DG-FET is not great with a low-κ dielectric, it may approach a factor of two for high κ. The limitations on depletion layer thickness also depend on the application and its sensitivity to subthreshold slope and tunneling leakage current. This sensitivity is the reason for the advantage of the DG-FET for low power applications, see Table 1. Augmenting this advantage is the fact that the two gates of the DG-FET may be controlled independently, extending circuit possibilities and allowing the use of one of the gates for power management.

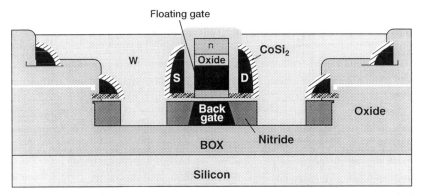

Figure 4. Schematic cross-section of an experimental planar DG-FET structure,[25] showing silicided sidewall self-aligned S/D, tungsten self-aligned S/D plugs, and isolated undercut back gate.

Research into DG-FETs has picked up pace in universities and the industry. Planar, fin, and pillar topologies have been used,[23] with the so-called FinFet geometry appearing the easiest to make and yielding excellent characteristics.[24] While the planar DG-FET achieves the best control of the SOI thickness, it is more difficult to make.[25] Figure 4 shows a cross-section of an experimental self-aligned planar DG-FET where the S/D regions are etched away leaving sidewalls, allowing the back gate to be under-cut.

7. Power management

Of all the issues that confront scaled integrated circuits, power dissipation is the most serious, simply because larger and larger systems are being built than ever before, due to cost reductions on scaling, and more function is crammed onto ever larger and denser chips. The power per function actually continues to decrease, but there is an increasing conflict between the demands for low dynamic power ($\sim CV^2$), which requires low voltages, and low standby power which requires higher V_T in order to be able to turn the transistors off. This trade-off leads to power supply optimization according to application,[3] as in Table 2, where one sees higher power supply voltages for lower power applications.

Power management technology is still in its early stages of development, since up until recently voltage reduction, improvements in cooling technology, and improvements in system architecture have kept the problem within bounds. The largest power-saving improvements are often realized at the highest levels of architecture.[26] Within the domain of devices, there are several techniques that could result in a better static *vs.* dynamic power trade-off, such as:

- multiple threshold voltages;

- multiple oxide thicknesses and power supplies;

- dynamically adjustable threshold voltages; and

- block switching.

Today, multiple threshold voltages are a standard offering for many technologies. Low-V_T devices are typically used for clocking, where duty factors are high, and speed-sensitive parts of the system. High-V_T devices are used especially in memory, where duty factors are very low. As technology progresses and gate oxides become thinner, the gate oxide leakage also becomes part of the power equation. Some technologies today have multiple oxide thickness offerings to be able to come with higher voltage requirements for output devices, and this approach could be applied to on-chip power management as well. A particularly clear case is between logic and memory, even within a memory function block. For instance, in a large static random access memory (SRAM), the cells themselves can be run at larger logic swings than the accessing logic since the internal cell capacitance is so small and the duty factor is small compared to the heavily loaded and high-duty-factor logic.[27] Thus, technologies with differing requirements on gate and subthreshold leakage vs. performance could coexist in the same functional block, and even more so on the chip-system.

A more radical approach to power management is block switching,[28] as illustrated in Fig. 5, where the power supply to a circuit block, or even an entire processor, is gated. If the gating switch consists of high-V_T thick-gate-oxide FETs, both the sub-threshold and gate leakage components of the standby power may be greatly reduced. An issue is the size, speed, and power needed to switch the gating transistor, which we will investigate below.

Consider a processor to be switched by a switching transistor (labeled with subscripts P and S, respectively). The power dissipation of the processor, $P_P = \alpha C_P V_P^2 f_P$, where α, C_P, V_P, and f_P are the duty factor, total capacitance, power supply voltage, and clock frequency, respectively. The resistance R_S of the switching transistor has to be low enough that only a small fraction δ_{PS} of V_P is dropped across the series switch.

Applying standard FET equations to calculate the resistance, one can derive the area ratio between the switch and the processor assuming area inefficiency

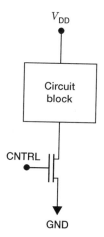

Figure 5. Switching of a circuit block using a series switch.

factors ($\beta \equiv$ total area/gate area), and a capacitance mulplier ($\gamma \equiv$ total capacitance/gate capacitance), taking wiring and parasitic capacitance into account. This approach gives:

$$A_S/A_P = (\gamma_P\beta_P/\gamma_S\beta_S) \times (C_{OP}/C_{OS}) \times (\alpha f_P \tau_S/\delta_{PS}) , \tag{1}$$

where the first factor accounts for area and capacitance overhead, the second factor (capacitance/unit area) reflects the different gate oxide thicknesses used, and the third factor covers the switching parameters of the switching transistor, where $\tau_S = L_S^2/(\mu_S V_{GS})$ is the transit time of the transistor with L_S being the gate length, μ_S the mobility, and V_{GS} the gate voltage. The switching time is determined by the time constant, $R_S C_P = k_{sat}\delta_{PS}/\alpha f_P$, where the extra factor $k_{sat} > 1$ accounts for the increase in switching time caused by velocity saturation. The ratio between gate power of the switching transistor and the processor power is

$$P_{GS}/P_P = (V_{GS}/V_P) \times (f_S\tau_S/\delta_{PS}) . \tag{2}$$

By following these equations, it can be appreciated that the area and gate-power penalties of the switching transistor are small. The area penalty is small mainly because of the small duty factor α of the average processor transistor, and also because of the small gate area utilization β_P. The switching-gate power penalty is low for switching frequencies less that $\sim f_P/10$, and the switching time constant is a small number of clock cycles, mainly because the requirement for small voltage drop δ_{PS} means the switching transistor must have good current drive. Note however that the switch should be designed to cope with the largest expected duty factor, which increases the area and power penalties. The above analysis illustrates the practicality of block switching and the plausibility of its increased use in future chip-systems.

If block switching can be done with high efficiency, it might pay to run the on-chip CPUs in *burst* mode, where the computation is done as quickly as possible and the CPU is gated-off during idle periods. This way minimizes the impact of the standby power of the ungated CPU.

8. Redundancy

A premise of the above power-management approach is that in future chip-systems, devices are cheap but power is expensive. At issue is whether devices are used to implement large numbers of processors on a chip, many of them idle because of power constraints, or whether the transistors are better used with fewer processors and a very large cache memory. Most likely a particular balance will be struck depending on the application. One should bear in mind that the availability of $\sim 10^{10}$ transistors in "end of road-map" chips allows for placement of hundreds of processors. Thus, innovations like embedded DRAM (or other dense on-chip memory technology) might mean that a large numbers of processors, each accompanied by a large cache, may fit onto a single chip.

Redundancy is already an essential part of memory design, including error detection and correction. Without it, memory could never achieve the integration levels available today. It is not an unwarranted speculation to assume that it will be equally important for future chip-systems. The increased susceptibility of scaled devices and circuits to soft errors, increasing parameter spreads caused by material fluctuations, and increasing numbers of devices point to the need for such techniques, while the increased on-chip computing power gives an increased capability of implementing them. Redundancy techniques have long been applied to fault-tolerant and highly reliable computing systems[29,30] and have been implemented in hardware and software and at high and low levels. Redundancy is an important area for research and a possible key for exploiting future systems to the fullest.

9. Conclusions

In summary, the data processing potential of future CMOS chips is seen to be tremendous. Even as we push up against ultimate limits, device designers will exploit structural possibilities to the utmost, continuing density and performance improvements for many generations to come. Indeed, the ultimate MOSFET is truly a "nanotransistor" competitive, as foreseen, with other nanochallengers. Even as device scaling approaches its logical conclusion, the future of the VLSI revolution will be as much or even more in the hands of the chip architect.

Acknowledgments

The author would like to acknowledge his many colleagues who provided the background for this presentation, including Max Fischetti, Dave Frank, Steve Laux, Ravi Nair, Jim Stathis, Yuan Taur, Philip Wong, and Nagui Halim.

References

1. M. V. Fischetti and S. E. Laux, "Monte Carlo simulation of electron transport in technologically significant semiconductors of the diamond and zinc-blende structures. Part II: Submicron MOSFETs," *IEEE Trans. Electron Dev.* **38**, 650 (1991).
2. Yukou Mochida, Takeshi Takano, and Hirohisa Gambe, "Future directions and technology requirements of wireless communications," *Tech. Digest IEDM* (2001), p. 15.
3. D. J. Frank, R. H. Dennard, E. Nowak, P. M. Solomon, Y. Taur, and H.-S. P. Wong, "Device scaling limits of Si MOSFETs and their application dependencies," *Proc. IEEE* **89**, 259 (2001).
4. S. Lai, "Tunnel oxide and ETOX flash scaling limitation," *Digest 1998 Intern. Nonvolatile Memory Conf.* (1998), p. 6.
5. Y. Yamashita, A. Endoh, K. Shinohara, M. Higashiwaki, K. Hikosaka, *et al.*, "Ultra-short 25-nm-gate lattice-matched InAlAs/InGaAs HEMTs within

the range of 400 GHz cutoff frequency," *IEEE Electron Dev. Lett.* **22**, 367 (2001).

6. M. Hong, J. N. Baillargeon, J. Kwo, J. P. Mannaerts, and A.Y. Cho, "First demonstration of GaAs CMOS," *Proc. IEEE Intern. Symp. Compound Semicond.* (2000), p. 345.

7. Richard Martel, Hon-Sum Philip Wong, Kevin Chan, and Phaedon Avouris, "Carbon nanotube field effect transistors for logic applications," *Tech. Digest IEDM* (2001), p. 159.

8. Ananth Dodabalapur, "Organic and polymer semiconductor devices," *Tech. Digest IEDM* (2001), p. 10.

9. P. Bunyk, K. Likharev, and D. Zinoviev, "RSFQ technology: Physics and devices," *Intern. J. High-Speed Electronic Syst.* **11**, 257 (2001).

10. D. J. Frank and H.-S. P. Wong, "Analysis of the design space available for high-κ gate dielectrics in nanoscale MOSFETs," *Superlatt. Microstruct.* **28**, 485 (2000).

11. Y. Taur, C. Wann, and D. J. Frank, "25nm CMOS design considerations," *Tech. Digest IEDM* (1998), p. 789.

12. J. Welser, J. Hoyt, S. Takagi, and J. Gibbons, "Strain dependence of the performance enhancement in strained-Si n-MOSFETs," *Tech. Digest IEDM* (1994), p. 373.

13. K. Rim, J. L. Hoyt and J. F. Gibbons, "Transconductance enhancement in deep submicron strained-Si n-MOSFETs," *Tech. Digest IEDM* (1998), p. 707.

14. M. V. Fischetti and S. E. Laux, "Band structure, deformation potentials, and carrier mobility in strained Si, Ge, and SiGe alloys," *J. Appl. Phys.* **80**, 2234 (1996).

15. M. V. Fischetti, D. A. Neumayer, and E. A. Cartier, "Effective electron mobility in Si inversion layers in MOS systems with a high-κ insulator: The role of remote phonon scattering," *J. Appl. Phys.* **90**, 4587 (2001).

16. B. Tavel, T. Skotnicki, G. Pares, N. Carrière, M. Rivoire, F. Leverd, C. Julien, J. Torres, and R.Pantel, "Totally silicided ($CoSi_2$) polysilicon: A novel approach to very low-resistive gate ($\Omega/\sigma\theta$) without metal CMP nor etching," *Tech. Digest IEDM* (2001), p. 825.

17. M. V. Fischetti and S. E. Laux, "Long-range Coulomb interactions in small silicon devices. Part I: Performance and reliability," *J. Appl. Phys.* **89**, 1205 (2001); M. V. Fischetti, "Long-range Coulomb interactions in small silicon devices. Part II: Effective electron mobility in thin-oxide structures," *J. Appl. Phys.* **89**, 1232 (2001).

18. A. Chatterjee, R. Chapman, G. Dixit, J. Kuehne, S. Hattangady, *et al.*, "Sub-100 nm gate length metal gate NMOS transistors fabricated by a replacement gate process," *Tech. Digest IEDM* (1997), p. 821.

19. J. W. Sleight, P. R. Varekamp, N. Lustig, J. Adkisson, A. Allen, *et al.*, "A high performance 0.13 μm SOI CMOS technology with a 70 nm silicon film and with a second generation low-κ Cu BEOL," *Tech. Digest IEDM* (2001), p. 245.

20. Robert Chau, Jack Kavalieros, Brian Doyle, Anand Murthy, Nancy Paulsen, *et al.*, "A 50 nm depleted-substrate CMOS transistor (DST)," *Tech. Digest IEDM* (2001), p. 621.

21. H.-S. P. Wong, D. Frank, and P. Solomon, "Device design considerations for double-gate, ground-plane,and single-gated ultra-thin SOI MOSFETs at the 25 nm channel length generation," *Tech. Digest IEDM* (1998), p. 407.

22. D. Esseni, M. Mastrapasqua, G. Celler, F. Baumann, C. Fiegna, L. Selmi, and E. Sangiorgi, "Low field mobility of ultra-thin SOI *n*- and *p*-MOSFETs: Measurements and implications on the performance of ultra-short MOSFETs," *Tech. Digest IEDM* (2000), p. 671; see also this volume.

23. H.-S. P. Wong, D. Frank, P. M. Solomon, H. J. Wann, and J. Welser, "Nanoscale CMOS," *Proc. IEEE* **87**, 537 (1999).

24. Yang-Kyu Choi, Nick Lindert, Peiqi Xuan, Stephen Tang, Daewon Ha, *et al.*, "Sub-20nm CMOS FinFET technologies," *Tech. Digest IEDM* (2001), p. 421.

25. K. W. Guarini, P. M. Solomon, Y. Zhang, K. K. Chan, E. C. Jones, *et al.*, "Triple-self-aligned, planar double-gate MOSFETs: Devices and circuits," *Tech. Digest IEDM* (2001), p. 424.

26. T. H. Meng, B. M. Gordon, E. K. Tsern, and A. C. Hung, "Portable video-on-demand in wireless communication," *Proc. IEEE* **83**, 359 (1995).

27. A. Fridi, P. Solomon, D. Frank, S. Reynolds, D. Pearson, and M. Elmasry, "A 0.22 μm CMOS 0.65 V, 500 MHz 64 Kb SRAM macro," unpublished, 1998.

28. S. Shigematsu, S. Mutoh, Y. Matsuya, Y. Tanabe, and J. Yamada, "A 1-V high-speed MTMOS circuit scheme for power-down application circuits," *IEEE J. Solid State Circ.* **32**, 861 (1997).

29. D. Siewiorek and R. Swarz, *Reliable Computing Systems, Design and Evaluation*, Burlington, MA: Digital Press, 1992.

30. B. Culbertson, R. Amerson, R. Carter, P. Kuekes, and G. Snider, "The Teramac custom computer: Extending the limits with defect tolerance," *Proc. IEEE Intern. Symp. Defect Fault Tolerance VLSI Syst.* (1996).

Driving Technology to Re-engineer Telecommunications

T. P. Smith III, H. A. Maudar, and J. W. Liu
Cable & Wireless, Vienna, VA, U.S.A.

S. J. Farrell
Cable & Wireless, Covent Garden, London, U.K.

1. Introduction

In today's Internet age, consumers have access to almost all types of information and services on line. In spite of the global growth in Internet use, telecom companies have been suffering large financial losses. Amongst the numerous bankruptcies, surviving players are still fighting to provide the most innovative and efficient services.

Technological advances have made bandwidth plentiful and intelligent edge devices ubiquitous. Networked computing technology has driven data traffic to levels comparable with voice. This parity has brought into question some of the basic design assumptions and architecture of traditional public switched telephone network (PSTN) businesses. Changes in network design are perceptively discussed in "The rise of the Stupid Network" by David Isenberg.[1] This groundbreaking paper brought to light problems with intelligent networks that are expensive, have scarce capacity, and focus on voice. Isenberg elucidated changing the basic network structure to a "stupid network" with intelligent user-controlled endpoints and only bit transport in the middle.

The PSTN physical networks and service-providing companies were built to operate and derive profit under currently outdated assumptions. Operating companies were structured to turn large capital investments into clearly defined products that required delivery to millions of customers in order to achieve margins. The engineering constructs of the global networks are changing in order to meet the new parameters. Under the new assumptions, functionality will be defined on an individual customer basis. Telecommunication companies must therefore re-engineer themselves to deal with customers one-on-one rather than *en masse*. Smart edge devices will allow customers to define their own services, and will require finer gradations of quality of service and flexibility from the network.

Peer-to-peer (P2P) computing is a new network computing architecture that will work alongside the traditional centralized client/server model and help drive the transition to "stupid networks."[2] Napster is a well known application using P2P computing. Interestingly, the industry is beginning to talk about the use of P2P computing for common business-to-business (B2B) applications, especially

B2B communications using emerging web-services technologies. These applications are designed to exploit a network of interacting intelligent devices and do not require additional network capabilities over and above transmission and quality of service. The adoption of P2P computing therefore reinforces Isenberg's arguments. What P2P applications do require, however, are new infrastructure capabilities to help them work and interact on the network. Network providers are well positioned to offer smart directories to authenticate and authorize users, locate resources, store and manage resource policies and profiles; content distribution to optimize performance; storage of data in the network; and network functions configurable by applications so that end users can govern network quality of service. The ultimate aim of "stupid networks" is to allow end users to create, offer, and use services unconstrained by the network.

New breeds of edge service providers such as content distribution network (CDN) providers and web-hosting providers have emerged. These edge providers are changing, and in some cases replacing, the relationship between network providers and customers, resulting in lost revenue to network providers. Oversupply, existing uneconomic hosting models, and the anticipated change in IP traffic mix have sparked an industry-wide consolidation and rationalization. IP network providers are well positioned to benefit from this transformation, and by providing new infrastructure services will build the application-based network of the future.

In this contribution, we will characterize and review major industry areas that have been the focus of activity and investment by telecommunication companies supporting the transition to stupid networks. We will highlight problems that have resulted through misalignment between these areas and why the industry should move forward in exploiting new technology to realign the industry and telecommunications services.

2.   The transition to "stupid networks"

The public switch telephony network (PSTN) design is composed of intelligent services at the core of the network that regulate distribution of resources and service quality. The PSTN is based on the long-held view of reality that circuit-switched capacity is an expensive and scarce resource. Within the telephony structure, traffic is predominantly voice and the network is highly controlled. Signaling systems have been created to provide call-establishment, billing, routing, and information-exchange functions. These systems were developed to provide intelligence that maximizes utilization and revenue generation from scarce resources. Intelligence at the core gives the network provider control of both the network and the services offered.

Abundant cheaper computing and resources are driving a transition towards packet-switched networks with intelligence at the edge rather than at the core. These new edge applications consider the network simply as demanding bandwidth and quality of service when required.

In support of the transition to packet-switched networks, the telecommunications industry has recently focused investment and activity in three major areas: core network capacity, IP services at the edge, and last-mile broadband connectivity.

3. Industry players

The companies driving the transition toward the stupid network can be characterized as having concentrated efforts and investment in numerous packet-switched network areas. These areas are the core, the edge, and the last mile. Core players provide the backbone structure of the network and are deploying massive increases in available capacity across and between continents. Edge players are companies building data caches and providing IP services such as authentication, authorization, service provisioning, indexing, web hosting, content distribution, and content filtering. Last-mile companies provide broadband connectivity from the edge of the network to the end user.

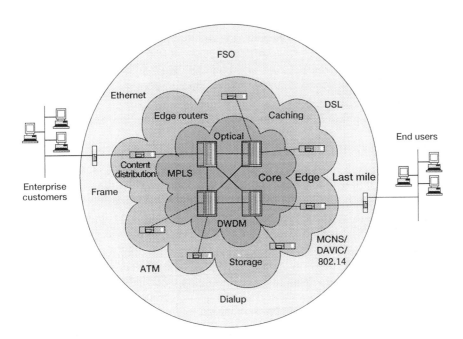

Figure 1. Major industry areas involved in the transition to the "stupid network".

4. Core players

There has been massive investment in global backbone networks. The objective was to dramatically increase network capacity to the levels thought necessary to meet projected data traffic volumes, reduce voice traffic operating costs, and maintain margins as a result of declining voice revenue. The build-out of additional network capacity has been executed largely successfully. However, lower than projected data traffic on abundant network capacity is driving prices down, presenting a serious threat to some companies.

Key features of core networks going forward will be higher bandwidth on existing fibers using enhanced dense wavelength division multiplexing (DWDM), optical switching at rates of 100-Gbit/s and above, and direct connections between DWDM equipment and routers giving rise to the optical Internet. The dominance of the synchronous optical network (SONET) approach for metropolitan area network (MAN) and wide area network (WAN) connectivity will be challenged through a combination of Ethernet and DWDM. Until recently, Ethernet could not compete with the range and network-management capabilities of SONET. However, DWDM equipment can now carry Ethernet signals over virtually any distance and the DWDM manufacturers are adding more administration and maintenance tools to their equipment. Advances in Ethernet interfaces and DWDM technologies combined will present a serious threat to the range and management capabilities of SONET. The use of optical fiber and Ethernet everywhere will result in a flat network structure. Ethernet could become a unifying transport protocol that will eliminate inefficient translation between different segments of the network and blur the distinctions between local area networks (LANs), MANs and WANs. The future network structure will provide high capacity, resilience, high availability, manageable capacity assignment, and reduced operating costs. But, contrary to popular myths, bandwidth will not be free.

The challenge of a network carrying IP traffic today is how to manage network performance and stability. IP is a connectionless protocol whereby each packet moves through the network to its destination along a path determined by a distributed set of routing tables and the current network topology. This mechanism provides flexibility but is difficult to manage to improve network predictability. Increasing network bandwidth to guarantee performance is not an efficient use of resources.

Multi-protocol label switching (MPLS) is a developing standard for improving performance, reliability, and scalability of connectionless IP networks by engineering deterministic route selection and providing a mechanism for traffic management. MPLS lets managers explicitly configure paths to send selected traffic along specific routes. MPLS provides an elegant traffic engineering mechanism termed routing for resource reservation (RRR). RRR also provides fast restoration of logical circuits during failure in an IP network. Future releases of MPLS will implement constraint-based routing to automatically establish explicit paths for balancing traffic loads that conform to specific policies.

5. Edge players

A huge industry has arisen at the edge of the network providing IP-based services to end customers. Two types of services can be distinguished: content and application services, and infrastructure services. Content service providers specialize in the creation, aggregation, indexing, taxonomy, personalization, branding, supply, and management of Internet-based information. Application service providers host, provide, deliver, and manage multi-tenancy Internet-based applications to the customer. These services are sold on a rental model, where the customer does not need to invest in IT infrastructure.

Infrastructure service providers that provide core IP services such as dial-up and direct Internet access, remote authentication dial-in services (RADIUS), domain name services (DNS), and dynamic host configuration protocol (DHCP) services are now also providing additional services such as web hosting and content distribution overlay networks for two primary reasons:

- To resolve performance issues with the core and last-mile tiers of the network by distributing and caching content closer to the point of request.

- To provide professional services and secure, robust, resilient, and scalable environments for hosting Internet content and applications in a largely unpredictable Internet marketplace.

A key technology common across many edge service providers is the directory. The directory has been rapidly adopted as the Internet-based repository for storing customer profiles and policies, and service configuration and provisioning information. Directories are even being used to store information about the network itself. Directories based on the International Organization for Standardization (ISO) and the International Telecommunication Union's (ITU) distributed directory specification, X.500, together with the lightweight directory access protocol (LDAP) are widely deployed because of the strong support for distribution and their excellent read access. Given the wealth of information stored within these directories across the Internet and the fact that they can be accessed using open standards, Internet-based applications are increasingly becoming directory-enabled.

6. The "last mile" players

Last-mile companies provide broadband connectivity for individuals and enterprises to the Internet specifically between edge routers and end-user devices. Execution of this strategy has been poor given initial projections. Last-mile technologies include frame relay, asynchronous transfer mode (ATM), digital subscriber line (DSL), Ethernet over optical and digital cable modems using multimedia cable network system (MCNS), digital audio visual council (DAVIC), data over cable system interface specification (DOCSIS), or Institute of Electrical and Electronics Engineers (IEEE) 802.14 standards.

In the U.S., the Telecommunications Act of 1996 was designed to deregulate the industry and spark competition. It was hoped that deregulation would attract new entrants so as to increase choice, lower costs, and build the necessary economic conditions to allow providers to offer broadband connections to areas that were difficult to connect. The reasons for the poor last-mile penetration are numerous. Those that create the most debate amongst industry analysts and state regulators center around the power exercised by incumbent local exchange carriers (ILECs) over the competitive local exchange carriers (CLECs) through their control of the physical infrastructure, as well as the outdated residential telephony subsidy that some believe ought to be used to subsidize broadband connectivity to remote areas.

7. Conflict through misalignment

Over the last few years, core providers have engaged in significant build-out of core network capacity and capability. This build-out has been driven both by traffic volume predictions and by a desire to reduce the long-term cost of supporting both packet switched and PSTN networks by migrating to voice over IP (VoIP) or voice over broadband (VoBB). Similarly, edge providers have invested in new points of presence, high bandwidth routers in anticipation of increased IP traffic, and new IP Services. Last-mile providers have been investing heavily in new access technologies, investments that were thought to be justified by a predicted surge in IP traffic.

The telecom industry has been suffering immense losses over the past year, for multiple reasons. One obvious cause was the liberalization of the telecom industry in 1996, partnered with the boom in network and computer equipment demand in the 1990's. This demand caused the creation of hundreds of network providers and start-ups that were all fighting to bring consumers the fastest network access with the goal of exploiting future Internet traffic.

Although the levels of demand at the core and edge are in synch, the last mile has not kept pace. Only 7–8 million of 120 million U.S. households have broadband access. Homes are still served predominantly by dial-up technology. Only 4–5% of enterprises connected to the Internet use broadband connections, despite existing infrastructure. According to a McKinsey–J. P. Morgan industry analysis, the U.S. has only 5% broadband last mile although there is enough core network infrastructure to provide for 40% penetration.[3] At the end-user level, the services provided by the core and edge are not achieving their predicted revenues due to insufficient access to end users. End users with dial-up connections have limited access to services due to lack of speed and clarity. For example, the viewing of websites with interactive streaming would be minimal. As a result, the critical mass of applications required to generate and sustain the transition to "stupid networks" has not materialized, which has led to the dot-com implosion.

It is clear that telecom companies, investors, and analysts misjudged the growth of this market. This overstated growth, coupled with lower utilization and in some cases poor management, resulted in oversupply. Telecom companies around the globe have been forced to downsize and even close down. Despite, and perhaps due to, deregulation in the U.S. market in 1996, rivals of incumbent local exchange carriers that control the last-mile infrastructure are being hit the hardest. Alleged "dirty ticks" by the incumbents together with inappropriately targeted last-mile subsidies have left competition for the last mile void and consequently, demand for IP services has been lower than projected. This reduced demand has led to oversupply of core network capacity that was installed predicated on much larger volumes of traffic than that witnessed today, driving down prices, and hitting core network providers and equipment manufacturers.

It is interesting to note that the Japanese market has largely managed to avoid these negative consequences. Japanese projections for broadband usage were more conservative and based on a good appreciation of the available technologies, limiting the mismatch between core network, edge, and last-mile build-out.

8.   The future of telecommunications: New technology is the key

The McKinsey–J. P. Morgan report forecasts that the current Internet traffic mix will move rapidly towards P2P, streaming, and multimedia.[3] This transition is completely dependent on increasing the penetration of broadband access technologies, especially for the business sector.

We believe that the last-mile problem will be resolved by a combination of technologies, industry initiatives, and government regulation and that it will take some time to for these measures to take effect. Technologies that can significantly reduce the cost and time scales commonly associated with the deployment of last-mile broadband connectivity will be attractive to business customers. These technologies include Ethernet over free space optics (FSO) and emerging 1–10 Gbps Ethernet interfaces coupled with DWDM technologies. Technologies that can be used by content and applications providers to reduce the impact of the last mile bottleneck will also be widely adopted. These technologies include web services, content distribution overlay networks, and network caching and hosting.

The adoption of open Internet standards will help drive down cost, increase interoperability between providers and equipment manufacturers, and fuel innovation. This mix will result in advances in broadband access technologies that may even leverage existing assets, but most importantly will create new services for customers, thereby increasing demand for last-mile, edge, and core resources.

Measures that government could take to increase competition include:

- removing the existing outdated telephony subsidy and using these resources instead to subsidize broadband penetration in 'hard to get at' areas; and

- enforcing swifter and harsher penalties on ILECs that abuse their power or position to drive out new entrants.

There is also an opportunity to increase penetration of last-mile broadband connectivity by encouraging last-mile companies to move beyond the simple objective of securing revenue from the broadband connection. Instead, these companies should look at a revenue-sharing model with network and content providers based on the increased traffic enabled through broadband connectivity.

IP traffic will continue to slow down over the next years, but the cumulative effect of its growth will still be substantial. McKinsey–J.P. Morgan predict a period of consolidation in the network-provider sector and a shift in market strength from start-ups and high-cost regional wholesale-based companies to traditional service providers with strong enterprise proficiency and structural cost advantages.[3] Companies who target enterprises versus individual customers also will be favored.

Technology companies are increasingly supporting peer-to-peer architectures that allow people and businesses directly to share files and computing resources. An example of a P2P service would be that provided by Napster, where individuals can access and download music directly off the PCs of users around the globe. P2P services enable people to find and collaborate with each other at will, thereby decentralizing control; promoting flexibility, diversity, and collaboration; and increasing efficiency. This evolution ultimately will break down the inherent barriers of geography and time, standardizing communication and access protocols.[2]

We believe that with the shift of intelligence to the edge and the emergence of P2P computing, successful network providers also will have to provide basic infrastructure services to support the content- and applications-based network of the future. One emerging standard for building P2P applications across the Internet is the web-services paradigm. Web services provide the opportunity for customers to invoke discrete functions advertised and offered anywhere on the Internet using open standards. Web services can themselves aggregate other web services. Customers could configure and combine web services to deliver business benefit without requiring huge investment in IT infrastructure. The network provider will most likely host a number of these services but its key added value is a smart directory that incorporates a number of capabilities presently available only from independent software vendors (ISVs). These capabilities will include personalization, multi-tenancy, delegated security domains, simple business process automation services that allow the combination of a number of web services by a distributed state-management engine, and a transactional infrastructure that extends the application server environments to cover web services from multiple providers. Underpinning all of these capabilities will be the more traditional functions of a directory server — location transparency, availability, interface publication, and so on.

Web services are based on a combination of two standards: simple object access protocol (SOAP) and universal description discovery and integration (UDDI). This emerging technology is backed by established industry players such as Microsoft and IBM and has the potential to transform the way in which network providers must address the edge and last-mile tiers.

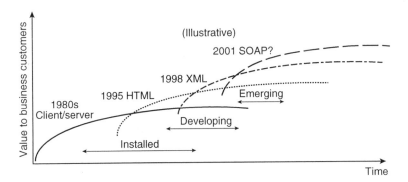

Figure 2. Recognizing the potential value of new technology to the telecom industry.

As seen in Fig. 2, hypertext mark-up language (HTML) made packet-switched networks a viable communications platform in 1995. In 1999, extensible mark-up language (XML) extended HTML and transformed these networks into a viable place to do business. In the next two years, SOAP will extend XML and transform packet-switched networks into a practicable computing platform. SOAP extends traditional component-based computing paradigms to the Internet. End customers will be able to offer business functionality and benefit across relatively low bandwidth connections by advertising fine granularity services.

The UDDI technology, an XML-based specification for a directory of businesses and the services they provide on the Internet, will allow users to locate and invoke these services. UDDI messages ride on top of the SOAP protocol. The standard is designed to enable software to discover services on the Web and automatically integrate with them by providing the necessary translations. UDDI capability depends on a network of UDDI servers, similar to the domain name system (DNS) network. Most current implementations are based on X.500 directory technology.

Edge providers will host UDDI directories as well as SOAP services of their own, adding significant value by providing carrier-grade aggregation, and translation and transactional services on top of the base services offered by customers. An interesting side note is the recent announcement by Microsoft of its intention to support the platform for privacy preferences (P3P) standard from the World Wide Web consortium (W3C) and to incorporate support for it into Internet Explorer version 6.[4] This type of preference-based service advertisement and usage is central to the concept of smart directory-enabled web services.

9. Change required to meet the new challenges

The problem facing network providers today is how to transform themselves to meet the challenges of the stupid network. The packet-switched network allows functionality to be defined on an individual customer basis. Smart edge devices will allow customers to define their own services and these services will require finer gradations of quality of service and flexibility from the network. Telecom companies must re-engineer themselves to deliver new applications and rich content to customers one-to-one rather than *en masse*. This change will require new network infrastructure and new business models, hence a change in the way network providers will operate and invest.

Network providers have the opportunity to generate revenue by tailoring and extending their management domains out beyond the edge and on to customer sites. They will need to manage and deliver the last mile and manage the enterprise user's own private network. This change may even require network providers to host the enterprise customer's web-services directory.

The network provider organization must shift to accommodate the new marketplace. Class of service will begin to apply not just to connectivity but also to web services and directories. Service-level agreements (SLAs) will also move up the value chain and apply to service availability, latency, and scalability. Connectivity will rely much more on partnership agreements and time-to-provide will become even more important.

In an environment where fine-grained web services become a major traffic type across the network, network providers will not be immune to the demand for a "customer of one" approach to service provision. Network providers will themselves need to take advantage of these emerging technologies to tailor services for individual customers on a low-cost and rapidly configurable basis. They will need to serve their own customers from a component-based set of services. In order to manage cost, it is essential that network providers architect their own building block services to promote reuse across customers while still presenting a unique solution for each customer.

Unsurprisingly, the boundary between provision of connectivity and provision of value-added services will blur even more than it has to date. Skill sets within network providers will need to move up to the application layer.

10. Vulnerability of the edge service providers

Assuming that over time, the last-mile problem will be solved by a combination of better access technologies and smarter use of bandwidth by applications, established network providers are well placed to win back ground from existing edge players. The majority of existing edge service providers make their margins by addressing performance, storage, reliability, and resilience issues associated with web-centric traffic. As the McKinsey–J.P. Morgan IP report shows, the future mix of traffic types will change in favor of higher value enterprise-to-enterprise, peer-to-peer, and multimedia and streaming traffic.[3] A smart network provider that acquires appropriate edge players today will be positioned to maximize the opportunity presented by the emerging last-mile solutions, particularly in the transition from CDN solutions that address static web pages to those that address streaming and multimedia. At present, although edge providers are able to add significant value to the Internet experience, we have not yet reached the critical mass of content and applications necessary for the Internet to transform the way individuals and enterprises work. Many edge service providers are experiencing lower than projected traffic and hence revenue. This situation has weakened their balance sheets, making them vulnerable to take-over by providers with deeper pockets, a longer-term view of the synergies to be expected between ownership of both core and edge, and the ability to create integrated services that sit at the edge and leverage the quality-of-service capabilities of the core.

11. Conclusion

The present economics of the industry imply a consolidation of network and edge service providers. The industry is refocusing investment toward higher value enterprise traffic and to accommodate the anticipated growth in multimedia and streaming IP traffic. Network providers with strong enterprise outsourcing skills that can build or acquire new infrastructure capabilities will be in a strong position to deliver end-to-end service for enterprise customers. Network providers must move to Internet-based open standards and embrace new developments such as web services and directories that will be key enabling technologies for building the application-based network of the future. Network providers must focus on early establishment of the key building blocks for this new era of directory-enabled applications. Components must be built that can be configured by the end customer but re-used across multiple users, with ubiquitous low-latency connectivity through the packet-switched backbone and granular directory-based integration. Network management domains must extend up into the application layer and out to the end user, providing seamless cross-domain knowledge of core, edge, and the last mile.

References

1.   David Isenberg, "The rise of the Stupid Network: Why the Intelligent Network was once a good idea, but isn't anymore. One telephone company nerd's odd perspective on the changing value proposition," *Computer Telephony*, August (1997), pp. 16-26; see also www.isen.com.
2.   Jonathan Hare, *P2P: Enabling Business Processes in a Global Marketplace*, Berkeley: Consilient Inc., 2000; see also www.consilient.com.

3. McKinsey–J. P. Morgan report, "Industry analysis – IP! How changes in the Internet are disrupting the telecom services industry", New York: J. P. Morgan Securities Inc, 2001.
4. W3C, P3P Specification 2001, see http://www.w3.org/P3P/.

Rare Earth Metal Oxides as High-κ Gate Insulators for Future MOSFETs

H. Iwai, S. Ohmi, S. Akama, A. Kikuchi, I. Kashiwagi,
C. Ooshima, J. Taguchi, H. Yamamoto, C. Kobayashi,
K. Sato, M. Takeda, K. Oshima, and H. Ishiwara
Tokyo Institute of Technology
4259 Nagatsuta Midori-ku, Yokohama 226-8502, Japan

1. Introduction

Development of a high-κ gate insulator is certainly the most urgent issue among the necessary technology items for the next generations of scaled CMOS, and many organizations in the world are now seriously engaged in this research. Among the high-κ dielectric materials, ZrO_2 and HfO_2 so far have been regarded as the most promising candidates. Excellent results of high-performance MOSFETs with small gate leakage current have been reported. However, problems such as interfacial layer growth and micro-crystal formation during the thermal process have been delineated as the materials come to be applied in the CMOS fabrication process. Although some good solutions to the problems have been shown,[1-3] it is now expected that a longer development period will be necessary.

On the other hand, rare earth metal oxides are possible alternatives to ZrO_2 or HfO_2 as shown in Fig. 1.[4] However, there have been only a few experimental reports of such oxides as the gate insulator in MOSFETs — La_2O_3,[5] Pr_2O_3,[6] and Gd_2O_3.[7] The purpose of this contribution is to report the first results of the examination of various rare earth metal oxides for future CMOS gate insulator applications.

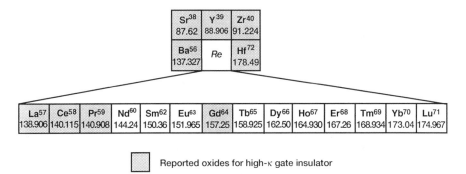

☐ Reported oxides for high-κ gate insulator

Figure 1. Candidates for metal oxide high-κ gate insulator.

2.   Experimental

Ultrahigh-vacuum physical vapor deposition (PVD) equipment with a base
pressure of 10^{-10} torr was used for the deposition of the high-κ films on n-type
Si(100) substrates, as shown in Fig. 2.   Pressure during the deposition was
10^{-9}~10^{-7} torr.   The ultrahigh vacuum is important to suppress the formation of
interfacial layers.   Four targets as evaporation sources of the high-κ material can
be installed for simultaneous deposition.   In these experiments, metal oxides are
used as the source.   PVD has advantages over CVD in terms of film purity.   The
metal oxide targets were heated by electron beam radiation for the evaporation.
The deposition by evaporation is advantageous over sputtering in terms of Si
substrate damage.

Figure 3 shows the sample fabrication flow for MOS capacitors.   Figure 3(a)

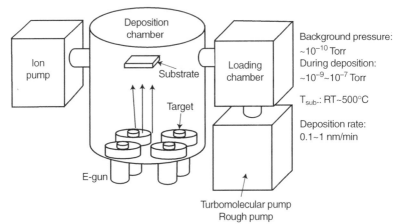

Figure 2.   Molecular beam deposition chamber equipped with 4 electron guns.

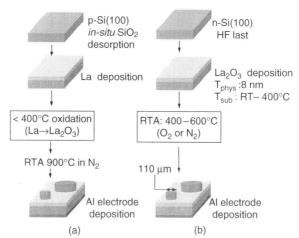

Figure 3.   Al/La$_2$O$_3$/Si(100) capacitor fabrication processes: method reported by
Chin *et al.*[5] (a); our method (b).

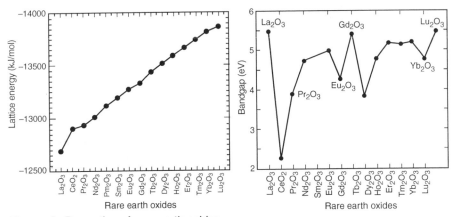

Figure 4. Properties of rare earth oxides.

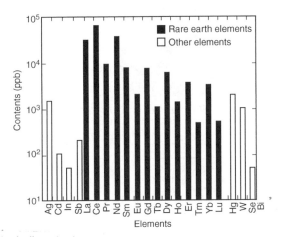

Figure 5. Earth shell contents.

shows the La_2O_3 thin film deposition method from the literature[5] and Fig. 3(b) shows our case for rare earth metal oxides. In the case reported in the literature, a thin La metal film was first deposited on a *p*-type Si substrate and then selectively oxidized at a low temperature (below 400 °C) to form amorphous La_2O_3. In our case, amorphous rare earth metal oxide films themselves were deposited on *n*-type Si. Most of the deposited films were subsequently annealed by rapid thermal annealing (RTA) at 200–600 °C in O_2 or N_2 for 5 min. Then, aluminum gate electrodes (110 μm) were formed by evaporation. We carried out *C–V*, *J–V*, AFM and cross-sectional TEM measurements to characterize these films.

Among the rare earth metals, oxides of La, Pr, Eu, Gd, Yb, and Lu were chosen as the first group for this experiment. These oxides have a variety of bandgap and lattice energy values as shown in Figs. 4(a) and 4(b), respectively. Thus, the oxides are expected to show quite different characteristics in both electrical and chemical properties such as leakage current and moisture absorption.

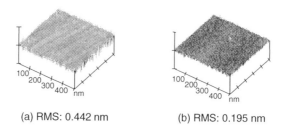

(a) RMS: 0.442 nm   (b) RMS: 0.195 nm

Figure 6. Typical AFM images for La_2O_3/n-Si(100) deposited at room temperature (a) and 250 °C (b). Scan area is 0.5 x 0.5 μm^2; z scale: 5 nm/div.

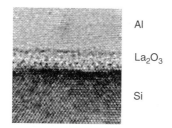

Al

La_2O_3

Si

Figure 7. Cross-sectional TEM image of La_2O_3/n-Si(100) deposited at 250 °C and annealed at 600 °C in O_2.

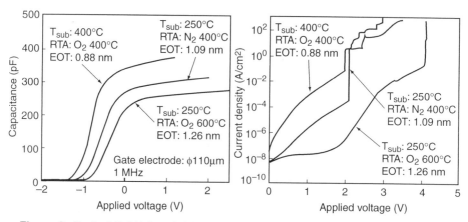

Figure 8. Typical $C-V$ (a) and $J-V$ (b) characteristics for La_2O_3/n-Si(100).

Figure 5 shows the comparison of earth shell contents of the rare earth metals to those of some other metals. It should be noted that the rare earth metals' abundances are sufficiently large compared with other important elements used in industry. In fact, they are larger than those of In, Bi, Sb, Cd, In, Hg, Ag, Se, Pt, and Au. Figure 6 shows the typical AFM images of films deposited at room temperature and 250 °C. It is found that the surface morphology is significantly improved by deposition at higher temperature. An RMS roughness of 0.195 nm is almost comparable to that of thermally grown SiO_2 on Si. Figure 7 shows a cross-sectional TEM image for La_2O_3/n-Si deposited at 400 °C followed by a 400-°C O_2 rapid thermal anneal. A smooth interface between the La_2O_3 and the Si is confirmed. No clear interfacial layer is observed. Figure 8 shows typical high-frequency C–V and J–V characteristics for La_2O_3 MOS capacitors. By changing the process conditions, leakage currents of 5.5×10^{-4} A/cm^2 (equivalent oxide thickness, EOT = 0.88 nm), 1.9×10^{-6} A/cm^2 (EOT = 1.09 nm), and 1.7×10^{-8} A/cm^2 (EOT = 1.26 nm) were obtained. Figure 9 shows the C–V frequency dependence. It should be noted that there is no frequency dependence at least to 1 MHz.

Figure 10 shows EOT $vs.$ J_G (gate leakage current) plots for MOS diodes obtained in our experiments (filled symbols) in comparison with published data

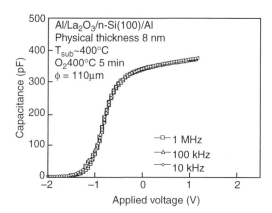

Figure 9. Frequency dependence of the C–V curve.

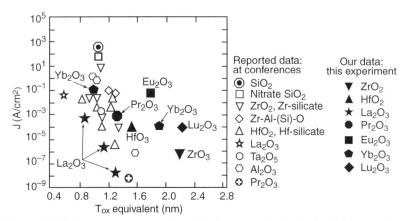

Figure 10. Gate leakage current density J_G $vs.$ equivalent oxide thickness T_{ox}.

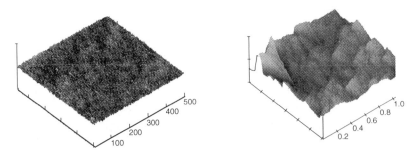

Figure 11. AFM images of as-deposited films at 250 °C: Eu_2O_3 (left; 0.5 x 0.5 μm², z scale: 5 nm/div) and Lu_2O_3 (right; 1.2 x 1.2 μm²; z scale: 5 nm/div).

Material	Temp. (°C)	Material	Temp. (°C)
La_2O_3	> 700	Yb_2O_3	>250
Pr_2O_3	>600	Lu_2O_3	>250
Eu_2O_3	>700		

Table 1. Crystallization temperature.

(open symbols) measured in the accumulation condition with $V_G = 1$ V. Among the rare earth metal oxides, La_2O_3 shows superior characteristics, suggesting little interface layer growth, good SiO_2-like properties of the interface layer, and small film roughness. In fact, the La_2O_3 data as shown in the figure are the best ones ever published. Regarding the EOT *vs.* J_G properties of other rare earth metal oxides, unfortunately they are not as good as those of La_2O_3 at this moment. However, there is a possibility that excellent results will be obtained in near future through optimization of the process conditions.

For example, Fig.11 shows AFM images for as-deposited Eu_2O_3 and Lu_2O_3 samples. Lu_2O_3 is relatively easily crystallized and forms rough polycrystalline films even at lower temperature, while Eu_2O_3 films maintain a smooth amorphous structure. Table 1 shows examples of the substrate temperature conditions for amorphous and crystal phases for different rare earth metal oxides. The temperature conditions are expected to change depending on the Si substrate conditions — whether the Si surface before the deposition is bare or covered with some oxide. Different optimizations for different rare earth metal oxides are necessary, but these refinements are still underway.

Moisture absorption and the resulting film degradation are the biggest concerns with the rare earth metal oxides at this time. Moisture absorption tests were carried out. Figures 12 and 13 show film characteristics before and after the test. The amount of degradation changes significantly among different rare earth

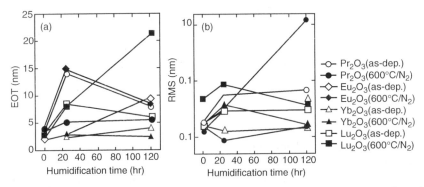

Figure 12. EOT (a) and leakage current density at +1 V (b) as a function of humidification time.

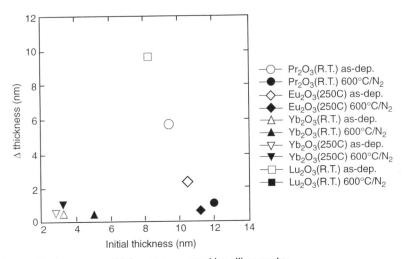

Figure 13. Increase in thickness measured by ellipsometry.

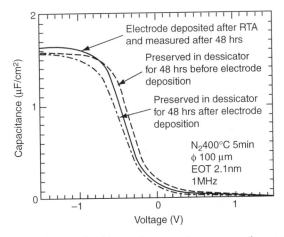

Figure 14. Pr₂O₃/p-Si(100) C–V dependence on the preservation conditions.

metal oxides. We found that prior annealing of the film considerably suppresses the degradation. Figure 14 shows that the degradation is almost completely suppressed for the film preserved in a low-pressure dessicator (~100 torr) or covered by a gate electrode.

3. Conclusions

Electrical and physical characteristics of rare earth metal oxides are at least comparable to those published for ZrO_2 and HfO_2. In the case of La_2O_3, the EOT vs. J_G characteristics were found to be even better. The degradation by moisture absorption is suppressed by gate electrode coverage of the films as well as by preserving the films in a dessicator. These results are only primitive-stage experiments. Rare earth metal oxides are certainly worth further investigation as high-κ gate dielectrics for future CMOS applications.

Acknowledgments

This work was partially supported by Semiconductor Technology Academic Research Center (STARC) and Grant-in-Aid for Scientific Research Priority Areas (A): Highly Functionalized Global Interface Integration. The authors would like to express sincere thanks to DENSO, Toshiba, Hitachi, NEC, and Sanyo for their technical support for the experiment. They are grateful to Drs. N. Nakayama, T. Arikado, J. Yugami, T. Kitano, and K. Fujita, Y. Tsunashima and K. Hara, Y. Horiike, and K. Masu for their useful discussion and advice for this research.

References

1. L. Manchanda et al., "Si-doped aluminates for high temperature metal-gate CMOS: Zr-Al-Si-O, a novel gate dielectric for low power applications," Tech. Digest IEDM (2000), p. 23.
2. C. H. Lee et al., "MOS characteristics of ultrathin rapid thermal CVD ZrO_2 and Zr silicate gate dielectrics", Tech. Digest IEDM (2000), p. 27.
3. B. H. Lee et al., "Characteristics of TaN gate MOSFET with ultrathin hafnium oxide (8–12 Å)," Tech. Digest IEDM (2000), p. 39.
4. J. R. Hauser, IEDM Short Course (1999).
5. A. Chin et al., "High quality La_2O_3 and Al_2O_3 gate dielectrics with equivalent oxide thickness 5–10 Å," Digest VLSI Tech. Symp. (2000), p.16.
6. H. J. Osten et al., "High-κ gate dielectrics with ultra-low leakage current based on praseodymium oxide," Tech. Digest IEDM (2000), p. 653.
7. J. A. Gupta, D. Landheer, J. P. McCaffrey, and G. I. Sproule, "Gadolinium silicate gate dielectric films with sub-1.5 nm equivalent oxide thickness," Appl. Phys. Lett. 78, 1718 (2001).

Ultra-Thin Single- and Double-Gate MOSFETs for Future ULSI Applications: Measurements, Simulations, and Open Issues

David Esseni
DIEGM, University of Udine, Udine, 33100 Italy

Claudio Fiegna
Dept. of Engineering, University of Ferrara, Ferrara, 44100 Italy

Marco Mastrapasqua
Agere Systems, Murray Hill, NJ 07974 U.S.A.

1. Introduction

CMOS technology represents the backbone of today's semiconductor industry and the enabler for the implementation of complex systems on single chips. The ITRS Roadmap is clearly pushing CMOS technology toward its limits to fulfill extremely demanding requirements even for near-term products.[1] The scaling of the MOS transistor to physical channel lengths below 70 nm is projected to occur by the years 2004–2005, despite the fact that the shrinkage of the bulk MOSFET to this gate-length range presents some issues that have no known solutions.[1]

In particular, in order to preserve satisfactory behavior of the bulk MOSFET, the ratio between the polysilicon gate length L_{poly} and the oxide thickness T_{ox} should not drop below approximately 50.[1-4] That constraint results in an equivalent T_{ox} around 1.0–1.5 nm for L_{poly} below 70 nm. Most probably such a small equivalent T_{ox} will not be attainable with standard silicon dioxide (SiO_2) because of excessive gate leakage and/or because of reliability limitations that are still actively debated.[5-7] Though new high-κ dielectrics could possibly replace SiO_2 as gate insulators in the future, oxide scaling below an equivalent T_{ox} of approximately 1.3 nm is presently an issue with no known solutions.[1]

Besides aggressive oxide scaling, the bulk MOSFET also demands channel doping concentrations larger than 10^{18} cm^{-3} to suppress short-channel effects (SCE) for L_{poly} below 100 nm. The statistical fluctuations of the doping in the channel introduce an appreciable dispersion of threshold voltage (V_T), which is an important fundamental limitation on bulk MOSFET design for extremely short channel lengths. Furthermore, a large channel doping increases the source-drain junction capacitances (that are comparable to the channel capacitance in scaled MOSFETs) and reduce the effective mobility (μ_{eff}), thus degrading the device performance for a given L_{poly}.[8]

Finally, in bulk MOSFETs the scaling of source-drain (S/D) junction depth is another essential ingredient to suppress SCE.[9] The roadmap predicts that S/D junctions shallower than approximately 30 nm will be necessary for L_{poly} around 70 nm, but the realization of reliable low-resistance junctions that shallow is again recognized as an issue with no known solutions. In a relatively long-term future it is possible that novel electron devices based on quantum-mechanical principles and featuring truly nanometer geometrical dimensions can challenge and take over CMOS. However, silicon technology has so far mainly relied on evolutionary solutions, hence device structures related to the conventional MOSFET concept deserve the maximum attention.

In this perspective, fully-depleted silicon-on-insulator (FD-SOI) transistors realized on ultra-thin (UT) silicon films are very promising candidates for the design of the ultimate MOSFET because, when the silicon thickness (T_{Si}) is reduced below 10 nm, SOI transistors exhibit excellent scalability even with essentially undoped channels.[10,11] Numerical simulations have shown that the double-gate (DG) MOSFET on UT silicon layers is probably the most scalable design for the MOS transistor.[12] In particular, UT-SOI MOSFETs can reduce the ratio between L_{poly} and the equivalent T_{ox}, thus relaxing the very critical requirement on oxide scaling of bulk MOS devices. For example, for $T_{Si} = 10$ nm a gate length of 50 nm seems feasible with $T_{ox} = 3$ nm.[13] Furthermore, undoped silicon films eliminate the issue of statistical doping fluctuations and result in lower effective fields[14] for a given inversion density, with possible μ_{eff} improvements over the bulk MOSFET. Moreover, T_{Si} inherently limits the depth of S/D junctions and the corresponding capacitances are greatly reduced.

It is apparent that in UT-SOI devices the silicon film thickness is a key design parameter for the scaling of the transistor. In this contribution we will discuss the dependence on T_{Si} of mobility, threshold voltage and device performance, presenting both experimental data and numerical simulations. We will also point out design and modeling challenges that should be addressed effectively to take full advantage of this very interesting transistor concept.

2.   Low-field mobility

Low-field effective mobility μ_{eff} is a very important electrical parameter for the current-drive capability of deep-submicron MOS transistors because a large mobility allows carrier velocity in the channel to exceed the bulk-silicon saturation value,[15,16] thus approaching the ballistic transport regime.[17–20] In this section, μ_{eff} of UT-SOI transistors is studied for T_{Si} down to approximately 5 nm.

- *Experimental procedure*
   The experimental determination of μ_{eff} is a relatively standard procedure performed on a long-channel transistor, consisting of the determination of

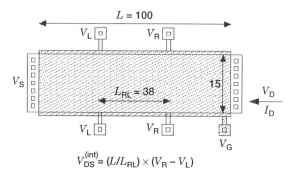

$$V_{DS}^{(int)} = (L/L_{RL}) \times (V_R - V_L)$$

Figure 1. Schematic top view of a UT-SOI MOSFET used for μ_{eff} measurements. The voltage probes V_R and V_L allow extraction of the intrinsic drain to source voltage $V_{DS}^{(int)}$ and conductance $g_D^{(int)}$ (dimensions L, L_{RL}, and W are in μm).

the inversion density N_{inv} (using gate-to-channel capacitance measurements) and of the intrinsic conductance $g_D^{(int)}$. Then the mobility is calculated simply as $\mu_{eff} = (L/W) \cdot g_D^{(int)}/qN_{inv}$ where L and W are device length and width, respectively. In UT-SOI transistors an issue arises for the determination of $g_D^{(int)}$ because when T_{Si} is very small and no elevated S/D structures are employed, the series resistance of the extremely shallow junctions can artificially reduce the device conductance. In order to circumvent this possible issue the μ_{eff} measurements discussed in this paper were obtained on the test structure illustrated in Fig. 1 that allows a determination of $g_D^{(int)}$ independent of series resistance.[21]

The T_{Si} values indicated in the remainder of this paper were experimentally determined using the characterization procedure based on $C–V$ measurements discussed in Ref. 22. A good correlation was found between the electrical T_{Si} determination and the values obtained using ellipsometric measurements.[21]

- *Single-gate SOI mobility*

Figure 2 illustrates electron μ_{eff} obtained from our samples compared to previously published bulk MOSFET data.[8] Mobility is plotted *vs.* the inversion density N_{inv} because for current drivability it is important to obtain the largest possible μ_{eff} for a given N_{inv}. From an applications viewpoint, it is interesting to compare μ_{eff} of different MOS designs at the same N_{inv}.

As can be seen, at large N_{inv} the mobility of SOI devices is largely independent of T_{Si} and consistent with μ_{eff} of the lightly-doped bulk MOSFETs. Since the heavily doped bulk transistor has a larger effective field than the UT-SOI device for a given N_{inv} due to a large depletion charge,[14] a remarkable μ_{eff} improvement can be obtained with SOI devices at large N_{inv}.

At lower inversion densities, however, μ_{eff} of SOI devices becomes more sensitive to T_{Si} and it is reduced with decreasing T_{Si}. The much larger μ_{eff}

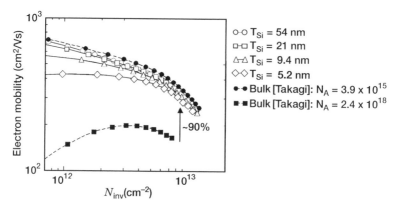

Figure 2. Measured room-temperature electron mobility versus inversion charge density N_{inv} for different T_{Si}. A comparison with previously published data for bulk transistors[8] is illustrated.

reductions for T_{Si} below 10 nm that were previously reported in the literature might be due to poor quality of the thin silicon films.[23] A similar qualitative behavior was also observed for the hole mobilities that are reported in Fig. 3, but a weaker μ_{eff} dependence on T_{Si} was found.

It also has been proposed that the T_{Si} dependence of electron μ_{eff} in UT-SOI transistors can be due to additional scattering with the back interface. When SOI wafers are obtained with the SIMOX process a poorer quality of the back interface may be expected, however the Smart-Cut wafers featuring a thermally grown back interface are used in this work. Figure 4 reports the mobility at the front interface for different back-gate voltages V_{BG} (open symbols) versus a properly defined effective field accounting for the back-gate bias.[23,24] In the same figure we also report μ_{eff} measured when the inversion layer is induced at the back interface (filled symbols) for two different values of the front-gate voltage V_{FG}. As can be seen, essentially the same mobility *vs.* effective field curve is observed at the two interfaces indicating that in our samples, back-interface quality should not be the dominant cause for μ_{eff} reduction at small T_{Si}.

Besides possible non-idealities of either the silicon film or the back interface, however, a μ_{eff} dependence on T_{Si} also is expected because carrier confinement in real space is known to enhance the phonon-scattering rate because of the corresponding delocalization in reciprocal k-space.[26,27] Figure 5 shows a comparison between the experimental μ_{eff} data of this work and the numerical simulations discussed in Ref. 25. In the simulations, the mobility dependence on T_{Si} at low N_{inv} is due to the above-mentioned increase of the phonon scattering. As can be seen, the experimental mobility degradation when T_{Si} is reduced from approximately 50 nm to 10 nm is reproduced fairly well by simulations.

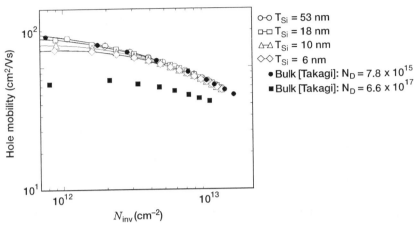

Figure 3. Measured room-temperature hole mobility versus inversion charge density N_{inv} for different T_{Si}. A comparison with previously published data for bulk transistors[8] is illustrated.

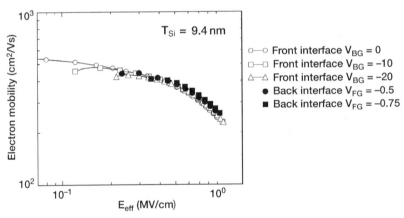

Figure 4. Measured room-temperature electron mobility versus the effective field E_{eff} at either the front interface (open symbols) or the back interface (filled symbols) for an SOI transistor with $T_{Si} = 9.4$ nm.

Furthermore, Fig. 6 reports μ_{eff} for $N_{inv} = 2\times10^{12}$ cm^{-2} as a function of the absolute temperature. In the explored temperature range between −50 and +100 °C, μ_{eff} exhibits an approximately $T^{-1.4}$ dependence that is close to the $T^{-1.7}$ dependence observed in bulk MOSFETs for phonon-limited mobility.[8]

Both the comparison with numerical simulations (Fig. 5) and the temperature dependence of mobility in the low N_{inv} range (Fig. 6) suggest that a modulation of the phonon-scattering rate could play a key role in the mobility dependence on T_{Si}.

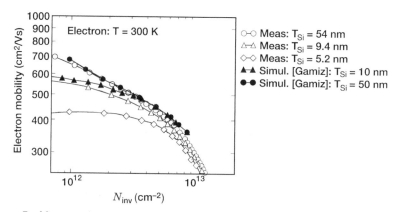

Figure 5. Measured room-temperature electron mobility *vs.* the inversion charge density N_{inv} compared with the numerical simulations reported in Ref. 25.

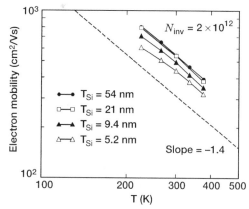

Figure 6. Measured electron mobility *vs.* temperature for different T_{Si} and for $N_{inv} = 2 \times 10^{12}$ cm^{-2}. A power law with an exponent of approximately -1.4 is found in the temperature range explored.

- *Double-gate SOI mobility*

Since Fig. 4 indicates that the back interface in our samples has a quality comparable to the front interface, it is very interesting to study μ_{eff} in a double-gate (DG) operation mode compared to the mobility in the single-gate (SG) mode considered so far.

In order to measure μ_{eff} in DG mode in our samples, which feature a back oxide that is much thicker than the front oxide (*i.e.* 400 nm for the former and 4.5 nm for the latter), we devised an experimental technique to determine the inversion charge in the transistor for arbitrary V_{FG} and V_{BG} values. Then, numerical simulations that account for subband quantization and poly-Si depletion are used to identify, for each V_{FG} value, the V_{BG} value that results in a symmetric electron density in the silicon film.

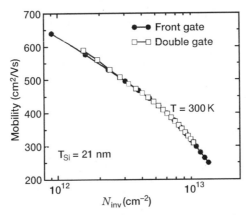

Figure 7. Measured electron mobility *vs.* inversion charge density N_{inv} in either SG or DG mode for T_{Si} = 21 nm. Virtually the same mobility is found in the two operating modes. In DG mode, N_{inv} is one half of the total inversion density.

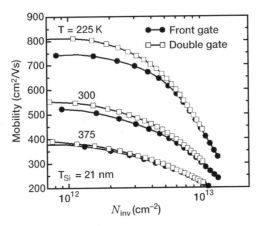

Figure 8. Measured electron mobility *vs.* inversion charge density N_{inv} in either SG or DG mode for T_{Si} = 9.4 nm. A clear improvement in μ_{eff} is observed in the DG mode at low N_{inv}. This effect is more pronounced for decreasing temperatures. In DG mode, N_{inv} indicates one half of the total inversion density.

Using the above procedure, we could measure μ_{eff} for the same device operated either in SG or DG mode, and the results are illustrated in Figs. 7, 8 and 9 for different silicon thicknesses (note that N_{inv} in the DG mode is one half of the entire inversion density). As can be seen for T_{Si} = 21 nm, essentially the same mobility is observed in DG and SG mode and we know from Fig. 2 that μ_{eff} is also consistent with lightly-doped bulk MOSFETs. On the other hand, for the two

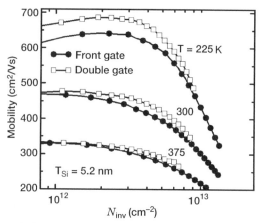

Figure 9. Measured electron mobility *vs.* inversion charge density N_{inv} in either SG or DG mode for T_{Si} = 5.2 nm. A clear improvement in μ_{eff} is observed in the DG mode at low N_{inv}. This effect is more pronounced for decreasing temperatures. In DG mode, N_{inv} indicates one half of the total inversion density.

thinnest silicon films and moderate inversion densities, a modest but systematic and unambiguous increase of μ_{eff} in the DG operation mode is observed. The increase for decreasing temperatures suggests a modulation of the phonon-scattering rate that has been predicted recently by numerical simulations.[28]

3. Threshold voltage and its dependence on silicon thickness

In Section 2 we discussed the dependence of the low-field carrier mobility in UT-SOI devices on the thickness of the silicon layer T_{Si}. Such a dependence is large in the case of low inversion-charge density (low-to-moderate gate bias) and may be ascribed to the modulation of phonon-scattering rates by T_{Si} as a result of increased carrier confinement at decreasing silicon thicknesses. Increased quantization at decreasing T_{Si} also affects the device threshold voltage V_T (the gate voltage needed to induce a given inversion-charge density) because it is strictly related to the 2D subband structure in the inversion layer and, in particular, to the energy eigenvalues in the inversion layer and to the occupation of the corresponding subbands.

Figure 10 reports experimental and simulated threshold voltages as a function of T_{Si} for the same devices considered in the previous section. The threshold voltage is too low for digital applications and it depends on T_{Si}. Regarding the control of V_T, it must be mentioned that an increase in silicon doping would have little effect, due to the limited amount of depletion charge that would be available, especially for very thin silicon layers. For this reason, in order to get acceptable

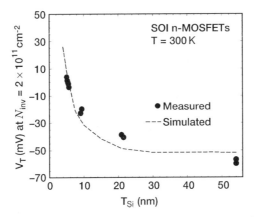

Figure 10. Dependence of threshold voltage on the thickness of the silicon layer for single-gate SOI MOSFETs (simulation shown by a dashed line). Threshold voltage is defined as the gate voltage necessary to obtain $N_{inv} = 2\times10^{11}$ cm^{-2}.

V_T values, doping concentrations well in excess of 10^{18} cm^{-3} would be required, leading to unacceptable mobility degradation. Thus, it will be necessary to control the threshold voltage by adopting unconventional gate materials with almost mid-gap Fermi levels.

It is evident from Fig. 10 that the experimental V_T increases as T_{Si} is reduced below approximately 10 nm. This result is confirmed by the self-consistent solution of the 1D Schrödinger and Poisson equations using the simulation tool adopted in Ref. 29. A detailed analysis of the simulation results indicates that the increase of V_T with decreasing T_{Si} is closely related to the confinement-induced increase of the energy corresponding to the lowest transverse (or primed) subband. This subband, in fact, has a lower effective mass in the direction of confinement than longitudinal (or unprimed) subbands ($m^* = 0.19m_0$ instead of $0.916m_0$) that results in a stronger dependence of the corresponding eigenvalues on the silicon film thickness. Furthermore, due to the larger multiplicity and the larger density-of-states effective mass, the lowest transverse subband contributes significantly to the inversion density when the device is biased at $V_G = V_T$. More precisely, the lowest transverse energy eigenvalue at threshold increases by about 58 meV when T_{Si} scales from 30 nm to 5 nm and the corresponding subband accounts for 42% of the total inversion density for $T_{Si} = 30$ nm and $V_G = V_T$.

The modulation of the energy eigenvalues by T_{Si} is particularly effective when the devices are biased below or close to the threshold condition. On the other hand, this modulation is expected to be suppressed as the gate bias is increased, reaching the strong inversion condition for which the confinement of inversion charge tends to be dominated by the gate-induced electric field, rather than by the width of the quantum well associated with the oxide-silicon-oxide system. Nonetheless, the dependence of V_T on T_{Si} will be a major issue for the implementation of UT-SOI technologies, because in order to suppress V_T fluctuations, very good control and reproducibility of T_{Si} will be required.

4. Implications for SOI device performance

In this section we report the results of device simulations in order to clarify the pros and cons of UT-SOI MOSFETs compared to more conventional bulk counterparts. To this purpose, the intrinsic performance of UT-SOI structures suitable for the realization of 70-nm gate length MOSFETs has been compared with that of bulk n-MOSFETs by means of hydrodynamic device simulations.[30] The main technological parameters used for the two transistor designs are summarized in Table 1. Series resistance effects have been deliberately minimized by assuming ideal contacts extended over the entire source/drain regions with the exception of the under-diffusion below the gate. A mid-gap Fermi level gate material is assumed for the thin-SOI devices in order to achieve an acceptable threshold voltage. Quantization effects are accounted for with the model of Van Dort et $al.$[31] The accuracy of this approach in reproducing the inversion charge of SOI devices was verified by comparing simulated to experimental C–V characteristics of the large-area transistors used for mobility characterization.

As for mobility, the model for μ_{eff} degradation at the Si-SiO$_2$ interface proposed in Ref. 32 has been adopted, with slightly modified parameter values in order to improve the fit to the experimental μ_{eff} of heavily-doped bulk devices[8] (N_A = 7.7×10^{17} – 2.4×10^{18} cm^{-3}) up to high N_{inv}. Furthermore, in the case of the 5.2-nm SOI device, the low-field maximum mobility was reduced to 1000 cm^2/V·s to match our experimental results. The results of these empirical adjustments are shown in Fig. 11, illustrating that the experimental mobility increase of SOI devices is quantitatively reproduced by simulations.

This simulation approach is believed to provide realistic results for MOSFETs with L = 70 nm because first-order non-stationary effects such as carrier heating and velocity overshoot are accounted for. Ballistic effects (occurring when the channel length becomes comparable to or shorter than the electron mean free path) are not accounted for in these simulations. However, this lack should not cause relevant errors even for the shortest simulated devices: in fact, the calculated velocities reported in the following do not exceed the upper limit set by quasi-equilibrium diffusion from the source in the ballistic regime.[17,20]

Figure 12 reports the dependence of threshold voltage on gate length for the

Parameter	T_{ox} (nm)	T_{Si} (nm)	T_{box} (nm)	N_{sub} (cm^{-3})	X_J (nm)
Bulk	1.5	—	—	2.4×10^{18}	20
SOI	1.5	5.2	400	1.0×10^{15}	5.2

Table 1. Main technological parameters of the SOI and bulk MOSFETs compared by means of 2D numerical simulations.

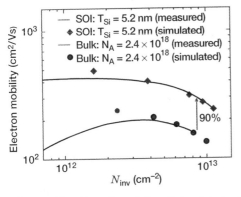

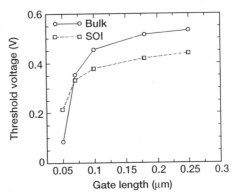

Figure 11. Experimental and simulated effective mobility *vs.* inversion-layer density for highly-doped bulk MOSFETs and thin-film single-gate SOI.

Figure 12. Dependence of V_T on gate length for single-gate (mid-gap gate work function) and conventional (highly doped *n*-poly gate) MOSFETs.

thin SOI and bulk MOSFETs; by adopting a thin lightly-doped silicon layer, the short-channel effects of SOI devices are lower compared to highly-doped bulk counterparts with the same gate oxide thickness.

The results of our simulations point out that, in order to compare the current drive of SOI and bulk structures, the effects of two counteracting factors must be considered. Figure 13 shows the simulated average electron velocity along the channel for the two devices biased at $V_{GS} = V_{DS} = 1.5$ V. The work functions of the two transistors were adjusted to give the same off current $I_{OFF} = 1$ nA/μm for $V_{DS} = 1.5$ V. As a consequence of the different 2D electric field patterns and different mobilities within the channel, different profiles of average electron velocity are obtained. In particular, the increased low-field μ_{eff} of SOI devices leads to a higher electron velocity near the source of the device. The lower velocity in the drain region, instead, is due to the smaller maximum longitudinal field at the drain end of the channel. However, the design of the SOI transistor with an essentially undoped silicon film implies that the centroid of the inversion layer is farther from the Si-SiO$_2$ interface than in the highly-doped bulk MOSFET. This electrostatic effect results in a degradation of the control capacitance of the SOI device for a given oxide thickness, which is illustrated in Fig. 14 and becomes progressively more severe when t_{ox} is reduced.

In the case of the 70-nm-long device simulated in this work the increase of the electron velocity at the source and the reduced control capacitance tend to cancel out. Figure 15 shows the simulated I–V characteristics of the two devices biased in the linear ($V_{DS} = 50$ mV) and saturation regimes ($V_{DS} = 1.5$ V); the comparison is made for the same off current ($I_{DS} = 1$ nA/μm at $V_{GS} = 0$) for both devices and both drain bias conditions. Thanks to a larger low-field mobility, the SOI MOSFET provides a larger linear current (especially at large gate overdrive — see Fig. 11). However, at high drain voltages, due to the large electric field in the channel, carrier transport deviates from the ohmic (linear) behavior, carrier

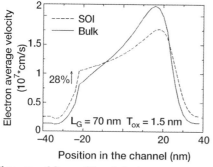

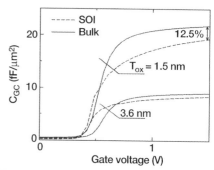

Figure 13. Simulated average electron velocity along the channel of thin-SOI and bulk $L_G = 70$ nm devices biased at $V_{GS} = V_{DS} = 1.5$ V (I_{OFF} is taken as 1 nA/μm for both transistors).

Figure 14. Simulated gate-to-channel capacitance C_{GC} for large thin-SOI and bulk devices; $T_{ox} = 1.5$ nm and 3.6 nm are representative of the 70-nm and 180-nm technology nodes, respectively.

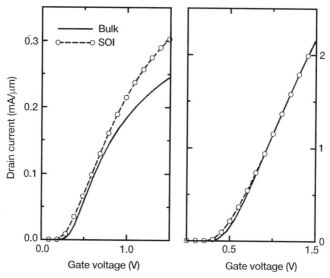

Figure 15. Transfer I–V characteristics for thin-film SOI and heavily-doped bulk MOSFETs with $L_G = 70$ nm at $V_{DS} = 50$ mV (left) and $V_{DS} = 1.5$ V (right). The comparison is made for the same $I_{OFF} = 1$ nA/μm at $V_{GS} = 0$ V for both devices and both bias conditions.

velocity is no longer simply proportional to low-field mobility, and essentially the same on current (*i.e.* I_D for $V_{GS} = V_{DS} = 1.5$ V) was found for the two structures. Therefore, within the predictive capabilities of the model, the maximum current drive capability of the two MOSFET designs turns out to be very similar.

5. Discussion and conclusions

In this paper, the main issues related to fully depleted SOI structures suitable for the implementation of high-speed circuits starting from the 100-nm technology node have been discussed on the basis of experimental results and numerical simulations. In particular, since the suppression of short-channel effects requires the adoption of extremely thin silicon layers that lead to relevant size-induced quantization effects, we investigated the impact of thinning the silicon layer to 10 nm and below.

The low-field mobility has been characterized in single- and double-gate UT-SOI MOSFETs. The results of accurate experimental characterization point out that at large inversion densities, both n- and p-type UT-SOI single-gate MOSFETs exhibit mobilities that are very similar to those of lightly-doped bulk transistors in the entire T_{Si} range. When compared to heavily-doped bulk devices suitable for deep sub-micron technologies, UT-SOI transistors can lead to a remarkable mobility improvement for a given inversion density because the negligible depletion charge of UT-SOI films results in a significantly smaller effective field. At relatively small N_{inv}, a systematic mobility reduction with decreasing T_{Si} is observed. The investigation of the mobility dependence on the back-gate bias suggests that in our samples the proximity of the back interface does not play a relevant role in the mobility reduction observed at the smallest T_{Si}. Instead, the experimental temperature dependence indicates that an increased phonon scattering due to spatial confinement in the silicon layer could be an important ingredient to explain the T_{Si} dependence of mobility, at least in the explored temperature range.

Operation of the devices in double-gate mode (symmetric inversion layer in the silicon film) has also been studied and the electron mobility of UT-SOI transistors operated in DG mode has been experimentally investigated for T_{Si} down to 5 nm and temperatures between 225 and 375 K. For $T_{Si} = 20$ nm, mobility is essentially the same in DG and SG modes. For thinner silicon films, however, at small N_{inv} we observed a clear μ_{eff} improvement in DG mode that increases for decreasing temperatures.

Experimental results and simulations indicate that the threshold voltage of UT-SOI MOSFETs strongly depends on T_{Si} due to the modulation of the subband structure in the inversion layer.

Conventional device simulation has been applied in order to estimate the impact on the on currents of the larger μ_{eff} in UT-SOI MOSFETs, compared to highly-doped bulk devices. In particular, the current drive of 70-nm MOSFETs was studied and the results indicate that, for a given I_{OFF}, the improved mobility and the degraded control capacitance of SOI transistors tend to compensate so that the saturation I_{ON} of SOI transistors could be approximately the same as for bulk MOSFETs.

In conclusion, based on the results of experiments and simulations of intrinsic MOSFETs, we expect that fully-depleted UT-SOI devices can provide significant improvements with respect to conventional bulk technologies in terms of short-channel effects and reduction of parasitic capacitances. On the other hand, the advantages in terms of low-field mobility are not likely to result in relevant I_{ON} improvements.

In order to make UT-SOI MOSFETs a viable solution for future ULSI, a few very important issues must be taken into consideration.

- Since the threshold voltage of UT single- and double-gate devices is almost independent of the doping in the silicon film, the control of the threshold voltage will necessarily require gate work-function engineering based on alternative gate materials with mid-gap Fermi levels.

- As for threshold voltage engineering and control, another important issue is given by the V_T dependence on silicon thickness. For silicon thicknesses below approximately 10 nm, accurate control and reproducibility of the silicon thickness is necessary to avoid unacceptable fluctuations of V_T.

- As for current drive capability, source-drain engineering is very critical for thin silicon films in order to achieve low series resistance. The use of non-planar elevated source/drain structures will most probably be necessary to access effectively the intrinsic transistor.

- From the point of view of modeling, the dependence of mobility on the thickness of the silicon layer is not correctly accounted for by available mobility models implemented in conventional device simulators used for technology development. For this reason, it will be necessary to develop a more accurate mobility model that consistently accounts for the effects of size-induced confinement that occurs in ultra-thin devices.

- Finally, given the relevance of quantization effects in the thin silicon film, the development of device-simulation tools able to account efficiently and consistently for these effects is mandatory.

Acknowledgments

The authors wish to acknowledge the collaboration with G. K. Celler, L. Selmi, and E. Sangiorgi.

References

1. International Technology Roadmap for Semiconductors: 1999. Austin, TX, SEMATECH, 1999.
2. R. H. Dennard, F. H. Gaensslen, L. Kuhn, N. Y. Yu, V. L. Ridout, E. Bassous, and A. LeBlanc, "Design of ion-implanted MOSFETs with very small physical dimensions," *IEEE J. Solid State Circuits* **9**, 256 (1974).
3. G. Baccarani, M. R. Wordeman, and R. H. Dennard, "Generalized scaling theory and its application to a 1/4 micrometer MOSFET design," *IEEE Trans. Electron Dev.* **31**, 452 (1984).

4. J. R. Brews, "The submicron MOSFET," Chapter 3 in: S. M. Sze, ed., *High-Speed Semiconductor Devices*, New York: Wiley, 1990, p. 139.
5. J. Stathis and D. J. DiMaria, " Reliability projection for ultra-thin oxides at low voltage," *Tech. Digest IEDM* (1988), p. 167.
6. B. E. Weir, M. A. Alam, J. D. Bude, P. J. Silverman, A. Ghetti, *et al.*, "Gate oxide reliability projection to the sub-2 nm regime," *Semicond. Sci. Technol.* **15**, 455 (2000).
7. R. Degrave, B. Kaczer, and G. Groeseneken, "Reliability: A possible showstopper for oxide thickness scaling?" *Semicond. Sci. Technol.* **15**, 436 (2000).
8. S. Takagi, A. Toriumi, M. Iwase, and H. Tango, "On the universality of inversion-layer mobility in Si MOSFETs. Part I: Effect of substrate impurity concentration," *IEEE Trans. Electron Dev.* **41**, 2357 (1994).
9. C. Fiegna, I. Iwai, T. Wada, M. Saito, E. Sangiorgi, and B. Ricco, "Scaling the MOS transistor below 0.1 μm: Methodology, device structures and technology requirements," *IEEE Trans. Electron Dev.* **41**, 941 (1994).
10. E. Suzuki, K. Ishii, S. Kanemaru, T. Maeda, T. Tsutsumi, T. Sekigawa, K. Nagai, and H. Hiroshima, "Highly suppressed short-channel effects in ultrathin SOI *n*-MOSFETs," *IEEE Trans. Electron Dev.* **47**, 354 (2000).
11. M. Jurczak, T. Skotnicki, M. Paoli, B. Tormen, J. Martins, J. L. Regolini, D. Dutartre, P. Ribot, D. Lenoble, R. Pantel, and S. Monfray, "Silicon-on-nothing (SON) – an innovative process for advanced CMOS," *IEEE Trans. Electron Dev.* **47**, 2179 (2000).
12. H. S. Wong, D. Frank, and P. M. Solomon, "Device design considerations for double-gate, ground-plane and single-gate ultra-thin SOI MOSFETs at the 25 nm channel length generation," *Tech. Digest IEDM* (1998), p. 407.
13. K. Suzuki, T. Tanaka, Y. Tosaka, H. Horie, and Y. Arimoto, "Scaling theory for double-gate SOI MOSFETs," *IEEE Trans. Electron Dev.* **40**, 2326 (1993).
14. A. G. Sabnis and J. T. Clemens, "Characterization of the electron mobility in the inverted <100> Si surface," *Tech. Digest IEDM* (1979), p. 18.
15. G. Baccarani and M. Wordeman, "An investigation of steady-state velocity overshoot in silicon," Solid State Electronics **28**, 407 (1985).
16. M. R. Pinto, E. Sangiorgi, and J. Bude, "Silicon MOS transconductance scaling into overshoot regime," *IEEE Electron Dev. Lett.* **14**, 375 (1993).
17. K. Natori, "Ballistic metal-oxide-semiconductor field effect transistor," *J. Appl. Phys.* **76**, 4879 (1994).
18. M. Lundstrom, "Elementary scattering theory of the Si MOSFET," *IEEE Electron Dev. Lett.* **18**, 361 (1997).
19. G. Timp, J. Bude, K. K. Bourdelle, J. Garno, A. Ghetti, *et al.*, "The ballistic nano-transistor," *Tech. Digest IEDM* (1999), p. 55.
20. F. Assad, Z. Ren, D. Vasileska, S. Datta, and M. Lundstrom, "On the performance limits for Si MOSFETs: A theoretical study," *IEEE Trans. Electron Dev.* **47**, 232 (2000).
21. D. Esseni, M. Mastrapasqua, G. K. Celler, F. H. Baumann, C. Fiegna, L.Selmi, and E. Sangiorgi, "Low field mobility of ultra-thin SOI *n*- and *p*-

MOSFETs: Measurements and implications on the performance of ultra-short MOSFETs," *Tech. Digest IEDM* (2000), p. 671.

22. J. Chen, R. Solomon, T.-Y. Chan, P.-K. Ko, and C. Hu, "A CV Technique for measuring thin SOI film thickness," *IEEE Electron Dev. Lett.* **12**, 453 (1991).

23. J.-H. Choi, Y.-J. Park, and H.-S. Min, "Electron mobility behavior in extremely thin SOI MOSFETs," *IEEE Electron Dev. Lett.* **16**, 527 (1995).

24. B. Mazhari and D. E. Ioannou, "Surface potential at threshold in thin-film SOI MOSFETs," *IEEE Trans. Electron Dev.* **40**, 1129 (1993).

25. F. Gamiz, J. A. Lopez-Villanueva, J. B. Roldan, J. E. Carceller, and P. Cartujo, "Monte Carlo simulation of electron transport properties in extremely thin SOI MOSFETs," *IEEE Trans. Electron Dev.* **45**, 1122 (1998).

26. P. J. Price, "Two-dimensional transport in semiconductor layers. I. Phonon scattering," *Ann. Phys.* **133**, 217 (1981).

27. B. K. Ridley, "The electron-phonon interaction in quasi-two-dimensional semiconductor quantum-well," *J. Phys. C* **15**, 5899 (1982).

28. M. Shoji and S. Horiguchi, "Electronic structures and phonon-limited electron mobility of double-gate silicon-on-insulator Si inversion layers," *J. Appl. Phys.* **85**, 2722 (1999).

29. C. Fiegna and A. Abramo, "Analysis of quantum effects in nonuniformly doped MOS structures," *IEEE Trans. Electron Dev.* **45**, 877 (1998).

30. *DESSIS 6.0 – User Manual*, ISE A.G., 1999.

31. M. J. Van Dort, P. H. Woerlee, A. J. Walker, C. A. Juffermans, and H. Lifka, "A simple model for quantization effects in heavily-doped silicon MOSFETs at inversion conditions," *Solid State Electronics* **37**, 411 (1994).

32. M. N. Darwish, J. L. Lentz, M. R. P. M. Zeitzoff, T. J. Krutsick, and H. H. Vuong, "An improved electron and hole mobility model for general purpose device simulation," *IEEE Trans. Electron Dev.* **44**, 1529 (1997).

Future Silicon-on-Insulator MOSFETs: Chopped or Genetically Modified?

F. Allibert
IMEP-ENSERG, BP 257, 38016 Grenoble Cedex 1, France and
SOITEC S. A., Chemin des Franques, Bernin, 38926 Crolles, France

J. Pretet
IMEP-ENSERG, BP 257, 38016 Grenoble Cedex 1, France and
STMicroelectronics, 38926 Crolles, France

A. Zaslavsky
CEA-LETI, 17 rue des Martyrs, 38054 Grenoble Cedex 9, France and
Div. of Engineering, Brown University, Providence, RI 02912, U.S.A.

S. Cristoloveanu
IMEP-ENSERG, BP 257, 38016 Grenoble Cedex 1, France

1. Introduction

A reasonable postulate in microelectronics is that the last generations of CMOS circuits will be made with silicon-on-insulator (SOI) technology. Two possible avenues are included in the roadmap: size shrinking or architectural modification of the conventional MOS transistor.[1] In this paper, we present new data and arguments for both approaches.

2. Narrow-channel devices

The shrinking of SOI MOSFETs is a process that involves all dimensions and is better described in terms of *minimum volume* rather than minimum length. The fabrication of sub-50-nm long transistors with reduced short-channel effects requires the use of extremely thin (<15 nm) SOI films. At the same time, very narrow MOSFETs are needed for low-power applications. In this section, we focus on the critical (and less explored) issue of narrow-channel effects. The relationship between narrow, short, and ultra-thin channels has been investigated for both fully- and partially-depleted LOCOS-isolated SOI *n*-MOSFETs. We first address the impact of the channel width reduction on the main transistor parameters, before describing the special behavior of floating-body effects (FBEs).

- *Typical narrow-channel effects*

Figure 1 shows the threshold voltage variation with channel width for various film thicknesses and two channel lengths. We note that the threshold voltage decreases with channel width in the ultra-thin 15-nm film, demonstrating the inverse narrow-channel effect. The normal effect is visible in 47- and 100-nm thick films, where V_T tends to increase with decreasing width. These contrasting trends suggest that the over-doping of the transistor sidewalls, which is used for V_T control, loses efficiency in ultra-thin films.[2]

For a 47-nm-thick channel, the threshold voltage roll-off between a long and a short channel does not vary with the channel width. There is no apparent correlation between short- and narrow-channel effects in LOCOS-isolated SOI MOSFETs. This lack of correlation is confirmed by the drain-induced barrier lowering (DIBL), which remains rather constant with width — see Fig. 2. It is noteworthy that DIBL is greatly reduced in the ultra-thin films below 15 nm.

Additional experiments reveal that the subthreshold slope (in the ohmic region) does not vary significantly with width, whereas the low-field mobility degrades in narrow devices (presumably reflecting a degradation of the Si crystal in the thin edges and an influence of the lateral over-doping).

- *Reduced floating-body effects in narrow-channel devices*

In ultra-thin films, the saturation subthreshold swing remains constant (75 mV/dec) with width, as shown by the 15-nm curves in Fig. 3. But in a thicker fully-depleted 37-nm film, the swing can drop below 60 mV/dec, due to the appearance of FBEs. This effect is more pronounced in short and wide devices, as can be seen in Fig. 3.

The reduction of FBEs in narrow MOSFETs is confirmed by the improvement of the breakdown voltage with decreasing width, illustrated in Fig. 4. The parasitic bipolar transistor, whose activation via FBEs causes premature breakdown in SOI devices, is clearly attenuated in narrow channels.

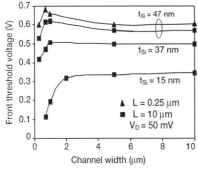

Figure 1. Threshold voltage *vs.* channel width for various transistors.

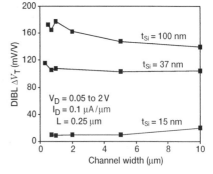

Figure 2. DIBL effect *vs.* film thickness and width.

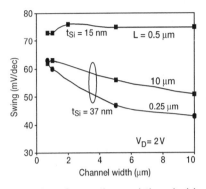
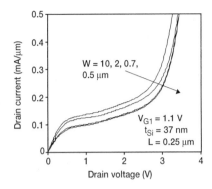

Figure 3. Saturation subthreshold swing *vs.* width for different L and t_{Si}.

Figure 4. Drain current breakdown characteristics *vs.* width.

The FBEs are related to the charging and discharging of the transistor body with majority carriers (generated by impact ionization or during the formation of depletion regions). Three main mechanisms may be invoked for the reduction of FBEs in narrow SOI MOSFETs:

- A thinner transistor (fully depleted) is less sensitive to majority carriers accumulated in the body.[3,4] TEM data show a lateral thinning of the film on the edges, under the LOCOS isolation. The overall effect is a reduced effective thickness in narrow channels.

- A degraded carrier lifetime on the edges increases the recombination rate. This degradation was established directly by carrier lifetime measurements using the method of drain current transients.[5,6] Both recombination (Fig. 5) and generation transients (Fig. 6) are shorter in the narrower devices, which proves that carrier lifetime is reduced on the edges, where the Si crystal has a lower quality.

- A more efficient removal of the majority carriers from the body, by the source/body junction, reduces the FBEs. The junction current increases along the sidewalls, due to boron out-diffusion and segregation into the isolation oxide: the doping level on the edges and the source/body barrier height are reduced, thus increasing the junction leakage. Boron out-diffusion was confirmed by measurements[7] and process simulations.

In conclusion, the fabrication of narrow MOSFETs is a natural solution to the critical problem of FBEs in SOI. All of the above coexisting mechanisms must be accounted for in the future development of small-volume devices, although the prevailing effect may depend on the specific isolation technology.

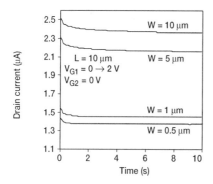

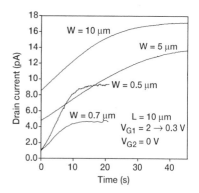

Figure 5. Drain current overshoot for different widths (t_{Si} = 100 nm).

Figure 6. Drain current undershoot for different widths (t_{Si} = 100 nm).

3. "Genetically modified" devices

It is clear that transistor length reduction, without geometry alteration, will eventually reach a limit. The question is how far can this limit be pushed by structural modifications of the device. SOI is a more flexible technology, hence SOI transistors are better prepared to undergo "genetic" modifications, infusing new functionality into extremely thin channels.

- *Ground plane*
  We first consider the problem of fringing fields in short-channel MOSFETs. The electric field induced by the drain penetrates the body of the transistor, lowering the threshold voltage (DIBL). Solutions to reduce DIBL consist of increasing the doping of the body (which unfortunately degrades the carrier mobility) or using thinner films. Below 15-nm thickness, DIBL is almost suppressed, as shown in Fig. 2, and no longer affected by the body doping,[8] which thus becomes unnecessary.

  But fringing fields can also penetrate through the buried oxide (BOX) and the substrate underneath. They modify the potential of the body as if the substrate itself were biased in weak inversion. This phenomenon, which dramatically lowers the threshold voltage, is known as the drain-induced virtual substrate biasing and it can be attenuated[9] by reducing the BOX thickness (so as to limit the field penetration) and then "grafting" a conductive layer underneath. This layer can be a ground plane, thus preventing the drain field from spreading into the substrate (Fig. 7).

  It is worth noting that a ground plane under an ultra-thin buried oxide is not an optimal solution, because it increases the BOX capacitance.[10] In particular, this increase affects the subthreshold swing S, which is a function of front oxide (C_{ox1}), Si film (C_{Si}) and BOX (C_{ox2}) capacitances. Ignoring the capacitances of front and back interface traps, we have:[4]

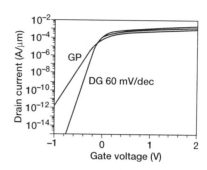

Figure 7. SOI MOSFET with a ground plane (black layer underneath the BOX).

Figure 8. Degradation of the subthreshold slope in an SOI MOSFET with a ground plane under a 2-nm BOX.

$$S = 2.3 \frac{kT}{q} \left(1 + \frac{C_{ox2}}{C_{ox1}} \frac{C_{Si}}{C_{Si} + C_{ox2}} \right). \tag{1}$$

It is obvious from this equation that if the BOX thickness is reduced then C_{ox2} increases and the slope degrades significantly — see Fig. 8. The strategy we promote is to use a thin and undoped film with high mobility, a reasonably thin BOX (50 nm to be compared to a 1–2 nm gate oxide) and a ground plane. The threshold voltage can be adjusted by using an appropriate gate material (mid-gap metal) and/or slightly biasing the ground plane.

- *Double gate*

Double-gate (DG) MOSFETs are unchallenged in terms of their potential for ultimate scaling and performance.[11-13] For a DG fabrication process to be efficient, it should allow for variable transistor widths and ultra-short channels, with a uniform Si film thickness, low series resistance, and an ultra-thin film (for volume inversion). In general, it is also described as a symmetrical device with mandatory self-aligned front and back gates (Fig. 9).

More or less technologically viable, albeit admirably sophisticated, grafting techniques for double-gate transistors have been proposed. A key difficulty is the fabrication of self-aligned double gates in a realistic fashion. To circumvent this problem, we have explored the possibility of a slight misalignment of the

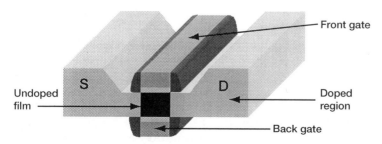

Figure 9. Schematic representation of a double-gate MOSFET on SOI.

Figure 10. Asymmetrical DG MOSFET with 50-nm front gate and 100-nm back gate. The back gate can be shifted towards the drain (a), centered (b), or towards the source (c).

gates,[14,15] compensated by an increase of the back gate length — see Fig. 10. We have simulated the effect of shifting one gate with respect to the other, using Silvaco's process (ATHENA) and device (ATLAS) modules. The three asymmetrical DG geometries are shown schematically in Fig. 10.

The simulated DG device had a 50-nm front (top) gate, a 100-nm back gate, a 12-nm thick Si film (*i.e.* negligible quantum effects), 2-nm gate oxides, *n*-poly Si gates, an undoped channel, and elevated source (S) and drain (D). The front gate is the one used for self-aligned S/D implantation and thus determines the effective S/D length of our transistors (diffused doped regions reach the front gate edges, as shown in Fig. 10). The difference in gate length allows for a 25-nm shift, in each direction, of one gate with respect to the other. This value corresponds to state-of-the-art e-beam overlay accuracy.

The asymmetrical devices of Fig. 10 are compared to the symmetrical self-aligned DG device with both gates of 50 nm (Fig. 9). All DG MOSFETs exhibit ideal subthreshold slopes (60 mV/dec at $T = 300$ K), always outperforming ground-plane transistors with ultra-thin BOX. Figure 11 shows a somewhat surprising result: asymmetric DG devices have a higher saturation current than their symmetric counterpart. This difference arises because the longer back gate overlaps the source and drain, thus creating an accumulation layer that decreases the series access resistance. The configuration of the devices is such that transistor (a) has a lower drain resistance, transistor (c) has a lower source resistance, whereas device (b) has both series resistances reduced. Since the source resistance is the most significant parasitic effect (it reduces both the effective V_G and V_D), it is logical that devices (b) and (c) yield the highest drain saturation current.

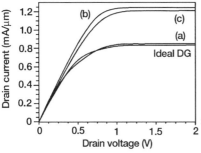

Figure 11. Drain saturation current for the various DG device structures.

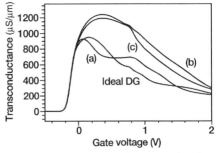

Figure 12. Transconductance for the various DG device structures.

The same advantages are visible in the transconductance curves in Fig. 12. In addition, the transconductance of the asymmetric devices shows either a wider peak or a plateau. This special feature is also explained by a series resistance effect: the transconductance peaks of the front and back channels do not appear at the same gate voltage and do not necessarily have the same magnitude. These two peaks can merge partially and result in the observed shapes.

4. Conclusions

The MOS/SOI transistor will be deservedly sanctified, but not before completing the terminal stages of its heroic evolution. A detailed analysis has been made of the narrow-channel effects and their relationship with the other dimensions (length and thickness). We have shown that the threshold voltage, the mobility, and the floating-body effects are strongly affected by the lateral isolation properties. As far as the implementation of innovative architectures is concerned, we expect the double-gate SOI transistor to be a leading competitor. Although we have demonstrated that some degree of asymmetry is acceptable for relaxing the process constraints, the device still needs to secure widespread sponsorship and an enthusiastic audience.

References

1. S. Cristoloveanu, T. Ernst, D. Munteanu, and T. Ouisse, "Ultimate MOSFETs on SOI: Ultra-thin, single gate, double gate, or ground plane," *Int. J. High Speed Elec. Syst.* **10**, 217 (2000).
2. J. Pretet, T. Ernst, S. Cristoloveanu, C. Raynaud, and D. Ioannou, "Narrow-channel effects in LOCOS-isolated SOI MOSFETs with variable thickness," *Proc. Int. IEEE SOI Conf.* (2000), p. 66.
3. J.-Y. Choi and J. G. Fossum, "Analysis and control of floating-body bipolar effects in FD submicrometer SOI MOSFETs," *IEEE Trans. Electron Dev.* **38**, 1384 (1991).
4. S. Cristoloveanu and S. S. Li, *Electrical Characterization of SOI Devices*, Boston: Kluwer (1995).
5. D. Munteanu, D. A. Weiser, S. Cristoloveanu, O. Faynot, J.-L. Pelloie, and J.G. Fossum, "Generation/recombination transient effects in partially depleted SOI transistors: Systematic experiments and simulations," *IEEE Trans. Electron Dev.* **45**, 1678 (1998).
6. D. E. Ioannou, S. Cristoloveanu, M. Mukherjee, and B. Mazhari, "Characterization of carrier generation in enhancement-mode SOI MOSFETs," *IEEE Electron Dev. Lett.* **11**, 409 (1990).
7. A. Ono, R. Ueno, and I. Sakai, "TED control technology for suppression of reverse narrow channel effect in 0.1 μm MOS devices," *Tech. Dig. IEDM* (1997), p. 227.

8. F. Allibert, T. Ernst, J. Pretet, N. Hefyene, C. Perret, A. Zaslavsky, and S. Cristoloveanu, "From SOI materials to innovative devices," *Solid State Electronics* **45**, 559 (2001).

9. T. Ernst, C. Tinella, C. Raynaud, and S. Cristoloveanu, "Fringing fields in sub-0.1 μm FD SOI MOSFETs: Optimization of the device architecture," *Proc. ULIS* (2000).

10. H.-S. P. Wong, D. J. Franck, P. M. Solomon, C. H. J. Wann, and J. J. Welser, "Nanoscale CMOS," *Proc. IEEE* **87**, 537 (1999).

11. J.-P. Colinge, *Silicon-on-Insulator Technology: Materials to VLSI*, 2nd ed., Boston: Kluwer (1997).

12. X. Huang, W. C. Lee, C. Kuo, D. Hisamoto, L. Chang, J. Kedzierski, E. Anderson, H. Takeuchi, Y. K. Choi, K. Asano, V. Subramanian, T. J. King, J. Bokor, and C. Hu, "Sub-50 nm *p*-channel FinFet," *IEEE Trans. Electron Dev.* **48**, 880 (2001).

13. H.-S. P. Wong, K. K. Chan, and Y. Taur, "Self-aligned (top and bottom) double-gate MOSFET with a 25 nm thick silicon channel," *Tech. Dig. IEDM* (1997), p. 427.

14. K. Suzuki, T. Tanaka, Y. Tosaka, H. Horie, and T. Sugii, "High-speed and low-power n^+-p^+ double-gate SOI CMOS," *IEICE Trans. Electron.* **E78-C**, 360 (1995).

15. J. P. Denton and G. W. Neudeck, "Fully depleted dual-gated thin film SOI P-MOSFETs fabricated in SOI islands with an isolated polysilicon backgate," *IEEE Trans. Electron Dev.* **17**, 509 (1996).

Current Transport Models for Engineering Applications

Tibor Grasser and Siegfried Selberherr
Institute for Microelectronics, TU Wien, A-1040 Wien, Austria

1. Introduction

Numerical simulation of carrier transport in semiconductor devices dates back to the famous work of Scharfetter and Gummel.[1] Since then the transport models have been continuously refined and extended to capture more accurately transport phenomena occurring in modern semiconductor devices. The need for refinement and extension is caused primarily by the ongoing feature size reduction in state-of-the-art technology. As the supply voltages cannot be scaled accordingly without jeopardizing the circuit performance, the electric fields inside the devices have increased. Large electric fields that rapidly change over small length scales give rise to non-local and hot-carrier effects that begin to dominate device performance. An accurate description of these phenomena is required and is becoming a primary concern for industrial applications.

Traditionally, the drift-diffusion model[2] has been used to describe carrier transport in semiconductor devices. However, the drift-diffusion model assumes equilibrium between carrier energy and electric field, which is no longer valid in modern devices. Extended models have been proposed that consider the carrier energy an independent solution variable.[3,4] These models are capable of describing non-local and hot-carrier effects to first order. Recent results, however, suggest that the average energy is in many cases not sufficient for accurate modeling. Both the transport models themselves and the models for the physical parameters are affected. In this article we review the most commonly used transport models and point out their most important limitations. In addition, an extended transport model based on six moments of the distribution function is presented, which seems to be a balanced trade-off between accuracy and complexity.

2. Boltzmann's transport equation

Transport equations used in semiconductor device simulation normally are derived from Boltzmann's transport equation, which provides a semiclassical description of carrier transport. For a general inhomogeneous material with arbitrary band structure it reads[5]

$$\partial f/\partial t + \mathbf{u}\cdot\nabla_r f + \hbar^{-1}\mathbf{F}\cdot\nabla_k f = C[f] . \tag{1}$$

Here, **u** is the group velocity, **F** the force exerted on the particles, and C the collision operator. For inclusion of quantum effects, equations based on the Wigner-Boltzmann equation have been considered.[6] Boltzmann's equation needs to be solved in the seven-dimensional phase space, which is prohibitive for engineering applications. Monte Carlo simulations have been proven to give accurate results but are restrictively time consuming. Furthermore, if the distribution of high-energy carriers is relevant, or if the carrier concentration is very low in specific regions of the device, Monte Carlo simulations tend to produce high variance in the results. Therefore, a common simplification is to investigate only some moments of the distribution function, such as the carrier concentration and the carrier temperature. These moments of the distribution function are typically defined as

$$\langle \phi \rangle = (1/4\pi^3) \int \phi f d^3 \mathbf{k} \,, \tag{2}$$

with a suitable weight function $\phi = \phi(\mathbf{k})$.

3. The drift-diffusion model

The drift-diffusion model is the simplest current transport model that can be derived from Boltzmann's transport equation by the method of moments[2] or from basic principles of irreversible thermodynamics.[7] It has been the workhorse in industrial applications for over thirty years. Within the drift-diffusion model, the well known continuity and current equations have to be solved. In their static form, these equations read

$$\nabla \cdot \mathbf{J} = qR \,, \tag{3}$$

$$\mathbf{J} = q\mu n\mathbf{E} + \mu k_B T_L \nabla n \,. \tag{4}$$

Here μ denotes the electron mobility, T_L the lattice temperature, **E** the electric field, **J** the current density, R the recombination rate, n the carrier density and k_B the Boltzmann constant. The average energy w can be estimated via the local energy balance equation. This method neglects the lag between the electric field and the average energy characterized by the energy relaxation time. One consequence of the lag is that the maximum energy can be much smaller than the one predicted by the local energy balance equation. Furthermore, this lag gives rise to an overshoot in the carrier velocity, because the mobility depends to first order on the energy and not on the electric field.

Therefore, modeling deep-submicron devices is becoming more and more problematic. Although successful reproduction of terminal characteristics of nanoscale MOS transistors has been reported,[8] the values of the material parameters used significantly violate basic physical principles. In particular, the saturation velocity v_S has to be set to more than twice the value observed in bulk measurements.

4.   The full hydrodynamic model

The full hydrodynamic model was first derived by Bløtekjær.[4] In its original form the first three moments of Boltzmann's transport equation were considered. Closure was obtained by applying a heuristic model for the heat flux using Fourier's law. Furthermore, the band structure was assumed to be parabolic and the tensor quantities were approximated by scalars. The resulting equation set reads[4]

$$\nabla \cdot \mathbf{J} = qR \,, \tag{5}$$

$$\mathbf{J} - (\tau_m/q)\nabla \cdot (\mathbf{J} \otimes \mathbf{J}/n) = q\mu n\mathbf{E} + \mu k_B \nabla(nT_n) \,, \tag{6}$$

$$\nabla \cdot \mathbf{S} = \mathbf{E} \cdot \mathbf{J} - n(w{-}w_0)/\tau_E + G_E \,, \tag{7}$$

$$\mathbf{S} = -(w + k_B T_n)\mathbf{J}/q - \kappa(T_n)\nabla T_n \,; \tag{8}$$

where τ_m is the momentum relaxation time. The additional parameters are the energy relaxation time τ_E, the thermal conductivity κ, the energy flux $\mathbf{S}$, the electron temperature T_n, the average energy in equilibrium w_0, and the generation rate G_E. For the thermal conductivity an empirical relation analogous to the Wiedemann-Franz law is used with a correction factor p:

$$\kappa(T_n) = (5/2 - p)(k_B/q)^2 \, q\mu n T_n \,. \tag{9}$$

This equation system is similar to the Euler equations of gas dynamics with the addition of a heat conduction term and the collision terms. Thus, the electron gas has a sound speed and the electron flow may be either subsonic or supersonic.[9] In the case of supersonic flow, electron shock waves will in general develop inside the device. These shock waves occur at either short length scales or at low temperatures. Furthermore, the traditionally applied Scharfetter-Gummel[1] discretization scheme and its extensions cannot be used for this type of equation, which makes handling the full hydrodynamic model quite difficult.[9, 10]

5.   The energy transport model

As the closure of the full hydrodynamic model has been shown to be problematic, the fourth moment of Boltzmann's equation is added to give a more accurate description for the energy flux $\mathbf{S}$. For the closure of the equation system a heated Maxwellian distribution is generally assumed.[11] Since the resulting equation system is difficult to handle, simplifications are generally considered. The four-moments energy-transport model is obtained by the simplification of the four-moments hydrodynamic model. The convective term,

$$(\tau_m/q)\nabla \cdot (\mathbf{J} \otimes \mathbf{J}/n) \,, \tag{10}$$

in the current relation is neglected, as are the corresponding convective term in the energy flux relation and the contribution of the kinetic energy to the total carrier energy:

$$w = mv^2/2 + 3k_B T_n/2 \approx 3k_B T_n/2 . \tag{11}$$

This simplification gives the four-moments energy-transport model which reads as follows:

$$\nabla \cdot \mathbf{J} = qR , \tag{12}$$

$$\mathbf{J} = q\mu n\mathbf{E} + \mu k_B \nabla(nT_n) , \tag{13}$$

$$\nabla \cdot \mathbf{S} = \mathbf{E} \cdot \mathbf{J} - 3nk_B(T_n - T_L)/2\tau_E + G_E , \tag{14}$$

$$\mathbf{S} = -(5\mu_S/2\mu) [(k_B T_n/q)\mathbf{J} + (k_B/q)^2 q\mu nT_n\nabla(nT_n)] . \tag{15}$$

Here the energy flux mobility μ_S appears instead of the thermal conductivity in Eq. (8). Considering the different definitions for the mobilities, the energy-transport model is equivalent to the energy-balance model proposed by Stratton.[3] See for example Ref. 12.

A comparison of the energy flux equation of the hydrodynamic model and the four-moments energy transport model shows that the correction factor p in the thermal conductivity has to be set to zero to obtain a consistent equation set. Furthermore, the ratio of the mobilities μ_S/μ is assumed to be unity in the hydrodynamic model.

6. Problems of the hydrodynamic and energy-transport models

During the derivation of the models given above various approximations of different severity have been employed. The most important approximations will be summarized in the following.

- *Closure*
  The method of moments transforms Boltzmann's equation into an equivalent infinite set of equations. One of the severest approximations is the truncation to a finite number of equations (normally three or four). The equation of highest order contains the moment of the next order, which has to be suitably approximated using available information, typically the lower order moments. Even though no form of the distribution function needs to be assumed in the derivation, an implicit coupling of the highest order moment and the lower order moments is enforced by this closure. One approach to derive a suitable closure relation is to assume a distribution function and calculate the fourth order moment, where a heated Maxwellian shape is almost exclusively used. Ramaswami and Tang[13] gave a comparison of different closure relations available in the literature.

- *Tensor quantities*

An issue that has been only vaguely dealt with is the approximation of the tensors by scalar quantities, such as the trace of the tensors. For example, the carrier mass and the carrier temperature are approximations introduced that way. One-dimensional simulations show[14] that the longitudinal temperature component is larger than the transverse temperature component. This result indicates that the distribution function is elongated along the field direction and thus that the normally assumed equipartition of the energy is invalid. A rigorous approach has been taken by Pejcinovic et al.[15] who model four components of the temperature tensor. They observed no significant difference between the scalar temperature and the trace of the temperature tensor for ballistic diodes and bipolar transistors but a 15% difference for aggressively scaled MOSFETs in the linear region of the transfer characteristics.

- *Drift energy versus thermal energy*

Another common approximation is that the contribution of the drift energy to the total carrier energy is neglected.[16] As has been pointed out by Baccarani and Wordeman,[17] the convective energy can reach values comparable to the thermal energy. The error introduced by this approximation can be significant in the beginning of the channel where the carrier temperature is still low and a velocity overshoot is observed. This effect has been studied in detail in Ref. 18.

- *Modeling of the physical parameters: mobility and impact ionization*

The relaxation times traditionally have been derived from homogeneous field measurements or Monte Carlo simulations. For homogeneous fields there is a unique relationship between the electric field and the carrier temperature via the local energy balance equation which can be used as a definition for τ_E. From Boltzmann's equation it is clear, however, that the relaxation times depend on the distribution function through the collision operator. Since the distribution function is not uniquely described by the average energy, models based solely on the average energy are bound to fail.

Two models for the energy dependence of the mobility are frequently used, the model after Baccarani and Wordeman,[17]

$$\mu(T_n)/\mu_0 = T_L/T_n , \qquad (16)$$

and the model after Hänsch,[19,20]

$$\mu(T_n)/\mu_0 = [1 - (3\mu_0/2\tau_E v_s^2)(k_B T_L/q + 2S/5J)]^{-1} . \qquad (17)$$

For homogeneous materials $(S/J) = 5k_B T_n/2q$, which can be used to simplify Eq. (17) to

$$\mu(T_n)/\mu_0 = [1 + (3\mu_0 k_B/2\tau_E v_s^2 q)(T_n-T_L)]^{-1} . \qquad (18)$$

A comparison of these three expressions with Monte Carlo simulation results for an n^+-n-n^+ test structure with channel length $L_C = 200$ nm is given in Fig. 1.

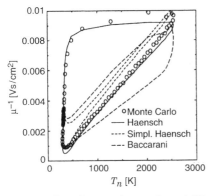

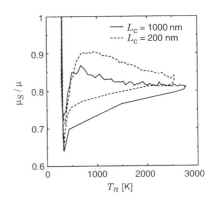

Figure 1. Comparison of mobility models with Monte Carlo data for an n^+-n-n^+ test structure with $L_c = 200$ nm.

Figure 2. Ratio of the mobilities obtained by Monte Carlo simulations for two n^+-n-n^+ test structures.

The analytical expressions were evaluated using the data from the Monte Carlo simulation. Note that the temperature dependence of the inverse mobility is frequently plotted because of the expected linear dependence of Eq. (16). The small hysteresis in the simplified Hänsch model and in the Baccarani model is due to the doping dependence of the zero-field mobility μ_0.

The ratio of the mobilities μ_S/μ as a function of the carrier temperature is shown in Fig. 2 for two n^+-n-n^+ test structures. To obtain comparable behavior the same doping profile has been used for both structures and the bias condition has been chosen to give a maximum electric field of 100 kV/cm. Note that in commercial device simulators the mobility ratio is normally assumed to be unity.

In Figs. 3 and 4 the error of the analytical models is shown for the two n^+-n-n^+ test structures. As has been pointed out in Ref. 14, Eq. (17) is the only expression that gives reasonable results in both increasing and decreasing field regions. However, at the beginning of the channel where the carrier temperature is still low, the mobility is considerably over- or underestimated. Furthermore, due to the quotient of the magnitudes of two vector quantities **S** and **J**, Eq. (17) is rather difficult to handle in a multidimensional device simulator.

As for impact ionization, it is poorly described by models that use the local average energy as the only parameter. In general, ionization rates obtained by local-energy models start rising too early and fall off too sharply. Furthermore, local-energy models considerably overestimate the ionization rates if not calibrated for the investigated device. In particular, local-energy models cannot capture impact ionization caused by hot electrons in the drain because there the cold carriers dominate the average energy, which is close to the equilibrium value.

Several non-local models have been proposed[21] which are, however, both difficult to implement in a conventional device simulator and difficult to justify on a theoretical basis, especially for multi-dimensional problems.

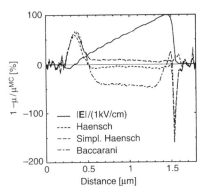

Figure 3. Error in the analytical mobility models for an n^+-n-n^+ test structure with $L_C = 1000$ nm.

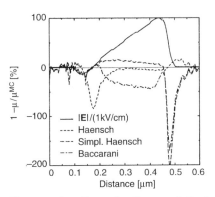

Figure 4. Error in the analytical mobility models for an n^+-n-n^+ test structure with $L_C = 200$ nm.

7. Possible solution: The six-moments model

One can include the kurtosis β_n of the distribution function in addition to the temperature T_n without making any assumption on the shape of the distribution function except that the diffusion approximation holds.[22] With the new variables (the kurtosis β_n and the kurtosis flux $\mathbf{K}$ along with its generation rate G_β) the static flux and balance equations of the six-moments model for electrons read

$$\nabla \cdot \mathbf{J} = qR, \tag{19}$$

$$\mathbf{J} = q\mu n\mathbf{E} + \mu k_B \nabla(nT_n), \tag{20}$$

$$\nabla \cdot \mathbf{S} = \mathbf{E} \cdot \mathbf{J} - 3nk_B(T_n - T_L)/2\tau_E + G_E, \tag{21}$$

$$\mathbf{S} = -(5\mu_S k_B^2/2\mu q)\,\mu[qEnT_n/k_B) + \nabla(nT_n^2\beta_n)], \tag{22}$$

$$\nabla \cdot \mathbf{K} = 2q\mathbf{E} \cdot \mathbf{S} - 15nk_B^2(\beta_n T_n^2 - T_L^2)/4\tau_\beta + G_\beta, \tag{23}$$

$$\mathbf{K} = -(35\mu_K k_B^3/4\mu q)\,\mu[qEn\beta_n T_n^2/k_B + \nabla(nT_n^3\beta_n^3)]. \tag{24}$$

Only one additional second-order partial differential equation for β_n is introduced.

The additional parameters are the kurtosis relaxation time τ_β and the kurtosis flux mobility μ_K. The solution variables are defined as

$$T_n = 3\langle\varepsilon\rangle/2k_B \quad \text{and} \quad \beta_n = 3\langle\varepsilon^2\rangle/5\langle\varepsilon\rangle^2. \tag{25}$$

For a heated Maxwell-Boltzmann distribution and parabolic bands $\beta_n = \beta_{MB} = 1$. When non-parabolicity is taken into account, the value of β_{MB} depends on the energy but stays close to unity.

The main difference to the energy-transport equations is that the kurtosis β_n appears in the equation for the energy flux. As a consequence, the energy flux equation cannot be written in the form frequently used for energy-transport models as proportional to the current density. This modification makes the coupled equation system difficult to solve and approximations have been used. Note that the six-moments model reduces to the standard energy-transport model when the equations for **K** are dropped and a value of unity is assumed for β_n. The kurtosis assumes values in the range 0.75–2.5 and gives the deviation from a heated-Maxwellian distribution. This deviation is vital and can be used to formulate more accurate models for the physical parameters.

8.   Application of the six-moments model: Substrate currents

For reliability issues and for the calculation of substrate currents accurate modeling of impact ionization is required. As shown in Ref. 23, the kurtosis can be used to obtain an analytical expression for the distribution function that goes beyond the Maxwellian shape approximation. With this distribution function, microscopic scattering rates traditionally used in Monte Carlo simulations can be incorporated into macroscopic device simulators. For the purpose of demonstration we considered two MOSFETs with gate lengths $L_G = 1.0$ μm and $L_G = 0.25$ μm. A comparison of the impact ionization rate predicted by a local-energy model and by a six-moments model is given in Fig. 5 for the short-channel device. also shown is the metallurgical junction (fat line). For the local-energy model the

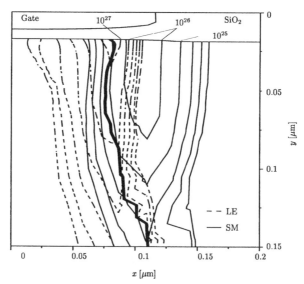

Figure 5. Comparison of the impact ionization rate as predicted by a local-energy model (LE) and by a six-moments model (SM). Also shown is the metallurgic junction (fat line).

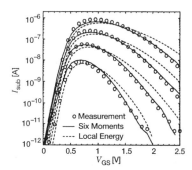

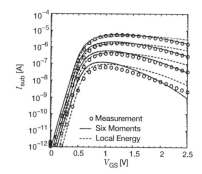

Figure 6. Comparison of modeled substrate currents and measurements for the long-channel device.

Figure 7. Comparison of modeled substrate currents and measurements for the short-channel device.

maximum occurs at the junction where the average energy rapidly decreases, because the hot carriers from the channel meet the large pool of cold carriers in the drain. In the case of the six-moments model, the maximum is inside the drain region, which is in agreement with Monte Carlo simulations.

Simulated substrate currents are given in Figs. 6 and 7 for the long-channel and short-channel devices, respectively. Both characteristics were calculated using the same parameter values. Note that the local-energy model had to be calibrated for these devices. Although the local-energy model delivers reasonable substrate currents, the calculated ionization profiles inside the devices are at the wrong position and have a wrong shape that requires individual calibration.

9. Problems of higher-order moment equations

Even though the vast majority of routinely performed device simulations still employ the drift-diffusion model, hydrodynamic and energy-transport models have been investigated thoroughly during the last ten years. Accurate results have been obtained for a large variety of devices. However, the values of the physical parameters used vary considerably.

Uncertainties are introduced by the approximation of the collision terms that are modeled via relaxation times or mobilities and by the derivation of closure relations. Expressions for the parameters normally are calibrated with homogeneous Monte Carlo simulations. As has been clearly shown, homogeneous Monte Carlo simulation data are not sufficient for the simulation of state-of-the-art devices as neither the relaxation times nor the closure relations are single-valued functions of the average energy. Unfortunately, data for inhomogeneous situations are difficult to extract from measurements due to the complex interaction between the various parameters. Therefore, Monte Carlo simulations of n^+-n-n^+ test-structures were performed to extract the desired data. However, the results obtained by available Monte Carlo codes differ significantly.[24] Impurity scattering is especially difficult to model[25] and any error in the mobility influences the simulated energy relaxation times where large differences are found in the published data.

It is particularly important to note that all models should be able to reproduce correctly the homogeneous limit. Unfortunately, model parameters calibrated to particular devices frequently do not fulfill this basic requirement, indicating that some of the underlying assumptions need to be reconsidered.

From a practical point of view it has to be pointed out that convergence problems are still an issue and inhibit the use of higher-order moment equations in everyday engineering applications. Unfortunately, simulation codes based on these equations have never reached a robustness comparable to the drift-diffusion model. One reason may be that a consistent discretization of the current equation, the energy flux equation, and the heat source term in the energy balance equation is difficult.[26] No generally accepted scheme like the Scharfetter-Gummel[1] scheme for the drift-diffusion equations exists.

Furthermore it has been shown that hydrodynamic and energy-transport models capture velocity overshoot only to first order. In general, the velocity overshoot is overestimated. Moreover, a spurious peak in the velocity at the drain end is observed which has so far not been eliminated with a unique parameter set. Comparison with Monte Carlo simulations indicate that this is a principal problem caused by the truncation of the infinite moment series.[27] In addition, inaccuracies in the physical parameters, such as the mobility, amplify the spurious peaks.[14]

With the ongoing reduction of feature size, the influence of quantum-mechanical effects such as confinement in the channel and tunneling currents is increasing.[28] Accurate modeling of these effects still requires large amounts of computation time. In particular, the influence of the surface still can not be modeled properly. For instance, the energy relaxation time is known to differ significantly from its bulk value.

10. Conclusions

Various transport models have been proposed so far. Apart from the drift-diffusion model, higher-order models based on either Stratton's or Bløtekjær's approach have been considered. Despite its well-known limitations, the drift-diffusion model is still predominantly used in engineering applications. The need for higher-order models is well understood and these models have delivered excellent results in carefully set-up simulations. However, handling of higher-order models still requires a lot of fine-tuning and a detailed understanding of the underlying physical phenomena.

References

1. D. Scharfetter and H. Gummel, "Large-signal analysis of a silicon Read diode oscillator," *IEEE Trans. Electron Dev.* **16**, 64 (1969).
2. S. Selberherr, *Analysis and Simulation of Semiconductor Devices*, Vienna and New York: Springer, 1984.
3. R. Stratton, "Diffusion of hot and cold electrons in semiconductor barriers," *Phys. Rev.* **126**, 2002 (1962).
4. K. Bløtekjær, "Transport equations for electrons in two-valley semiconductors," *IEEE Trans. Electron Dev.* **17**, 38 (1970).
5. D. Ferry, *Semiconductors*, New York: Macmillan, 1991.
6. C. Gardner, "The classical and quantum hydrodynamic models," in: *Proc. Intern. Workshop Computational Electronics*, Univ. of Leeds (1993), pp. 25–36.
7. G. Wachutka, "Rigorous thermodynamic treatment of heat generation and conduction in semiconductor device modeling," *IEEE Trans. Computer-Aided Design* **9**, 1141 (1990).
8. J. Bude, "MOSFET modeling into the ballistic regime," in: *Proc. Simul. Semicond. Dev. Processes Conf.*, Seattle (2000), pp. 23–26.
9. C. Gardner, "Numerical simulation of a steady-state electron shock wave in a submicrometer semiconductor device," *IEEE Trans. Electron Dev.* **38**, 392 (1991).
10. E. Fatemi, J. Jerome, and S. Osher, "Solution of the hydrodynamic device model using high-order nonoscillatory shock-capturing algorithms," *IEEE Trans. Computer-Aided Design* **10**, 232 (1991).
11. T.-W. Tang, S. Ramaswamy, and J. Nam, "An improved hydrodynamic transport model for silicon," *IEEE Trans. Electron Dev.* **40**, 1469 (1993).
12. T.-W. Tang and H. Gan, "Two formulations of semiconductor transport equations based on spherical harmonic expansion of the Boltzmann transport equation," *IEEE Trans. Electron Dev.* **47**, 1726 (2000).
13. S. Ramaswamy and T.-W. Tang, "Comparison of semiconductor transport models using a Monte Carlo consistency check," *IEEE Trans. Electron Dev.* **41**, 76 (1994).
14. S.-C. Lee and T.-W. Tang, "Transport coefficients for a silicon hydrodynamic model extracted from inhomogeneous Monte-Carlo calculations," *Solid State Electronics* **35**, 561 (1992).
15. B. Pejcinovic, H. Tang, J. L. Egley, L. Logan, and G. Srinivasan, "Two-dimensional tensor temperature extension of the hydrodynamic model and its applications," *IEEE Trans. Electron Dev.* **42**, 2147 (1995).
16. R. Cook and J. Frey, "An efficient technique for two-dimensional simulation of velocity overshoot effects in Si and GaAs devices," *COMPEL* **1**, 65 (1982).

17. G. Baccarani and M. Wordeman, "An investigation of steady-state velocity overshoot in silicon," *Solid State Electronics* **28**, 407 (1985).

18. M. Stettler, M. Alam, and M. Lundstrom, "A critical examination of the assumptions underlying macroscopic transport equations for silicon devices," *IEEE Trans. Electron Dev.* **40**, 733 (1993).

19. W. Hänsch, *The Drift Diffusion Equation and Its Application in MOSFET Modeling*, Vienna and New York: Springer, 1991.

20. W. Hänsch and M. Miura-Mattausch, "The hot-electron problem in small semiconductor devices," *J. Appl. Phys.* **60**, 650 (1986).

21. C. Jungemann, R. Thoma, and L. Engl, "A soft threshold lucky electron model for efficient and accurate numerical device simulation," *Solid State Electronics* **39**, 1079 (1996).

22. T. Grasser, H. Kosina, M. Gritsch, and S. Selberherr, "Using six moments of Boltzmann's transport equation for device simulation," *J. Appl. Phys.* **90**, 2389 (2001).

23. T. Grasser, H. Kosina, C. Heitzinger, and S. Selberherr, "An impact ionization model including non-Maxwellian and nonparabolicity effects," *Appl. Phys. Lett.* **80**, 613 (2002).

24. A. Abramo, L. Baudri, R. Brunetti, *et al.*, "A comparison of numerical solutions of the Boltzmann transport equation for high-energy electron transport silicon," *IEEE Trans. Electron Dev.* **41**, 1646 (1994).

25. H. Kosina and G. Kaiblinger-Grujin, "Ionized-impurity scattering of majority electrons in silicon," *Solid State Electronics* **42**, 331 (1998).

26. W.-S. Choi, J.-G. Ahn, Y.-J. Park, H.-S. Min, and C.-G. Hwang, "A time-dependent hydrodynamic device simulator SNU-2D with new discretization scheme and algorithm," *IEEE Trans. Computer-Aided Design* **13**, 899 (1994).

27. T. Grasser, H. Kosina, and S. Selberherr, "Investigation of spurious velocity overshoot using Monte Carlo data," *Appl. Phys. Lett.* **79**, 1900 (2001).

28. Z. Yu, R. Dutton, and R. Kiehl, "Circuit/device modeling at the quantum level," *IEEE Trans. Electron Dev.* **47**, 1819 (2000).

Advanced Physically Based Device Modeling for Gate Current and Hot-Carrier Phenomena in Scaled MOSFETs

Pierpaolo Palestri, Luca Selmi,
Alberto Dalla Serra, Antonio Abramo, and Enrico Sangiorgi
DIEGM, Univ. of Udine, Via delle Scienze 208, 33100 Udine, Italy

Maura Pavesi and PierLuigi Rigolli
Dept. of Physics, Univ. of Parma, Parco Area delle Scienze 7/A, Parma, Italy

Frans Widdershoven
Philips Research Leuven, Kapeldreef 75, B-3001, Leuven, Belgium

1. Introduction

Gate oxide scaling is a key issue for further progress down the semiconductor roadmap towards ultimate MOSFET performance.[1] Reliability and leakage currents compete as limiting factors of this aspect of the scaling process.[2,3] Whatever the bottleneck will be, detailed calibrated physically-based models of carrier transport through silicon-dielectric-silicon stacks are becoming increasingly important to our understanding of leakage and degradation mechanisms and to meaningful reliability projections. Significant progress has been made recently in this direction that allows for important advances in the understanding of thin dielectrics.[3-6]

In this paper, we present a comprehensive physically-based model of carrier injection in the gate oxide and of hot-carrier processes in MOS devices, as well as application of this model to tunneling MOS capacitors. In light of this model, we revisit the anode hole injection (AHI) and photon emission experiments of Ref. 7 and we demonstrate that conclusions quite different from those of Ref. 7 can be drawn from the same experiments with the help of a more detailed physical picture of the processes involved. Implications of this analysis for future MOSFETs are also briefly discussed.

The paper is organized as follows: Sections 2 and 3 describe the model and its verification, respectively. Section 4 discusses applications to MOS capacitors, focusing on the interpretation of the experiments of Ref. 7. Considerations on thin oxides are given in Section 5. Conclusions are drawn in Section 6.

2. Simulation procedure and model description

Figure 1 illustrates the tunneling and hot-carrier phenomena included in the model (a positive gate voltage condition is assumed). Electrons tunnel from the inversion layer into the gate (anode), where they generate electron-hole pairs by impact ionization (II). The injected electrons and the generated holes may emit photons. Some of the generated holes have enough energy to overcome the Si-SiO$_2$ barrier and to be injected back into the substrate, giving a first contribution ($I_{SUB,II}$) to the total majority carrier substrate current (I_{SUB}). The photons emitted with energy higher than the gap can be absorbed in the substrate, thus generating e-h pairs; holes give an additional contribution to the substrate current ($I_{SUB,PH}$), while some of the minority electrons can be collected by a reverse-biased junction (I_{COLL}).

Figure 2 illustrates the simulation procedure used to evaluate the tunneling current, anode hole injection, and the majority and minority carrier substrate currents. The key features of our model are the following:

1. The contribution of each quantized level to the gate tunneling current, $I_G(E)$, is computed as described in Ref. 8 from self-consistent Schrödinger-Poisson solutions of the Si-SiO$_2$-Si stack incorporating polysilicon quantization and a semi-classical expression of the quasibound-state lifetime.

2. The gate current $I_G(E)$ is fed to a bipolar full-band Monte Carlo simulator of silicon devices (FBMC),[9,10] suitably extended to include the calibrated SiO$_2$ transport model of Ref. 11. The silicon band structure and the phonon and the impact ionization scattering rates (SR$_{II}$) have been extended well above 10 eV by considering 8 conduction bands, instead of the usual 4 in the frame of non-local pseudopotential theory.[12] This

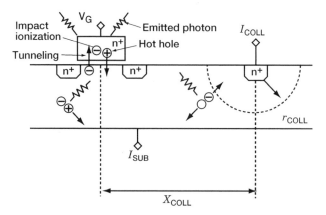

Figure 1. Schematic representation of tunneling, impact ionization, and photon emission-absorption processes in an *n*-MOS capacitor under positive gate bias, and definition of some terms and symbols used in the text.

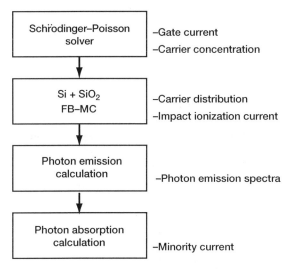

Figure 2. Flowchart of the simulation procedure.

extension provides accurate modeling of very high-energy carriers emerging from the SiO$_2$ into Si.[10] Particles inside the Si and SiO$_2$ are run in parallel, thus eliminating the need for multi-step simulations.

3. Carriers are injected through and above the image-force-lowered interfacial barriers with an emission probability (T_P), computed following the transfer matrix approach and using the same parameters as in Ref. 13. As a result, cold electrons injected into the anode generate hot holes that can be injected back to the cathode. These holes contribute to the majority carrier substrate current ($I_{SUB,II}$) and generate cathode hot electrons that are eventually injected back towards the gate.[9]

4. The model[12] for photon emission in Si is an extension of those in Refs. 14 and 15: direct (D) and phonon-assisted (PA) spontaneous emission rates for inter- and intra- conduction and valence band transitions (c-v, c-c, and v-v) were computed employing the same band structure, wavefunctions, and phonons of the FBMC. The spectral distribution of the emitted light intensity is then computed at each grid point based on the FBMC carrier distributions.

5. The generation rate due to photon absorption is computed with a simple two-dimensional propagation model. The absorption coefficient for Si is taken from Ref. 16. Zero emission and absorption are assumed in SiO$_2$. Reflections at the Si-SiO$_2$ interfaces are negligible, due to the large ratio between the photon wavelength and oxide thickness.

6. The upper bound for the majority carrier substrate current due to photon emission/absorption processes ($I_{SUB,PH}$) is estimated by assuming that half of the photons with an energy higher than the band gap energy ($E_G = 1.12$ eV) generate an e-h pair and that all generated holes are collected by the substrate contact.

7. It is assumed that all minority carriers generated within an effective diffusion length of the remote junction (r_{COLL}) have a chance to be collected.[17] The minority carrier substrate current (I_{COLL}) thus is calculated by integrating the generation rate over the region surrounding the reverse-biased junction within a distance r_{COLL}. The domain for photon absorption calculations can be arbitrarily larger than that for FBMC simulations.

8. The FBMC and the photon emission and absorption blocks are two-dimensional (2D). Therefore, if the potential of the Schrödinger-Poisson solver is replaced with that of a 2D drift-diffusion simulator, the same simulation procedure can be used to simulate photon emission in MOSFETs or n^+-n-n^+ devices.[18]

3. Model verification

Each block of the model was validated separately by comparison with experimental data. Transport models in silicon were calibrated by the usual means (comparison with velocity-field curves and impact ionization data). As demonstrated in Ref. 10, the coupled Si and SiO$_2$ FBMC model reproduces average electron energy data in SiO$_2$[11] and impact ionization quantum yield (QY) data[19,20] over the whole range of available oxide thicknesses.

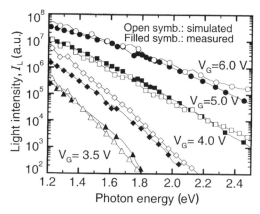

Figure 3. Comparison between simulated and measured[22] emission spectra for n^+-n-n^+ devices.

Photon emission spectra for template Maxwellian and Gaussian distributions[12,21] are in good agreement with those in Refs. 14 and 15; furthermore, calculated spectra compare very well also with the experimental data on n^+-n-n^+ diodes of Ref. 22 (Fig. 3).

As for the absorption model, Fig. 4 reports the simulated minority carrier collection efficiency (η) as a function of X_{COLL} and r_{COLL} (see Fig. 1) for the same junction geometry as in Ref. 7. For higher values of X_{COLL} (plotted in Ref. 21, but not here) the collection efficiency follows an exponential decay with characteristic length of approximately 700 µm, in agreement with Refs. 7 and 17.

4. Application to MOS capacitors

Recent experimental results on the correlation between I_{SUB} and I_{COLL} in 4–9 nm SiO_2 layers led researchers to question the validity of the AHI-based interpretation of I_{COLL},[7] and to propose that the majority carrier substrate current measured in tunneling n-MOS capacitors at $V_G > 0$ could actually be the result of photon emission processes.[23,24] If confirmed, this result would have important consequences on the validity of AHI models of device degradation, hence on reliability assessment procedures, which are currently based on the assumption that the oxide damage is due to backward injection of anode hot holes. The interpretation of these results, however, is based on comparison with simple analytical models,[25] hardly applicable to the problem at hand.

In this section we show that the detailed physically based model of the processes described in Sections 2 and 3 can lead to a different view, and to the possibility of reconciling the experiments of Ref. 7 with AHI models.

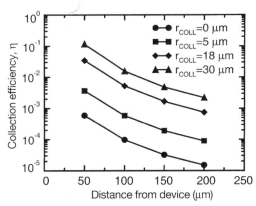

Figure 4. Collection efficiency $\eta = I_{COLL}/I_{SUB,PH}$ for reverse-biased junctions[7] as a function of the distance from the gate of a tunneling MOS capacitor. Different minority carrier diffusion lengths are assumed. The spectra are emitted by a MOS capacitor with $t_{ox} = 7.7$ nm and $V_G = 10$ V.

As a first step we consider n-MOSFETs. Since in these devices c-c transitions represent the dominant photon emission process,[14,22] we compare the efficiency of this mechanism in generating e-h pairs via photon re-absorption to that of electron impact ionization. To this purpose, Fig. 5 reports half of the total photon emission scattering rate for c-c transitions with $E_{PH} > E_G$ ($SR_{PH}/2$, so that the corresponding photon absorption process generates pairs in the substrate only) with the electron impact ionization scattering rate (SR_{II}). We observe that $SR_{PH}/2SR_{II}$ has a slightly decreasing trend with an average value of ~3×10^{-5} over a broad energy range. The small $SR_{PH}/2SR_{II}$ ratio implies that quantum yield experiments (gate injection in an n^+-poly p-MOSFET[19,26]) maintain their validity as a means to test and calibrate impact ionization models even in the presence of pair generation by photon emission/absorption. Furthermore, this result suggests a physically based explanation of the weakly drain-voltage–dependent ratio of approximately 3×10^{-5} between I_{COLL} and I_{SUB}, repeatedly observed in experiments where an efficient minority carrier collector is available (*i.e.* when the collection efficiency, η, is close to unity[17,27,28]).

Unlike the MOSFET case, the dominant photon emission process in tunneling MOS capacitors has not been clearly established yet.[24,29] Let us consider first emission at the anode side. The following emission mechanisms for photons with $E > E_G$ can be foreseen:

- Conduction to conduction band (c-c) transitions of the hot electrons injected by tunneling;

- Valence to valence band (v-v) transitions of the holes generated by impact ionization at the anode;

- Conduction to valence band (c-v) transitions involving the injected electrons and the anode holes.

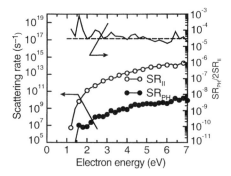

Figure 5. Scattering rate for electron impact ionization SR_{II} and for photon emission SR_{PH} with $E_{PH} > 1.12eV$ through direct and photon-assisted c-c transitions (the dominant emission mechanism in n-MOSFETs), left axis. $SR_{PH}/2SR_{II}$, right axis. A value ~3×10^{-5} is found, in excellent agreement with Refs. 27, 17, and 28.

The c-c and v-v transitions can be evaluated directly from the simulated FBMC carrier distributions, and c-c transitions dominate over v-v ones because of the larger electron energy. The c-v processes, instead, are more complicated to estimate, because a population of cold holes not accounted for by the Schrödinger-Poisson simulations, and difficult to estimate accurately with MC tools, piles up at the interface as a result of the balance between impact ionization of injected electrons and recombination processes.[25] These cold holes can contribute significantly to emission of near-bandgap photons. In our simulation procedure, the cold-hole concentration is evaluated by solving the continuity and drift-diffusion equation for holes. The carrier generation rate is taken from the FBMC simulations. Differently from Ref. 25, we account for nonzero field in the gate, which we found to have a dramatic impact on the concentration profile, hence on the results of emission calculations.

Figure 6 compares the photon emission intensity due to c-v transitions with that due to c-c and v-v transitions as a function of the hole lifetime in the gate. The c-v processes dominate over the c-c one only for unphysically large lifetimes; a reasonable value of $\tau = 10^{-8}$ s for highly doped polysilicon corresponds to much less emission from c-v than from c-c processes.

Essentially the same procedure is used to calculate photon emission in the cathode, except that holes and electrons exchange their respective roles.[18] The relative importance of these emission processes on the gate and substrate side of a typical capacitor biased in the tunneling regime are shown in Fig. 7. Since the oxide is essentially transparent in this photon energy range, these curves suggest that c-c transitions should dominate photon emission in MOS capacitors also. The bias dependence of these curves is easily explained by noting that photon emission in the substrate depends almost linearly on the AHI current, while photon emission in the gate depends almost linearly on the gate current.

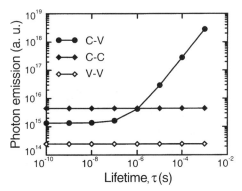

Figure 6. Contribution of c-c, c-v, and v-v transitions inside the gate as a function of the minority carrier lifetime for an n-MOS capacitor with $t_{ox} = 5.5$ nm, $V_G = 8$ V.

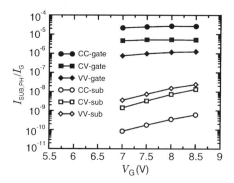

Figure 7. Efficiency of c-c, c-v and v-v transitions in an *n*-MOS capacitor with t_{ox} = 5.5 nm as a function of the gate voltage.

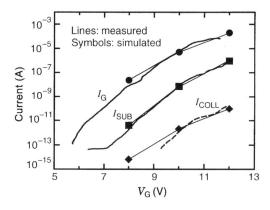

Figure 8. Direct comparison between simulated gate, substrate, and collector currents and the measurements in Ref. 7 (t_{ox} = 7.7 nm).

Having established the dominant emission mechanisms in MOS capacitors, we can compare them with AHI in order to interpret the experiments of Ref. 7. Figure 8 compares measured and simulated gate, substrate, and collector currents for a 7.7-nm oxide in Ref. 7; a good agreement is obtained for all the currents, thus confirming the validity of the models for tunneling, AHI, and photon emission/absorption. Figure 9 shows I_{SUB} normalized to the gate current for 6- and 7-nm oxides from a different technology; as can be seen in this case also the model is able to reproduce the measurements with a very good accuracy.

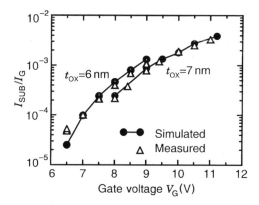

Figure 9. Measured and simulated I_{SUB}/I_G in n-MOS capacitors. Good agreement is observed using the same parameter set as in Ref. 9.

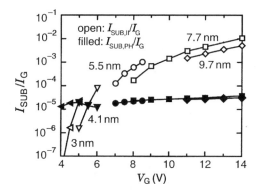

Figure 10. Maximum contribution of the reabsorption of photons emitted by c-c transitions ($I_{SUB,PH}/I_G$) and the contribution of impact ionization $I_{SUB,II}/I_G$ to the total simulated I_{SUB}/I_G for different values of t_{ox}. Note the totally different dependence on V_G (and hence on F_{OX}) of the two terms.

A direct comparison of the contributions of AHI and photon emission/absorption to I_{SUB} is given in Fig. 10 for different oxide thicknesses and gate voltages. Clearly, if we consider the same relatively thick oxides of Ref. 7 (5.5, 7.7 and 9.7 nm), we see that $I_{SUB,II}$ is a few orders of magnitude higher than $I_{SUB,PH}$. Furthermore, $I_{SUB,PH}/I_G$ also has a much weaker dependence on F_{OX} and t_{ox} than $I_{SUB,II}/I_G$, a dependence that does not match that of the experiments in Fig. 9. This weak dependence clearly is a direct consequence of the fact that the impact ionization quantum yield is almost constant at high gate voltages[20,10] and of the small and almost constant ratio SR_{PH}/SR_{II} (Fig. 5 and Refs. 17, 27, and 28).

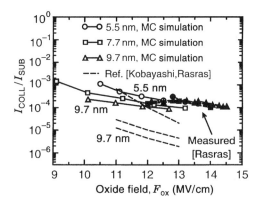

Figure 11. Ratio of the minority carrier current I_{COLL} collected by two junctions 100 µm away from the capacitor (as in Ref. 7) and the substrate current I_{SUB} as a function of F_{ox} for different oxide thickness.

Figure 11 compares the ratio I_{COLL}/I_{SUB} simulated with our model to the measurements in Ref. 7: aside from a shift in F_{OX} values (likely due to inaccurate F_{OX} extraction and imprecise knowledge of the device doping), a comparable dependence on F_{OX} and t_{OX} is observed in measurements and simulations. Therefore, differently from the conclusions of Ref. 7, the model suggests that a strong correlation between I_{COLL} and I_{SUB} can exist even if I_{SUB} is due to AHI.

In order to understand this result in more detail we express ratio I_{COLL}/I_{SUB} as

$$\frac{I_{COLL}}{I_{SUB}} = \frac{\eta I_{PH}}{I_{SUB,II} + I_{SUB,PH}} \cong \frac{\eta \alpha_{PH} I_G}{\alpha_{II} T_P I_G + \alpha_{PH} I_G} , \qquad (1)$$

where α_{PH} and α_{II} are the efficiencies of photon emission and impact ionization, and T_P is the injection probability for holes.

If I_{SUB} were dominated by photon emission/absorption, I_{COLL}/I_{SUB} would be given by the collection efficiency of the remote junction (η). For a distance of 100 µm between the emission region and the collecting junction, a ratio I_{COLL}/I_{SUB} = 10^{-4} was measured in Ref. 7. As can be seen in Fig. 4, this value is obtained only by assuming a negligibly small value for the minority carrier diffusion length. A reasonable value of a few tens of microns,[17] instead, results in $\eta \sim 10^{-2}$.

From the previous considerations, we know that both the ratio α_{PH}/α_{II} and the efficiency η are almost constant. As a result, I_{COLL}/I_{SUB} depends on F_{OX} and t_{OX} through T_P. If calculated with the simple analytical model of Ref. 25, T_P is expected to exhibit a rather strong F_{OX} and t_{OX} dependence (approximately a factor of 10^3 in the 5.5–9.7 nm range — see dashed lines in Fig. 11), whereas only a very weak one is found in experiments. Note, however, that T_P is the effective injection probability of the entire hole population, and therefore depends strongly

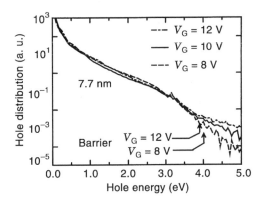

Figure 12. Energy distribution of holes impinging upon the anode interface according to the FBMC model (t_{ox} = 7.7 nm). Due to multiple scattering, the distribution is weakly dependent on F_{OX} around the image-force-lowered barrier height ($E = (\phi_{Bh} - \Delta\phi) \sim 3.9$–4 eV).

on the hole energy distribution at the anode interface. The results of Figs. 5 and 7 indicate that in order to have $I_{SUB,PH} > I_{SUB,II}$, T_P has to be lower than about 3×10^{-5}; much larger values are typically extracted from simulations of 5–10 nm oxides.

Figure 12 reports the anode hole energy distribution and illustrates the physical reason why in our model, consistently with the experiments, T_P has a much weaker dependence on F_{OX} and t_{ox} than expected.[21] Clearly, the assumption of a Maxwellian distribution whose effective temperature is directly related to F_{OX} and t_{OX}[25] is no longer valid for the holes at the Si/SiO$_2$ interface at high gate voltages, because secondary impact ionization is very strong and a large number of holes is generated by secondary carriers (see the quantum yield value > 1 obtained in Ref. 20). As a consequence, a large portion of the distribution, extending to energies comparable to the hole barrier height, does not vary too much with the gate bias (Fig. 12). Because scattering makes this large portion of the distribution lose its memory of oxide transport conditions, a weaker variation of T_P with voltage results.

We note that in order to detect unambiguously the possible F_{OX} and t_{OX} dependence of I_{COLL}/I_{SUB} measurements over a broad range of oxide fields would be needed, which is difficult to achieve because I_{COLL} soon becomes undetectable as F_{OX} is reduced.

5.   Implications for thin oxides

The results in the previous section apply to relatively thick oxides (such as those measured in Ref. 7) biased at high voltages (as for a reliability study). The voltage dependencies of $I_{SUB,II}$ and $I_{SUB,PH}$ in Fig. 10 suggest that for low voltages these two currents will become comparable and possibly exchange their roles. This exchange happens for the 3-nm and 4.1-nm oxides, where at low voltages $I_{SUB,PH} > I_{SUB,II}$. It has to be remarked, though, that $I_{SUB,PH}$ represents only an approximate upper estimate of the actual contribution of photons to the majority carrier substrate current, and that the real current would be lower than the simulated one. Furthermore, the gate voltage range where $I_{SUB,PH} > I_{SUB,II}$ can be probed only by means of very thin oxide devices, in which trap-assisted tunneling components could complicate significantly the overall physical picture.

6. Conclusions

In summary, we described a physically-based comprehensive model of tunneling currents, AHI, and photon emission/absorption processes in MOS structures. Besides reproducing many different experiments, the model explains some relevant features of minority carrier substrate currents in n-MOSFETs, such as the weakly bias dependent ratio $I_{COLL}/I_{SUB} \sim 2$–3×10^{-5} observed in Refs. 17, 27, and 28. Furthermore, the model provides a better understanding of the origin of substrate currents in tunneling MOS capacitors, reconciling the experimental results in Ref. 7 with the AHI model.

In particular, the model suggests that for thick oxides the majority carrier current is due to AHI while the minority carrier current is consistent with a photon emission/absorption mechanism. The two currents appear to be strictly correlated because of the relatively flat ratio between impact ionization and photon emission scattering rates, and because in thick oxides at very large fields the hole transmission probability is much less sensitive to oxide thickness and gate bias than predicted by simple analytical expressions.

The same model suggests that photon emission/absorption processes may become the dominant substrate current generation mechanism for tunneling MOS capacitors at low gate bias, because anode holes have insufficient energy to overcome the Si-SiO$_2$ barrier. However, this range of gate voltages can be probed only by means of very thin oxide devices in which trap-assisted tunneling components could complicate significantly the overall physical picture.

Acknowledgments

This work was supported in part by the Italian CNR (5% Microelectronics and Madess II projects) and MIUR (PRIN project).

References

1. International Technology Roadmap for Semiconductors, 2000.
2. J. Stathis and D. J. DiMaria, "Reliability projection for ultra-thin oxides at low voltage," *Tech. Digest IEDM* (1988), p.167.
3. M. A. Alam, J. Bude, B. Weir, P. Silverman, A. Ghetti, D. Monroe, K. P. Cheung, and S. Moccio, "An anode hole injection percolation model for oxide breakdown: The 'doomsday' scenario revisited," *Tech. Digest IEDM* (1999), p. 715.
4. R. Degraeve, G. Groeseneken, R. Bellens, M. Depas, and H. E. Maes, "A consistent model for the thickness dependence of intrinsic breakdown in ultra-thin oxides," *Tech. Digest IEDM* (1995), p. 863.
5. J. Bude, B. Weir, and P. Silverman, "Explanation of stress-induced damage in thin oxides," *Tech. Digest IEDM* (1998), p.179.
6. M. A. Alam, B. Weir, J. Bude, P. Silverman, and D. Monroe, "Explanation of soft and hard breakdown and its consequences for area scaling," *Tech. Digest IEDM* (1999), p. 449.
7. M. Rasras, I. DeWolf, G. Groeseneken, B. Kaczer, G. Degraeve, and H. E. Maes, "Photo-carrier generation as the origin of Fowler-Nordheim induced substrate hole current in thin oxides," *Tech. Digest IEDM* (1999), p. 465.
8. A. Dalla Serra, A. Abramo, P. Palestri, L. Selmi, and F. Widdershoven, "A comparison between semi-classical and quantum-mechanical escape-times for gate current calculations," *Proc. ESSDERC* (2000), p. 340.
9. P. Palestri, L. Selmi, E. Sangiorgi, M. Pavesi, and F. Widdershoven, "Cathode hot electrons and anode hot holes in tunneling MOS capacitors," *Proc. ESSDERC* (2000), p. 296.
10. P. Palestri, L. Selmi, M. Pavesi, F. Widdershoven, and E. Sangiorgi, "Coupled Monte Carlo simulation of Si and SiO_2 transport in MOS capacitors," *Proc. SISPAD* (2000), p. 38.
11. D. Arnold, E. Cartier, and D. J. DiMaria, "Theory of high-field electron transport and impact ionization in silicon dioxide," *Phys. Rev. B* **49**, 10278 (1994).
12. M. Pavesi, P. L. Rigolli, M. Manfredi, P. Palestri, and L. Selmi, "Spontaneous hot-carrier photon emission rates in silicon: Improved modeling and applications to metal oxide semiconductor devices," to be published in *Phys. Rev. B* (2001).
13. A. Ghetti, C.-T. Liu, M. Mastrapasqua, and E. Sangiorgi, "Characterization of tunneling current in ultra-thin gate oxide," *Solid State Electronics* **44**, 1523 (2000).
14. J. Bude, N. Sano, and A. Yoshii, "Hot-carrier luminescence in Si," *Phys. Rev. B* **45**, 5848 (1992).
15. S. Villa, A. Lacaita, and A. Pacelli, "Photon emission from hot electrons in silicon," *Phys. Rev. B* **52**, 10993 (1995).
16. E. D. Palik, *Handbook of Optical Constants in Solids*, New York: Academic Press, 1985.

17. S. Tam and C. Hu, "Hot-electron induced photon and photocarrier generation in silicon MOSFETs," *IEEE Trans. Electron Dev.* **31**, 1264 (1984).
18. A. Dalla Serra, P. Palestri, and L. Selmi, "Can photon emission/absorption processes explain the substrate current of tunneling MOS capacitors?" *Proc. ULIS* (2001), p. 117.
19. S. Takagi, N. Yasuda, and A. Toriumi, "Experimental evidence of inelastic tunneling and new *I–V* model for stress-induced leakage current," *Tech. Digest IEDM* (1996), p. 323.
20. Y. Kamakura, I. Kawashima, K. Deguchi, and K. Taniguchi, "Monte Carlo simulation of quantum yields exceeding unity as a probe of high-energy hole scattering rates in Si," *Tech. Digest IEDM* (1999) p. 727.
21. P. Palestri, M. Pavesi, P. L. Rigolli, L. Selmi, A. Dalla Serra, A. Abramo, and E. Sangiorgi, "Impact ionization and photon emission in MOS capacitors and FETs," *Tech. Digest IEDM* (2000), p. 97.
22. L. Selmi, M. Mastrapasqua, D. M. Boulin, J. D. Bude, M. Pavesi, E. Sangiorgi, and M. R. Pinto, "Verification of electron distributions in silicon by means of hot-carrier luminescence measurements," *IEEE Trans. Electron Dev.* **45**, 802, (1998).
23. D. J. DiMaria, E. Cartier, and D. A. Buchanan, "Anode hole injection and trapping in silicon dioxide," *J. Appl. Phys.* **80**, 304 (1996).
24. E. Cartier, J. C. Tsang, M. V. Fischetti, and D. A. Buchanan, "Light emission during direct and Fowler-Nordheim tunneling in ultra thin MOS tunnel junctions," *Microelectronics Eng.* **36**, 103 (1997).
25. K. Kobayashi, A. Teramoto, M. Hirayama, and Y. Fujita, "Model for the substrate hole current based on thermionic electron tunneling in *n*-channel metal-oxide semiconductor field effect transistors," *J. Appl. Phys.* **77**, 3277 (1995).
26. A. Ghetti, M. A. Alam, J. Bude, and F. Venturi, "Assessment of quantum yield experiments via full band Monte Carlo simulations," *Tech. Digest IEDM* (1997), p. 873.
27. P. A. Childs, R. A. Stuart, and W. Eccleston, "Evidence of optical generation of minority carrier from saturated MOS transistors," *Solid State Electronics* **26**, 685 (1983).
28. L. Selmi and D. Esseni, "A better understanding of substrate enhanced gate current in MOSFETs and flash cells, Part I and II," *IEEE Trans. Electron Dev.* **46**, 376 (1999).
29. K. DeKort and P. Damink, "The spectroscopic signature of light emitted by integrated circuits," *Proc. ESREF* (1990), p. 45.

2 THE FUTURE BEYOND SILICON: SEMICONDUCTORS, SUPERCONDUCTORS, PHASE TRANSITIONS, DNA

Contributors

2.1 M. Dorojevets

2.2 I. Lagnado and P. R. de la Houssaye

2.3 M. J. Kelly

2.4 F. Chudnovskiy and S. Luryi

2.5 R. Zia, A. Rakitin, and J. M. Xu

2.6 M. Cahay, Y. Modukuru, J. Thachery, *et al.*

FLUX-1: Designing the First Generation of 20-GHz Superconductor RSFQ Microprocessors in 1.75-μm Technology

Mikhail Dorojevets
Dept. of Electrical and Computer Engineering
State University of New York at Stony Brook, Stony Brook, NY 11794, U.S.A.

1. Introduction

This paper presents the first single-chip superconductor FLUX-1 microprocessor designed in rapid single flux quantum (RSFQ) logic and fabricated using 4-kA/cm^2 1.75-μm Nb/AlO$_x$/Nb Josephson junction technology as a result of the collaboration between SUNY-Stony Brook and TRW, Inc. A FLUX-1 chip represents an 8-bit microprocessor with a parallel architecture and a target clock rate of 17–20 GHz. It contains ~65,800 Josephson junctions on a 10.6×13.2 mm^2 die with flip-chip packaging. Fabrication began in June 2001 with the first FLUX-1 chips available for testing in August 2001. We will discuss the FLUX-1 architecture and organization as well as challenges in designing the first 20-GHz superconductor processor.

2. Design goals

The FLUX superconductor microprocessor project became possible due to several advances in superconductor technology and design made during the last few years. Theoretical studies and experimental chip designs based on RSFQ logic demonstrated the great potential of this logic for ultra-fast superconductor digital circuit implementation.[1,2] Quite powerful circuit-level software design and testing tools were developed to cope with numerous challenges of practical medium-scale RSFQ chip design and testing.[3–6] A new 4-kA/cm^2 1.75-μm Nb/AlO$_x$/Nb Josephson-junction technology developed by TRW, Inc. allowed reproducible fabrication of chips containing up to 10^5 junctions with tolerable variations of their technological parameters.[7] The final contribution came from the architectural and design studies of the SPELL RSFQ processors for a petaflops-scale system within the hybrid technology multithreaded (HTMT) architecture project.[8,9]

Taking into account these design opportunities, we concluded that design and fabrication of the first general-purpose single-chip RSFQ microprocessor prototype called FLUX[10] should be our first step on the way to a full-fledged superconductor digital system. Six major goals of the FLUX project were formulated as follows: VLSI-scale complexity, high performance with a processor clock rate close to 20 GHz, chip-to-chip communication at a 5-GHz rate, design scalability, testability, and tolerance to process imperfections. We will show how and to what extent these objectives were achieved in the FLUX-1 microprocessor design when presenting its architecture and implementation in the following sections.

3. Superconductor technology impact

A good match between the FLUX architecture, the superconductor technology, and RSFQ logic is crucial to achieving our performance and complexity goals. Despite being perhaps the state-of-the-art superconductor integrated circuit technology in 2001, TRW's fabrication process has its limitations that have a strong impact on the design of the FLUX-1 processor. Among key technological factors are the critical current density, number of metal layers, chip area, circuit speed, and interconnection density.

The complexity of the design depends on the number of Josephson junctions (JJs) that can be laid out in a given die area. In order to use TRW's current fabrication techniques and get an acceptable die yield, the maximum area of a FLUX chip was limited to ~140 mm^2. With the 1.75-μm technology providing the critical current density of 4 kA/cm^2, each junction has to be overdamped with an external shunt resistor of ~3.8 Ω in order to have its Stuart-McCumber parameter be approximately equal to 1. Due to fabrication requirements, Josephson junctions and shunt resistors, while occupying different layers, cannot overlap each other. The ratio of junction to shunt-resistor areas is ~1:5 for the current technology. A dramatic increase in chip density is expected in future technologies featuring ~0.3-μm junctions and critical current density j_C exceeding 150 kA/cm^2, when junctions become "naturally" overdamped without any external shunt resistors.[11]

Even though shunt resistors take up such a large area compared to junctions, it is Nb wires (more exactly, passive microstrip and strip transmission lines), not shunt resistors, that consume most of the FLUX-1 chip area because of line widths and a relatively small number of metal layers available. There are three wiring layers plus a ground plane in TRW's technology.[7] All FLUX gates use at least two metal layers called TRIW and WIRA plus the ground plane, thus leaving only one metal layer above them (WIRB) unused. This WIRB level alone, however, cannot be used for wiring signals over any gate. In order to be reliable, all gate-to-gate communication has to be done via microstrip and strip transmission lines with one and two grounded planes, respectively. While the minimum wire widths and spacings required by the TRW topological design rules are 1.5 μm, the actual FLUX lines are much wider: 5 μm for TRIW strip lines (over the ground plane and under grounded WIRA), and 21 μm for WIRB microstrip lines (over grounded WIRA). The reason for this sizing is the necessity for all microstrip and strip transmission lines to have the same characteristic impedance of ~4.6 Ω, which makes the WIRB lines much wider than the TRIW ones. When necessary, these "narrow" strip and "wide" microstrip lines can cross each other with very low signal attenuation and dispersion, being safely shielded from mutual interference by grounded planes. Special transceivers work as "impedance bridges" between FLUX gates with their 1.9 Ω output resistance and transmission lines with a ~4.6 Ω impedance. While the drivers' output resistance and the line impedance are matched to each other, it is not possible to provide a "perfectly matched load" for transmission lines at the receiver's (termination) side due to the nonlinear resistance of Josephson junctions.[12] This nonlinearity results in partial reflections off the receiver's connection to a transmission line, which limits the maximum data transfer rate over passive transmission lines and their length. However, even with these limitations, internal FLUX signals (represented by ~0.5-mV high and ~4-ps wide pulses) can be transmitted ballistically with very low attenuation and dispersion at a rate of 20 GHz over relatively long (up to 3–4 mm) Nb transmission lines at speeds of 90–130 μm/ps, depending on the layer. Figure 1 shows the layout of a 4×4-bit section of the FLUX instruction memory featuring both types of the lines.

4. RSFQ logic

The low-loss ballistic propagation of signals over transmission lines together with extremely high speed and low power consumption are the most attractive features of superconductor RSFQ circuits. On the other hand, the use of pulses rather than voltage levels to code signal values prevents RSFQ design elements from being connected to each other by buses with shared wires. Instead, binary signal distribution trees built of splitters have to be used, which leads to quite a large area being consumed by interconnect in RSFQ designs (*e.g.*, in the FLUX-1 memory whose fragment is shown in Fig. 1).

Another intrinsic feature of RSFQ logic is that almost all logical functions can be done only with clocked gates. Theoretically, the maximum clock frequency of

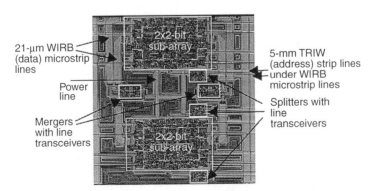

Figure 1.   Photograph of a 4×4-bit instruction memory fragment layout. Each 0.27-mm × 0.16-mm 2×2 sub-array consists of 118 Josephson junctions.

a pipelined processor can be reached with only one logic gate (in other words, one level of logic) per pipeline stage. (Current CMOS microprocessors feature from 8 to 12 levels of logic per stage.) Thus, the maximum rate at which these clocked gates can operate imposes the upper limit on a FLUX-1 processor clock rate. For any clocked gate used in FLUX-1, the reciprocal of the sum of the gate's set-up and hold times gives the maximum operating frequency (f_{max}) of the gate. Maximum speed is not the only requirement in gate design. RSFQ circuits must be operational with a given bit error rate and have relatively high tolerance to noise and technological process parameter variations. Table 1 shows key timing characteristics (accurate to ~10%) of major RSFQ gates, with the design of each of them optimized for a 20-GHz clock rate and noise margins of 30–35%.[13] Gate latencies do not depend on wire lengths because of the ballistic propagation of pulses over transmission lines in FLUX-1.

5.   The FLUX-1 architecture

The analysis of these technology- and logic-related factors led us to the conclusion that if built following a traditional design approach with a centralized memory, register file, and functional units, a FLUX microprocessor could become a chip with scarce "islands" of RSFQ logic surrounded by wires, where most of the time had to be spent on internal data transfer instead of processing. Moreover, long

Library element	JJs	Delay (ps)	Set-up (ps)	Hold (ps)	f_{max} (GHz)
JTL stage	1	3.0	—	—	—
Line driver	2	3.0	—	—	—
Line receiver	2	4.0	—	—	—
Splitter	3	6.5	—	—	—
Merger	5	4.5	—	—	—
D flip-flop (DFF)	6	7.5	6.5	7.5	71
Inverter	8	10.5	8.0	11.5	51
DFFC	9	14.0	10.0	17.5	36
D²FF	10	9.5	3.0	9.5	80
NDRO					
Read/write	10	10.0	17.5	25.5	23
Read only	10	10.0	—	25.5	39
XOR	11	9.5	8.0	8.0	62
AND	13	15.0	0	18.0	55

Table 1. Timing characteristics of major FLUX-1 RSFQ gate types (simulated with PSCAN[3]).

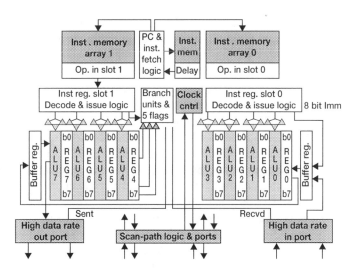

Figure 2. FLUX-1 block diagram. Each ALU shown here is a bit-stream ALU consisting of 8 single-bit pipelined ALUs that can work in parallel on data bits from adjacent registers.

transmission latencies would made it very difficult for us to reach a target clock rate of 20 GHz without using special design tricks unavoidably hurting overall processor performance. Thus, a new architectural design requiring localized data communication between processing elements and registers has been developed in order to address these realities of the current superconductor technology and RSFQ logic.

Figure 2 shows the block diagram of a FLUX-1 microprocessor. The major design units are: a 16-word instruction memory with embedded program counter (PC) and instruction fetch logic; a branch unit with five condition flags; the instruction register and dual decode/issue logic; eight 8-bit general-purpose integer registers (R0–R7) and eight bit-stream arithmetic logic units (ALUs); two 8-bit 5-GHz I/O ports; a clock controller; and built-in scan-path circuitry.

The FLUX "minimalist" instruction set consists of ~25 30-bit instructions, including conditional/unconditional branch, add/subtract, shift left/right, swap, logical or, invert, load an immediate value, *etc.* From a programmer's point of view, each bit-stream ALU operation has a two-register format, with the requirement that its source registers are to be adjacent (*i.e.*, their numbers differ by one only) and its destination register can be one of the source registers. After translation into binary code, ALU operations are represented by their codes consisting of a 2-bit ALU identifier and a 10-bit horizontal code field specifying the ALU operation to be performed on registers surrounding the ALU.

An 8-bit FLUX-1 microprocessor implements a new parallel architecture that features a synchronous dual-op long-instruction-word (LIW) architecture extended to support stream-like processing of data bits in integer registers. There are eight pipelined bit-stream ALUs interleaved with eight integer registers. An 8-bit word (signed/unsigned 8-bit value) in any of the registers is treated as a vector of eight single-bit elements starting from the least significant bit 0 (LSB) and ending with the most significant bit 7 (MSB). Each register bit has its own read and write ports. Each bit-stream ALU is built with eight single-bit ALUs. A single-bit ALU (bALU) has a 3-stage execution pipeline whose inputs and outputs are connected to the corresponding read and write bit ports of the registers located to the left and right of the bALU by relatively short (~ 500 µm) wires. Due to the short distances between ALUs and their source/destination registers, read and pre-process as well as write microoperations on each bit can be completed within one 50-ps cycle.

Thanks to the FLUX-1 dual-op LIW architecture, two integer (or a pair of one integer and one control) operations can be issued to two of the eight bit-stream ALUs (or to one ALU and the branch unit) and two other operations completed each cycle. In more detail, two operations of an instruction fetched from the instruction memory are placed into two slots of the instruction register. After two cycles spent in the first level of the pipelined instruction decoder, an ALU operation from each slot arrives at the ALU decoder specified by the operation's ALU-id field. Two cycles later, the pipelined ALU decoder initiates register read and one cycle later ALU operations by issuing properly delayed control signals to its least significant bALU and two registers surrounding it. Pipeline control logic embedded in each register bit and single-bit bALU transmits control, carry and data signals from one bit-stage to another until the completion of the operation. Four cycles (one for register read and three for execution) after the decoded integer operation reaches the target ALU, the result's first bit (LSB) is calculated. This start-up delay of four cycles would be the same even if we had 16 or 32 bits per register, thus making the FLUX-1 processing core design perfectly scalable. During the next eight cycles, 8 bits of the result (starting from LSB) will be written bit by bit into its destination register (located to the left or right of the ALU) by the eight single-bit bALUs comprising the target ALU. As a side effect of the operation on register R4, three control flags can be calculated and used in branch operations later.

What makes FLUX-1 different from any bit-sequential processor is that any operation that is dependent on data to be calculated by an operation in progress may start reading the data as early as its bit 0 (LSB) is ready, *i.e.*, not waiting for the operation in progress to complete. This type of processing resembles classical vector processing with so-called vector chaining, with two key differences:

- Operation chaining in FLUX-1 is implemented at the level of bits of each integer register rather than words in vector registers.

- Each bit-stream ALU in FLUX-1 can overlap execution of up to 12 independent or 3 serially-dependent instructions working with the same 8-bit integer register(s), while a classical vector processing unit cannot start execution of a new operation until it completes its current operation in progress.

Feature	Parameter
Architecture	2-way LIW with a processing-in-registers datapath and bit-level operation chaining
Instruction set size	~ 25 (branches, add/sub, logical, I/O, *etc.*)
Register data size, bits	8
Number of registers	8
Number of functional units	8 bit-stream arithmetic-logic units 1 branch unit
I/O port width, bits	8 data + 2 control
Instruction length, bits	30 (+ 2 reserved for future extensions)
Inst. memory size, bits	16×32
Operations issue rate	2 operations/cycle
Peak performance	40 billion 8-bit operations/second (with a 20-GHz clock rate)

Table 2. Architectural characteristics.

In fact, such overlapped execution of several data-dependent instructions makes FLUX-1 perhaps the first computer with multiple-instructions-single-data (MISD) processing.

Synchronous LIW architectures often suffer from an increase in code size (so-called code inflation) caused by the no-operation (NOP) instructions. This problem is solved in FLUX-1 by suppressing the NOP instructions using a 4-bit fetch delay field in each instruction. With a corresponding encoding of this field, the processor can postpone fetching of the instruction following the current one for the specified number of cycles (up to four), during which the FLUX hardware creates and inserts the NOP instructions into the processor pipeline.

Table 2 presents architectural and performance parameters of the FLUX-1 microprocessor.

6.   20-GHz clock frequency target

The major design challenge in FLUX-1 is an aggressive clock frequency target of 20 GHz. In order to achieve this goal, FLUX-1 blocks are ultra pipelined with a typical pipeline stage consisting of one clocked gate (serving as a pipeline latch simultaneously with performing its Boolean function), a few non-clocked elements such as splitters/mergers (the latter performing non-clocked *or*-like functions), and line drivers/receivers. Interconnect delays in critical paths are made relatively small (~10–12 ps, *i.e.* 20–25% of the cycle time) thanks to the localized communication model discussed above.

A different approach, wave pipelining, is used to implement the instruction memory, whose read latency is approximately three times larger than the clock cycle time of 50 ps. This technique is used in modern CMOS design to improve processor cycle time by pipelining functional units and caches without using intermediate pipeline latches with their additional synchronous clocking overhead resulting from flip-flop delay, set-up time, and clock jitter/skew. The FLUX instruction memory is pipelined across three stages with three waves of pulses propagating simultaneously one by one over non-clocked splitters/mergers and transmission lines in the memory address decoder. This design allows us to provide the instruction memory bandwidth of one (sequential) dual-op instruction per cycle.

Besides the high-bandwidth memory, a 20-GHz processor with an instruction fetch and issue rate of one instruction per cycle (in the absence of branches) must have very fast instruction-select logic associated with its program counter (PC). Each cycle, the FLUX-1 hardware has to be able to:

1) choose which address (the current PC value, delayed PC, or branch address) is to be sent to the instruction memory,

2) calculate the next PC address, and

3) read the most significant bit (the fetch delay flag for the next cycle) of the chosen instruction, and, if the flag is zero, postpone the start of the fetch operation for the next instruction by not writing its address into the PC until the required number of cycles has passed.

In order to fit this latency into one 50-ps cycle, the PC and its update circuitry were "fused" with the instruction memory word (row line) decoder. This move allows us to cut the latency of two steps: conversion from binary instruction addresses into memory word line (row) addresses, and next PC-address calculation. A 16-bit PC is implemented as a linear shift register holding such a pre-decoded instruction word address with additional inputs from the branch unit and scan-path logic. As a result, the next PC calculation can be done very quickly by shifting the contents of the PC by one bit. For each conditional/unconditional branch operation, the branch unit converts its binary 4-bit target address encoded in its instruction into the linear 16-bit address with only one non-zero bit corresponding to the position of the target instruction in the memory arrays (so-called "one-hot" encoding). The instruction-select stage is pipelined at an extremely fine-grain level, with all the critical path work done by only three non-clocked gates (two *or* and one *and* operations implemented by mergers and NDRO, respectively).

7. Chip design

Figure 3 shows a FLUX-1 chip micrograph. The chip is designed with the use of a full-custom design flow. In particular, the first task was the development of a new

RSFQ gate library[14] designed for TRW's technological process. This task is accomplished with the use of SUNY CAD tools, such as the Josephson junction circuit simulator PSCAN,[3] circuit optimizer Cowboy,[4] and quasi-2D inductance matrix calculator Lmeter[5] with a back-to-PSCAN annotator Lm2sch, as well as TRW libraries, all integrated with the Cadence CAD tools.

During the next step, the gate-level FLUX-1 design was developed in a way that allowed it to be partitioned into a small set of elementary design elements, such as a one-bit ALU and register block, a 2×2 memory sub-array, a one-bit shift register, *etc.* Each of these components is designed as a custom circuit macro optimized for high performance and small footprint. Placement and wiring of all elements (chip layout) is done manually with neither synthesis nor automatic place-and-route tools used.

All FLUX-1 blocks operate synchronously with their clock supplied by the clock controller. The controller has several test access ports (10 I/O pads total) connected to the external test equipment, allowing the processor clock frequency to be varied in a range from fractions of a hertz in a single-step mode, up to several hundreds of MHz using an external clock, and more than 20 GHz in "normal" mode using an on-chip clock generator with a programmable "run length" (the number of clock pulses to be generated). This clock is distributed throughout the chip using multiple-stage binary trees built of splitters.

Figure 3. FLUX-1 chip micrograph. There are ~66K Josephson junctions on a 140-mm² die. Photo courtesy of Lynn Abelson and George Kerber, TRW, Inc.

Even if these trees were perfectly balanced (which is not the case for FLUX-1), clock skew caused by thermal fluctuations and to a lesser extent technological process variations in these splitters could be quite large in any RSFQ processor with thousands of clocked gates. To solve the problem, FLUX-1 incorporates circuit-level clock delay adjustment capabilities. The clock controller drives four clock-domain buffers associated with each of the four major FLUX blocks: memory with its instruction-select logic and PC, the branch unit, instruction decoders, and the processing core including all ALUs, registers, and input/output ports. These clock-domain buffers are used for custom inter-domain deskewing of clock signals. Each clock-domain buffer represents a Josephson transmission line (JTL) consisting of 20 JJs in series. The bias currents of these JTLs are controlled from the outside, allowing the delays of individual junctions in these clock-domain buffers to be changed in order to create the desired clock skew on the order of ~5–7 ps. The elimination of signals generated by remote clocked gates from the FLUX-1 critical path allows us to keep the contribution from the data and clock skew relatively low (~8–10% of the cycle time).

The dc bias currents used for powering FLUX-1 RSFQ gates are in the range of ~0.2 mA for a transmission line transceiver and 0.4 mA for a splitter up to ~1.3 mA and ~1.7 mA for an AND gate and TRS flip-flop, respectively. With the average bias current of ~0.1 mA per Josephson junction, the total bias current is ~7 A. The 8 dc pads available in a FLUX-1 chip distribute this current to eight FLUX-1 "power regions" each with an almost equal number of junctions, limiting

Feature	Parameter
Technology	4 kA/cm^2 Nb-trilayer
Min. junction size	1.75-μm (in diameter)
Metalization	4 Nb layers (including one ground)
Circuit family	Rapid single flux quantum (RSFQ)
Max. clock rate	17–20 GHz
Max. I/O data rate	5 GHz via 8-bit I/O ports
Power dissipation	15 ± 20% mW at 4.2K
JJ count	65,759
Chip size	10.6 mm × 13.2 mm
Packaging	Flip-chip (solder bumps): 20 5-GHz I/O pads 56 low frequency test pads 8 dc pads
MCM	1.25 inch × 1.25 inch 2 Nb metal layers 56 signal + 16 DC pads

Table 3. FLUX-1 chip physical data.

the maximum bias current to be carried into the cryostat to ~1 A per copper lead. The power dissipation of a FLUX-1 chip (without the power dissipated in the dc copper leads carrying the bias current) is the sum of the static power dissipated in bias registers (~ 14 mW at 2 mV) and the dynamic power dissipated in shunt resistors when junctions switch. The value of the dynamic power is data- and program-dependent. With less than one-fourth of all FLUX junctions switched in each cycle and with power of ~0.06 mW dissipated per switching event, the dynamic power consumption does not exceed 1 mW at 20 GHz. Thus, the maximum power dissipation of a FLUX-1 chip is on the order of 15 mW. The additional power dissipated in the dc copper wires and BeCu spring fingers contacting the chip in the Z-cryostat's probe[14] is on the order of ~ 0.1–0.4 W.

FLUX-1 has full built-in testing (scan-path) capabilities, giving access to the contents of all registers and memory during testing. The chip has 10 test regions each with separate scan-path logic connected to I/O test ports. Within these regions, each bit of any register/memory word has additional inputs and outputs connected to adjacent bits of the same or other registers/memory words, forming a sequential scan path of the region. Each scan path operates as a long shift register whose data and control terminals are connected to four low-speed test pads: *data-in*, *external clock-in*, *data-out*, and *external clock-out*. Through these scan paths, data can be shifted bit by bit into registers/memory when loading their initial values and then shifted out when checking the results of computation under control of clock signals generated by the test equipment. There are 56 low-speed test pads in FLUX-1 used for scan path testing, clock deskewing, and I/O port tweaking. Table 3 summarizes physical characteristics of the FLUX-1 chip.

A cycle-accurate simulator with other programming tools was developed for the FLUX architecture verification, and VHDL simulation was used for partial functional verification of the FLUX-1 units. The full circuit-level simulation of a FLUX-1 chip with ~66K JJs was not possible due to the current state of superconductor CAD tools and inadequate computational resources available at SUNY. Instead, a multiple-chip approach with design, fabrication, and testing of several intermediate chips with increasing complexity was used. Eleven such chips containing individual FLUX gates and components were designed and tested before the first complete FLUX-1 chip went into fabrication after full layout-versus-schematics (LVS) and design rule checking (DRC) in June 2001.

8.   Current status

The first FLUX-1 "niobium" was delivered in August 2001. Currently, the processor is under testing at TRW, Inc. During testing, a cryogenic probe with a FLUX-1 chip flip-chip bonded to a Nb MCM is placed into a cryostat mounted on top of a liquid-helium storage dewar.[14] The multi-channel Octopux system[15] connected by shielded cables to the cryogenic probe is used to apply test vectors, read out, and analyze test results, as well as control differential dc current sources. Our general testing strategy is to check first that the SFQ/dc and dc/SFQ converters in the low-speed I/O test ports as well as all scan paths are fully operational before going on to the testing of individual FLUX blocks, and finally to low- and high-speed testing of a full FLUX-1 processor with a set of programs and test vectors generated by the FLUX software development and simulation tools.

Acknowledgments

This work is a result of the collaboration between the architecture, implementation and fabrication teams at SUNY Stony Brook and TRW, Inc. with management support from the JPL (NASA). I thank the current members of our project, namely Paul Bunyk (circuit-level design including the final layout of the chip), Quentin Herr (high data rate drivers), Mike Wire (MCM), Lynn Abelson and John Spargo (fabrication) from TRW, Inc, Arnold Silver and Alan Kleinsasser (project management) from the JPL (NASA) for their key contributions into the FLUX-1 design project. I also thank the former members of our SUNY team, Kostya Likharev and Dmitri Zinoviev, for their contributions to the gate library design and interconnection issues. This work would be impossible without support from our sponsoring agency, DoD, and in particular Doc Bedard.

References

1. K. K. Likharev, and V. K. Semenov, "RSFQ logic/memory family: A new Josephson-junction digital technology for sub-terahertz–clock-frequency digital systems," *IEEE Trans. Appl. Supercond.* **1**, 3 (1991).
2. W. Chen, A.V. Rylyakov, V. Patel, and K. K. Likharev, "Rapid single-flux quantum T-flip-flop operating at 770 GHz," *IEEE Trans. Appl. Supercond.* **9**, 3212 (1999).
3. S. V. Polonsky, V. K. Semenov, and P. N. Shevchenko, "PSCAN – personal superconductor circuit analyzer," *Supercond. Sci. Technol.* **4**, 667 (1991).
4. S. Polonsky, P. Shevchenko, A. Kirichenko, *et al.*, "PSCAN'96: New software for simulation and optimization of complex RSFQ circuits," *IEEE Trans. Appl. Supercond.* **7**, 2685 (1997).
5. P. I. Bunyk, and S. V. Rylov, "Automated calculation of mutual inductance matrices of multilayer superconductor integrated circuits," *Abstracts Int. Supercond. Electronics Conf.*, NIST, Boulder, CO (1993).
6. K. Gai, Q.P. Herr, V. Adler, *et al.*, "Tools for the computer-aided design of multigigahertz superconducting digital circuits," *IEEE Trans. Appl. Supercond.* **9**, 18 (1999).
7. G. K. Kerber, L. A. Abelson, M. L. Leung, *et al.*, "A high density 4 kA/cm^2 Nb Integrated Circuit process," *IEEE Trans. Appl. Supercond.* **9**, 1061 (2001).
8. M. Dorojevets, P. Bunyk, D. Zinoviev, and K. Likharev, "Superconductor electronic devices for petaflops computing," *FED J.* **10**, 3 (1999).

9. M. Dorojevets, "COOL Multithreading in HTMT SPELL-1 Processors," *Int. J. High-Speed Electronics Systems* **1**, 247 (2000).

10. M. Dorojevets, P. Bunyk, and D. Zinoviev, "FLUX chip: Design of a 20-GHz 16-bit ultrapipelined RSFQ processor prototype based on 1.75-μm LTS technology," *IEEE Trans. Appl. Supercond.* **9**, 326 (2001).

11. Y. Naveh, D. Averin, and K. Likharev, "Physics of high-j_C Josephson junctions and prospects of their RSFQ VLSI applications," *IEEE Trans. Appl. Supercond.* **9**, 1056 (2001).

12. P. Bunyk, P.-M. Koenig, and D. Zinoviev, "Passive interconnects: a revolutionary approach to RSFQ system design," *Abstracts 8th ISEC*, Osaka, Japan (2001).

13. P. Bunyk, K. Likharev, and D. Zinoviev, "RSFQ technology: Physics and devices," *Int. J. High-Speed Electronics Systems* **1**, 257 (2001).

14. M S. Wire, D. J. Durand, A. H. Silver, and M. K Wagner, "A multiple pin, flip-chip system for microwave and gigabit per second device testing at variable temperatures," *Rev. Sci. Instrument.* **2**, 1542 (2001).

15. D. Zinoviev, and Yu. Polyakov, "Octopux: An advanced automated setup for testing superconductor circuits," *IEEE Trans. Appl. Supercond.* **7**, 3240 (1997).

Silicon … Beyond Silicon:
Beginning of the End or End of the Beginning?

I. Lagnado and P. R. de la Houssaye

SPAWAR Systems Center San Diego, San Diego, CA 92152, U.S.A.

1. Introduction

In the early 1960's, at the beginning of the electronics revolution, silicon integrated circuits built their current dominance fundamentally and pervasively on tailor-made materials, starting at the atomic level. Early development in thin-film deposition techniques, particularly chemical vapor deposition (CVD) and molecular beam epitaxy, led the way to accurate control over material constituents in "atomic amounts," in order to form the active part of high-performance devices.

In the early 1970's, our team at Space and Naval Warfare Systems Center, San Diego, focused its vision to achieve a structure based on CMOS, an unknown and unproven technology at that time. One important attribute of that vision was the affordable implementation using devices with ever-increasing speed, wider bandwidth and lower power dissipation. During the eighties and nineties, CMOS became the mainstay technology and workhorse of the electronics revolution. It remains the dominant circuit configuration in today's systems because it offers low power, the largest signal-to-noise ratio, process simplicity, and flexibility in design. In this context,

- while new materials and highly engineered novel structures, such as silicon-germanium (SiGe) and silicon-on-insulator (SOI) for computing devices and processors, have demonstrated an incremental improvement in performance,

- our pursuit of the integration of SiGe on sapphire substrates has led to truly outstanding device performances that could not otherwise be obtained. The SiGe-on-sapphire configuration has helped integrate for the first time, without loss of performance, both analog and digital functions.

We have, therefore, projected increased levels of integration, increased performance, and lower cost for future integrated systems-on-a-chip. These systems will continue to fuel the exponential improvements predicted by Moore's Law, *e.g.* sub-50-nm devices, many tens of GHz speeds (f_T, f_{max} in excess of 200 GHz for both carriers) and billions of transistors per chip by the year 2010. These systems also will enable new applications to be realized at the high end of the performance spectrum. The driving force for the continuing explosive growth of the semiconductor industry, which doubles in size every 3.5 years (a corollary of Moore's Law) is continued process improvement. That was also demonstrated; CMOS on SOS reduces process complexity, as well as parasitic capacitance loading. Moreover, the introduction of Ge into the silicon wafer offers novel structures in strained material (tensile Si or compressive SiGe layers, 3–10 monolayers thick), fixes some existing device problems, and enhances performance without major complexity or cost penalties.

The advances resulting from the developments carried out in the context of our first visionary technological push have led to the investigation of SiGe on sapphire, a new vision for the next decade. Therefore, we will discuss the "past" results, as a prelude to the "present," with compelling reasons to pursue our new vision of CMOS on SiGe on sapphire.

2. The past

Silicon-on-sapphire (SOS) has all the advantages of other SOI technologies as well as many others specifically relevant to microwave circuits. These advantages include reduced self-heating effects (due to sapphire's higher thermal conductivity, 0.46 W/cm·K, compared to 0.014 W/cm·K in SiO_2), reduced device parasitic capacitances, radiation hardness, reduction of latch-up in CMOS structures, higher packing density, and improved isolation. SOS also has lower minority carrier lifetimes (~1 ns) that result in a higher source-drain breakdown voltage and a reduced parasitic bipolar gain. Another SOS characteristic to note is that the silicon film is under compressive stress. This stress splits the light and heavy hole valence bands leading to increased hole mobilities over that of bulk silicon. At the same time, this stress also causes lower electron mobilities as compared to bulk, similar to the effects seen in SiGe on Si. In SOS, p-MOS and n-MOS devices are more closely matched than in other CMOS variants.

Sapphire, and more generally polycrystalline sapphire (alumina), is known to have excellent dielectric properties. The dielectric constant, dielectric loss tangent, and resistivity of sapphire are ε_r = 9.39, tan δ < 0.0001 at 3 GHz, ρ = 10^{14} Ω·cm. Hence, in addition to the desirable traits of other SOI technologies, sapphire wafers make excellent microwave substrates for passive elements such as transmission lines and inductors, allowing on-chip integration with active devices.

The development of SOS started with Manasevit et al.,[1] who described a technology to achieve single-crystal silicon on a sapphire substrate in 1964. Since then, as evidenced from several well documented reviews,[2,3] the technological advances achieved for SOS can be traced, in parallel, with the evolutionary establishment of the silicon VLSI industrial infrastructure from p-MOS to n-MOS to n-MOS E/D and to the current workhorse CMOS. In the late 1970's, Lau et al.[4] demonstrated that the crystalline quality of SOS films could be improved by utilizing a silicon implant to create a buried amorphous layer followed by a thermal anneal, which causes regrowth of an improved film from the surface downward. This process is labeled "solid phase epitaxy" (SPE).

The application of the technique has caused a resurgence in the interest in SOS technology for its use in high-speed low-power CMOS circuitry implemented with sub-micrometer minimum device feature size. Almost concurrently, two research teams[5,6] applied the SPE (or double-SPE, DSPE) process to achieve device-quality crystal in 200-nm thin-film silicon on sapphire. G. Garcia and R. Reedy[7,8] extended their work to 20-nm thin films, commensurate with a 80–100 nm minimum device feature size. Three main observations were made. First, no aluminum out-diffusion from the substrate was observed, as shown from the SIMS profiling (Fig. 1). The SIMS data give a uniform background concentration for ^{27}Al of approximately $1-3 \times 10^{15}$ cm^{-3}, within the minimum detection margin of error for both the bulk Si and the DSPE-improved SOS samples.[6] Secondly, higher carrier mobilities ($\mu_e = 500$, $\mu_p = 200$ cm^2/V·s) were noted, along with a drastic reduction (albeit not elimination) of deep traps ($<10^{11}$ cm^{-2}) associated with the Si/sapphire interfacial region. The essential steps of the DSPE technique are deep amorphization by a 170 keV ^{28}Si implant at a dose of 1×10^{15} cm^{-2} followed by a thermal anneal of 550 °C for 2 hours, then 1050 °C for 1 hour. The top surface of the Si film is then rendered amorphous by a 100 keV Si implant at a similar dose of 1×10^{15} cm^{-2} followed by the same anneal cycle. Thinning the film is accomplished by growth of an oxide at 875 °C in steam. The film thickness is measured by interferometry and spreading resistance.

The 2.2 MeV ^{4}He$^+$ channeled RBS data and the high resolution TEMs of both "as-received" (non-implanted or non-improved) SOS and DSPE-improved SOS samples are shown in Fig. 2. The virtual elimination of "twin" defects and the ^{4}He$^+$ de-channeling in the DSPE-improved SOS wafers coming very close to that obtained for the bulk silicon demonstrate the substantial improvement achieved.

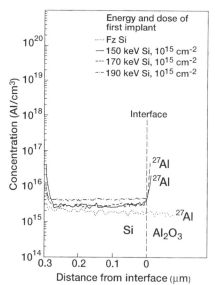

Figure 1. SIMS profile showing absence of Al in Si layer.

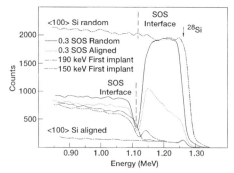

Figure 2. Channeled RBS data of as-received and DSPE-improved SOS samples for two different implant energies (150 and 190 keV).

3. The present

The measured characteristics of a front-end receiver at 2.4 GHz[9] and frequency dividers[10] at >20 GHz illustrate a real technological renaissance for silicon-on-sapphire. The demonstrated performance, unachievable in bulk silicon technologies, of a transmit/receive switch, mixer, and low noise amplifier[9] make possible the integration of an rf front end and VLSI digital circuits on the same substrate. The resulting system-on-a-chip implementation can reduce cost and increase overall performance for functions targeted at wireless/satellite communications, radar, image processing and others in the military, industrial, scientific, and commercial sectors.

A specific requirement for the function of the transmit/receive (TR) switch, needed to connect the antenna to the receiver chain (receive mode) or the transmitter chain (transmit mode), is that of very low insertion loss. It is important not to degrade the signal-to-noise ratio prior to the low-noise amplifier on the receive side, or reduce the efficiency on the transmit side.   In addition, nonlinearity must be small to accommodate large output power and isolation from one side to the other must be high. The measured data demonstrates the excellent suitability to this combination of requirements of the switch implemented in DSPE-improved SOS. The insertion loss is around 1.7 dB and the isolation is greater than 30 dB at 2.4 GHz. At 5 GHz, the insertion loss increased from 1.7 dB to only 2.0 dB and the isolation remained greater than 25 dB; while the isolation is the same, the insertion loss for SOS is three times lower than for SIMOX. The switch also displays excellent linearity and power handling capability. The two-tone measurements (f_1 = 2.4 GHz, f_2 = 2.425 GHz) show an input-referred third order intercept point (IP3) of 18 dBm.

The mixer, used in up- and/or down-conversion of the carrier frequency (rf or 2.4 GHz) to the intermediate frequency (IF or 250 MHz) has good isolation of rf and the local oscillator (LO at 2.15 GHz, 0.7 dBm power), as well as high IP3 (5 dBm) and low noise figure. Unlike III-V HBTs, the selected SOS design targets high IP3 (18 dBm) with little gain (the SOS mixer has 5 dB conversion loss), as it is easier to increase the gain in the receiver chain than to increase IP3.

The one-stage low-noise amplifier (LNA), a critical element of the receive path, has very low noise as well as high gain and output IP3 (OIP3) compared to a 2-stage HEMT LNA. Spiral inductors and MIM capacitors were used to match the FETs to the minimum noise figure at Γ_{opt}. The LNA was operated at V_{DS} = 1.5 V and V_{GS} = 0.7 V. The dc current consumption at low input power levels is 8.8 mA (while it rises to 9.9 mA for an input power corresponding to the 1 dB compression point). This corresponds to an average power dissipation of 14 mW. Using the spiral inductors, the LNA is well matched for gain (10 dB) and noise figure (2.8 dB) at 2.4 GHz. Two-tone (f_1 = 2.4 GHz, f_2 = 2.425 GHz) linearity measurements of the LNA were performed.   The output referred 1-dB compression point and third order intercept (OIP3) are 4 dBm and 14 dBm, respectively.

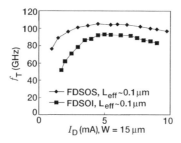

Figure 3. Transition frequency f_T vs. I_D for fully-depleted SOS and SOI.

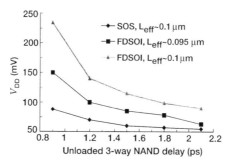

Figure 4. Unloaded 3-way NAND delay (ps) for SOS and fully-depleted SOI.

These results represent the highest-frequency CMOS low-noise amplifier reported to date.[11,12] The noise characteristics of this LNA are similar to Si bipolar circuits and satisfy wireless requirements. On the basis of simulation, even lower noise figure would be possible if the metal thickness used to realize the on-chip inductors were greater. The excellent linearity observed may be attributed to the good turn-off and low output conductance of the SOS FETs, together with the absence of the body effect. The combination of low noise figure, high OIP3, and low power consumption is desirable for wireless receiver circuits.

With further scaling, fully depleted devices with 100-nm gate length on 30–40 nm thin DSPE-improved SOS were characterized and were compared to similar devices implemented on SOI and bulk Si substrates. The 100-GHz f_T data for SOS are shown in Fig. 3; the heating effect in SIMOX devices is responsible for the lower value of f_T.

Very fast unloaded CMOS switching speeds have been observed (<10 ps delay). For three-way NAND ring oscillators, SOS devices are faster than SIMOX

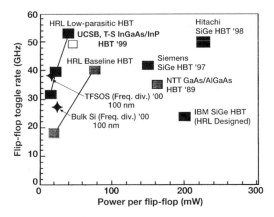

Figure 5. Flip-flop toggle rate *vs.* power per flip-flop in different technologies.

ones because of the higher hole mobility and lower threshold voltage (Fig. 4). Circuit functions such as a frequency divider[10] at 26 GHz have the highest frequency of operation for similar Si functions. Figure 5 compares several material implementations of the frequency divider.

4.   The future

Recent advances in SiGe epitaxial growth indicate that SiGe-based strained-layer MODFETs may be promising alternatives to III-V MESFETs and HEMTs for future high-speed analog communications applications. Electron and hole mobilities well in excess of bulk Si mobilities can be realized in tensile-strained Si quantum wells (QWs)[13] and compressive-strained SiGe[14] or pure Ge QWs,[15] respectively. These improvements have led to *n*-MODFETs with f_T = 62 GHz[16] and f_{max} = 120 GHz[17] and *p*-MODFETs with f_T = 70 GHz and f_{max} = 85 GHz.

Höck[18] and Hammond *et al.*[19] have both investigated a Ge QW grown on a $Si_{0.4}Ge_{0.6}$ relaxed buffer layer and showed the impressive performance of the Ge-channel MODFETs. However, a significant drawback of these devices is the high Ge content (60%) of the relaxed buffer layer which leads to increased surface roughness and defect density. In addition, 60%-Ge-content relaxed buffer layers are not well suited for integration with strained Si channel *n*-MODFETs.

We have observed that the *p*-MODFETs dc and rf performance figures are comparable to those of *n*-MODFETs suggesting the possibility of very high-speed complementary operation,[20] a capability not available in current III-V technology. Furthermore, *p*-MODFETs may actually be preferable to *n*-MODFETs for unipolar applications due to their comparable performance and the fact that they tend to be simpler to fabricate, particularly for self-aligned T-gated devices. Channels grown on buffer layers with $x < 40\%$ are much more attractive from a technological point of view, because they have better surface quality and are compatible with n-MODFETs. In order to address this issue, a new "composite-channel" layer structure design that utilizes a 30–35%-Ge-content buffer layer was developed. This heterostructure is still capable of producing hole mobilities as high as 1300 cm^2/Vs. In our work, we have applied the knowledge of material properties to the growth of strained layers of Si$_{0.2}$Ge$_{0.8}$ on relaxed Si$_{0.7}$Ge$_{0.3}$ layers on thin-film SOS for the fabrication of SiGe p-MODFETs.

Losses due to the conductivity of the Si substrate used in standard SiGe MODFET designs is another serious obstacle for applications in the microwave frequency regime. Insulating substrates, such as sapphire, provide an optimum solution. We have demonstrated the epitaxial growth of high-mobility modulation-doped composite-channel heterostructures on SOS substrates, and described the resultant outstanding rf characteristics of ≤ 100 nm T-gate p-MODFETs fabricated on these layers.

The composite-channel heterostructure device[21] on SOS has the following basic structure. A Si$_{1-x}$Ge$_x$ graded layer is first grown where the Ge alloy composition is increased from $x = 0$ to 30%. This layer is followed by a constant-alloy-composition Si$_{0.7}$Ge$_{0.3}$ layer, a 3.5-nm Si$_{0.7}$Ge$_{0.3}$ boron-doped supply layer, a 5-nm Si$_{0.7}$Ge$_{0.3}$ offset layer, an 8.5-nm Si$_{0.2}$Ge$_{0.8}$ QW, and finally a 9-nm Si$_{0.7}$Ge$_{0.3}$ capping layer. The quantum well is grown by nucleating a thin (<1 nm) Ge seed layer, followed by the 80% channel, thereby creating a "composite-channel" layer structure. This layer structure was grown on an SOS substrate as well as on a bulk Si control wafer. The devices had a gate length of 100 nm.

The room-temperature Hall mobility and sheet carrier density[21] of this composite-channel layer structure grown on an SOS wafer were 800–1200 cm^2/V·s and $3.1–2.5 \times 10^{12}$ cm^{-2} at room temperature, respectively. The room-temperature output (transfer) characteristic for devices with 100-nm gate length showed practically no difference between SOS and Si. Experimental data has also shown that the best SOS transistor has a maximum extrinsic transconductance of 377 mS/mm which is, to the authors' knowledge, the highest ever reported for alloy-channel p-MODFETs. The device has a corresponding output conductance of 25 mS/mm leading to a maximum dc voltage gain of 15. The only apparent degradation of the device performance due to the SOS substrate is a roughly one-order-of-magnitude higher gate leakage current compared to the Si monitor, a result that we again attribute to the increased defect density of the SOS wafers.

Figure 6 shows frequency-dependent plots[22] of the forward current gain ($|h_{21}|^2$) and the maximum unilateral gain (MUG) for a 0.1×50 μm^2 p-MODFET on SOS. Values of $f_T = 49$ GHz and $f_{max} = 116$ GHz, were obtained after de-embedding the contact pads; the latter value being the highest f_{max} ever reported for a SiGe p-MODFET. Figure 7 illustrates the fact that f_T and f_{max} saturate at a very low bias voltage; f_{max} reaches 100 GHz at $V_{DS} \sim -0.6$ V and is sustained over a wide bias range. Figure 8 further indicates low noise at high frequencies. This performance level should enable an entirely new technology. One of the characteristics of SOS technology is the higher hole mobility, as the transport of carriers is through "light" holes. Therefore, a primary driving factor in pursuing SOS investigation has been to increase the performance of the p-channel FET through higher hole mobility, to enhance CMOS circuitry performance. In this context, our objective is to increase p-channel mobility to a level close or equal to n-channel mobility in CMOS circuitry by use of a SiGe heterostructure MODFET.[23-25] By confining the holes to an undoped channel away from the Si/SiO_2 interface in a strained layer where the hole mobility is enhanced, a significant improvement in hole transport can be realized. More recently, 3–4× increases in both hole mobility and electron mobility using SiGe films with high levels of germanium (>70%) have been

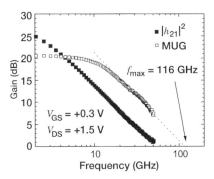

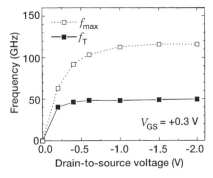

Figure 6. Highest reported f_{max} = 116 GHz for a p-FET in any material system (0.1 × 50 μm^2 composite-channel SiGe p-MODFET on SOS).

Figure 7. The f_T and f_{max} saturate at a very low bias voltage; f_{max} reaches 100 GHz at V_{DS} ~ –0.6 V.

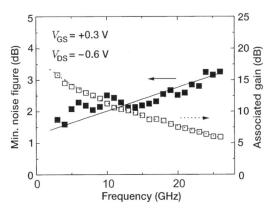

Figure 8. Minimum noise figure and the associated gain *vs.* frequency for a 0.1×90 μm^2 p-MODFET on SOS.

reported.[26] Performance of both polarity devices can be enhanced in the same set of layers using tensile strained Si on relaxed SiGe for the *n*-channel devices and compressively strained high-germanium-content SiGe layers for the *p*-channel devices. This structure requires, however, thick SiGe graded buffer layers, which are incompatible with ultrathin silicon-on-insulator technology. Epitaxially grown silicon-on-sapphire (SOS) has micro-twin densities (after improvement) in the 10^4 cm^{-2} range, resulting in an equal density of large faceted pits on ultra-high vacuum chemical vapor deposition (UHV-CVD) grown SiGe *p*-MODFET wafers. We should expect similar issues with deposited SiGe layers on sapphire, as discussed above. Although individual devices have been fabricated and tested on such wafers, the high density of pits makes these substrates unsuitable for ICs. Additionally, it is necessary first to grow a strain-relieved (600–800 nm thick) SiGe buffer layer[27,28] underneath the active layers, reducing the advantages of using a sapphire substrate.

However, the technique of bonding a thin (<100 nm) relaxed layer of $Si_{1-x}Ge_x$ ($x < 35\%$) provides a greater potential for an affordable reliable substrate, compatible with the deposition of strained $Si_{1-x}Ge_x$ ($x > 70$–75%) on the defect-free relaxed $Si_{1-x}Ge_x$ ($x < 35\%$) layer. It is well known that wafer-bonding methods rely on van der Waals attractive forces between the mirror-polished surfaces. The use of an interfacial medium between them, subsequent high temperature annealing, and a complete understanding of the underlying physical, chemical and electrical bonding processes are a prerequisite to achieve a permanently reliable bond of a very thin crystalline semiconductor to a supporting, preferably insulating, substrate.

A case in point is the successful "Smart-Cut" approach from SOITEC[29] which is, incidentally, not suitable for rf mixed-signal functions (integrated on a single chip) due to the high conductivity of the Si substrate. Other attempts to bond a thin Si layer on sapphire,[30–32] with and without an intermediary compliant SiO_2 medium, have met with various degrees of success. The presence of a thin oxide layer (>20 nm) prevents the void formation that is induced by poor water absorption by the native oxide. Due to the ready availability of sapphire substrates with a "bonding" finish, development focuses on materials integration alternatives and issues in order to achieve a stable interface at high temperatures (>550–850 °C). The next two figures[33] illustrate the evolution of this approach, from direct bonding of Si or SiGe on sapphire, to the use of SiO_2 as an integration medium that functions as a "sponge" for bond reaction products (H_2O, OH, H_2) to avoid the formation of bubbles at the interface during annealing.

Figure 9[33] illustrates the direct room-temperature bonding of a 92.5-nm thin silicon layer to sapphire, resulting from thinning the original bulk Si wafer. The bonding area is close to 75% of the 100-mm diameter of the substrate. Subsequent to the thinning, the annealing at 550 °C for 15 minutes results in a blistered film delaminated due to the expansion of trapped vapor at the interface. Also, the (004) high-resolution x-ray diffraction (HRXD) rocking curve data are compared to the theoretical calculation of the rocking curve from a defect-free Si layer. There is

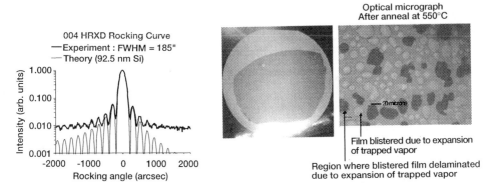

Figure 9. High-resolution x-ray (004) rocking curve data recorded from a 92.5-nm Si layer directly bonded to a sapphire wafer at room temperature, compared to the theoretical calculation of a rocking curve from a defect-free Si layer (a). There is good agreement between experiment and calculation. The result of thermally cycling the directly-bonded SOS structure to 550 °C (b).[33]

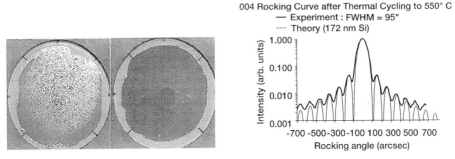

Figure 10. Photographs comparing the results of thermally cycling two SOS structures to 550 °C (the left wafer has no intermediary dry oxide at the bonded interface, whereas the right wafer has a 350-nm dry oxide). An x-ray diffraction (004) rocking curve from a bonded structure with an intermediate dry oxide after thermal cycling to increasingly higher temperatures (right). The rocking curve is compared to a theoretical calculation corresponding to a defect-free Si layer.[33]

agreement between experiment and calculation. The bonded silicon layer is relaxed.

Figure 10[33] demonstrates the results of thermally cycling two SOS structures to 550 °C. The structure on the left has no intermediary dry oxide, whereas the one on the right has a 350-nm dry oxide at the bonded interface. No additional blisters are observed if the sample on the right is subsequently cycled up to 850 °C. The dry oxide absorbs the interfacial OH groups that would otherwise form blisters. The experimental (004) rocking curve is shown for the wafer structure on the right, after thermally cycling the bonded structure to increasingly higher temperatures. The data is compared to a theoretical calculation of the rocking curve from a defect-free Si layer. The bonded structure can be thermally cycled to 550 °C without the introduction of enough structural defects to broaden the rocking curve. Also, as was observed in Fig. 9, the bonded structure with the intermediate dry oxide is relaxed.

5.   Conclusions

The incorporation of thin-film SiGe or Si on sapphire (either through SPE or bonding) leading to strained SiGe heterostructure n- and p-FETs characterized by superior transport carrier properties, high dynamic performance ($f_T, f_{max} > 200$ GHz), and low noise (~1–1.5 dB at 20 GHz) at high frequencies will enable an entirely new technology. This technology combines low-power high-speed analog and digital CMOS functions capable of very large scale integration, comparable to (if not higher performance than) III-V MESFETs and HEMTs, and at a greatly reduced cost due to compatibility with the Si CMOS IC manufacturing infrastructure. This enhanced silicon technology will enable the implementation of truly single-chip systems. These highly integrated systems will cost less to produce, occupy less volume, and draw less power than the same system realized with several chips. The incorporation of sapphire, an optimum rf substrate, will provide the required isolation between the front end and the noisy digital back-end functions, thus eliminating the need for III-V low-noise amplifiers. The resulting impact of the combined thin-film SOS and SiGe technology on the marketplace, both nationally and internationally will be quite revolutionary, as no other material can provide a complementary technology as efficiently, both technically and economically.

References

1.   H. M. Manasevit and W. S. Simpton, "Single-crystal silicon on sapphire substrate," *J. Appl. Phys.* **35**, 1349 (1964).

2.   G. W. Cullen and C. C. Wang, eds., *Heteroepitaxial Semiconductors for Electronic Devices*, Berlin: Springer-Verlag, 1978.

3.   S. Cristoloveanu, "Silicon films on sapphire," *Rep. Prog. Phys.* **50**, 327 (1987).

4.   S. S. Lau, S. Matteson, J. W. Mayer, P. Revesz, J. Gyulai, J. Roth, T. W. Sigmon, and T. Cass, "Improvement of crystalline quality of epitaxial Si layers by ion-implantation techniques," *App. Phys. Lett.* **34**, 76 (1979).

5.   T. Yoshii, S. Taguchi, T. Inoue, and H. Tango, "Improvement of SOS device state devices," *Jpn. J. Appl. Phys. Part 1* **21**, 175 (1982).

6.   R. E. Reedy, T. W. Sigmon, and L. A. Christel, "Suppressing Al outdiffusion in implantation amorphized and recrystallized silicon on sapphire films,"

Appl. Phys. Lett. **42**, 707 (1983).

7. G. A. Garcia and R. E. Reedy, "Electron mobility within 100 nm of the Si/sapphire interface in double-solid-phase epitaxially regrown SOS," *Electronics Lett.* **22**, 537 (1986).

8. G. A. Garcia, R. E. Reedy, and M. L. Burgener, "High-quality CMOS in thin (100 nm) silicon on sapphire," *IEEE Electron Dev. Lett.* **9**, 32 (1988).

9. R. Johnson, P. R. de la Houssaye, C. E. Chang, P. Asbeck, and I. Lagnado, "Advanced thin film silicon on sapphire technology: Microwave circuit applications," *IEEE Trans. Electron Dev.* **45**, 1047 (1998).

10. M. Wetzel, L. Shi, K. Jenkins, P. R. de la Houssaye, Y. Taur, P. M. Asbeck, and I. Lagnado, "A 26.5 GHz silicon MOSFET 2:1 dynamic frequency divider," *IEEE Microwave Guided Wave Lett.* **10**, 421 (2000).

11. A. N. Karanicolas, "A 2.7 V 900 MHz LNA and mixer," *Digest IEEE Intern. Solid-State Circuits Conf.* (1996), p. 50.

12. D. K. Shaeffer and T. H. Lee, "A 1.5-V, 1.5-GHz CMOS low noise amplifier," *IEEE J. Solid-State Circ.* **32**(5), (1997).

13. S. F. Nelson, K. Ismail, J. O. Chu, and B. S. Meyerson, "Room-temperature electron mobility in strained Si/SeGe heterostructures," *Appl. Phys. Lett.* **63**, 367 (1993).

14. K. Ismail, J. O. Chu, and B. S. Meyerson, "High hole mobility in SiGe alloys for device applications," *Appl. Phys. Lett.* **64**, 3124 (1994).

15. G. Höck, M. Glück, T. Hackbarth, H.-J. Herzog, and E. Kohn, "Carrier mobilities in modulation doped $Si_{1-x}Ge_x$ heterostructures with respect to FET applications," *Thin Solid Films* **336**, 141, (1998).

16. S. J. Koester, J. O. Chu, and R. A. Groves, "High-f_T n-MODFETs fabricated on Si/SiGe heterostructures grown by UHV-CVD," *Electronics Lett.* **35**, 86 (1999).

17. M. Zeuner, T. Hackbarth, G. Hock, D. Behammer, and U. Konig, "High-frequency SiGe n-MODFET for microwave applications," *IEEE Microwave Guided Wave Lett.* **9**, 410 (1999).

18. G. Hock, T. Hackbarth, U. Kohn, and U. Konig, "High performance 0.25 μm p-type Ge/SiGe MODFET," *Electronics Lett.* **34**, 1808 (1998).

19. R. Hammond, S. J. Koester, and J. O. Chu, "High transconductance 0.1 μm $Ge/Si_{0.4}Ge_{0.6}$ Schottky-gate p-MODFETs," *Proc. 57th IEEE Dev. Res. Conf.* (1999), p. 173.

20. M. A. Armstrong, D. A. Antoniadis, A. Sadek, K. Ismail, and F. Stern, "Design of Si/SiGe heterojunction complementary metal-oxide-semiconductor transistors," *Tech. Digest IEDM* (1995), p. 761.

21. S. J. Koester, R. Hammond, J. O. Chu, J. A. Ott, P. M. Mooney, L. Perraud, K. A. Jenkins, I. Lagnado, and P. R. de la Houssaye, *Proc. 7th Intern. Symp. Electron Dev. Microwave Optoelectronic Applications (EDMO-99)*, King's College, London, England (1999).

22. S. J. Koester, R. Hammond, J. O. Chu, P. M. Mooney, J. A. Ott, C. S. Webster, I. Lagnado, and P. R. de la Houssaye, *Proc. 58th IEEE Dev. Res. Conf.* (2000), pp. 31–32.

23. S. Verdonckt-Venderbroek, E. F. Crabbe, B. S. Meyerson, D. L. Harame, *et al.*, "High-mobility modulation-doped SiGe-channel *p*-MOSFETs," *IEEE Electron Dev. Lett.* **12**, 447 (1991).

24. T. P. Pearsall and J. C. Bean, "Enhancement- and depletion-mode *p*-channel Ge_xSi_{1-x} modulation-doped FETs," *IEEE Electron Dev. Lett.* **7**, 308 (1986).

25. D. Nayak, J. C. S. Woo, J. S. Park, K. L. Wang, and K. P. MacWilliams, "Enhancement-mode QW Ge_xSi_{1-x} PMOS," *IEEE Electron Dev. Lett.* **12**, 154 (1991).

26. A. Sadek and K. Ismail, "Si/SiGe CMOS possibilities," *Solid State Electronics* **38**, 1731 (1995).

27. P.M. Mooney, J. O. Chu, and J. A. Ott, "SiGe MOSFET structures on silicon-on-sapphire substrates grown by ultra high vacuum chemical vapor deposition," *J. Electronic Mater.* **29**, 921, (2000).

28. P. M. Mooney and J. O. Chu. "SiGe Technology: Heteroepitaxy and high-speed microelectronics," *Ann. Rev. Mater. Sci.* **30**, 335 (2000).

29. A. J. Auberton-Herve and M. Bruel, "Why can Smart-Cut change the future of microelectronics?" in: Y.-S. Park *et al.*, eds., *Frontiers in Electronics: From Materials to Systems*, Singapore: World Scientific, 2000, pp. 129-146.

30. T. Abe, K. Okhi, and A. Uchiyama, "Dislocation-free silicon on sapphire by wafer bonding," *Jpn. J. Appl. Phys.* **33**, 514 (1994).

31. P. T. Baine, H. S. Gamble, A Armstrong, B. M. Armstrong, and S. J. N. Mitchell, *Proc. 9th ECS Int. Symp. SOI Technol. Dev.* (1999), Vol. 3, p. 367.

32. Final communication to US Navy, Contract with Silicon Genesis.

33. Final report/communication to US Navy/SPAWARSYSCEN, SD-Contract N66001-00-M-1165 with the University of Wisconsin (Prof. T. Kuech and Dr. P. Moran, Principal Investigators).

Taming Tunneling

Michael J. Kelly

School of Electronics, Computing and Mathematics
University of Surrey, Guildford GU2 7XH, U.K.

1. Introduction

The gap between one-off device demonstration in the laboratory and volume manufacture is getting ever greater as device features become ever smaller. The challenges of achieving low-cost manufacture for tunnel devices sets the context for even greater challenges confronting the eventual manufacture of any active mesoscopic devices.

It is sobering to compare the ever greater number of new device ideas and prototypes emerging from the research laboratory, with the declining fraction that evolve into volume manufacture. The moves at the research frontiers towards mesoscopic electronics and beyond that into bioelectronics should have in their wake programmes that aim to take yesterday's discoveries into tomorrow's production. Without such programmes, a credibility gap between the research community and industry will open into an unbridgeable chasm. In this paper we describe some results that aim to take a semiconductor device the relies for its operation on quantum mechanical tunneling through a thin barrier layer from its prototype performance towards low-cost manufacture. It is sobering to recall that the original realisation of the concept, namely asymmetric $I–V$ characteristics from tunneling through thin semiconductor heterostructure barriers, is now thirty years old.[1] Although this paper focuses on a discrete device, the comments above and below apply *a fortiori* to any production involving integrated circuits.

2. Low cost manufacturing

Devices in volume production have a number of characteristics that are ($\checkmark$) or are not ($\times$) shared by heterojunction tunnel devices at this time:

superior performance ($\checkmark$)	high yield ($\times$)
design-tolerant ($\times$)	reverse engineered ($\times$)
right-first-time design ($\times$)	reproducible ($\times$)
reliable ($\checkmark$)	low cost ($\times$).

These qualities are interrelated, and if one could achieve a very high degree of reproducibility from a right-first-time design capability, the others would follow.

Workarounds that try to live with wide variations in performance from device to device drive up costs and seriously erode margins that are needed in the market place. Systems engineers typically want all devices to have less than ±10–15% absolute current variation about a specified value. They expect to use all the superior performance of new devices to improve the overall system capability, and not to be forced to trade some of that extra performance to cope with uncontrollable variability in performance between devices.

3.   The ASPAT diode

A thin AlAs tunnel barrier in an asymmetric doping environment in GaAs (the ASPAT diode) was first demonstrated a decade ago with a combination of superior properties for microwave detection — larger dynamic range, lower sensitivity to ambient temperature, lower noise — when compared with other devices such as the Schottky diode or the planar-doped-barrier diode.[2,3] The device consists of a 10-monolayer 100% pure AlAs barrier buried in a GaAs n-i-n multilayer structure that has an asymmetric doping profile from the top surface to the substrate: n^+ (750nm), n (40 nm), i (5nm), AlAs(2.85 nm), i (200nm), n (20 nm), n^+ (750 nm).

Early simulations drew attention to the extreme sensitivity of performance to the barrier parameters: ~360% change in current density per monolayer thickness change, and 30% increase in current density per missing 1% of Al in AlAs.[4] Later studies showed that electrical data could be very sensitive also to spatial fluctuations in the barrier properties.[5] Using commercial suppliers of both epitaxial wafers and materials qualification, a previous project to establish manufacturability was an abject failure: a 400% variation was achieved in current density at a fixed forward bias, between devices grown by MBE and MOCVD to the identical specification; a greater than 10% variability across single 2" wafers with opposite radial trends between MBE and MOCVD, and major run-to-run variations.[4,6]

4.   The present work

We have focused on MBE growth using a rapid *ex situ* calibration scheme to improve greatly on previous results, achieving standard deviations in current density of less than 6% within wafers by using a machine capable of handling 6" wafers, and 30% variation between wafers grown months apart. There is still some scope for further refinement to be able to meet engineering constraints. Only a right-first-time design capability is required now for low-cost manufacture.

We grow a sacrificial layer consisting of a superlattice and a doping staircase that can be assessed quickly by x-ray diffraction and C-V profiling as a result of which the growth conditions are fine-tuned. We then grow the layer of interest. (In production, one such layer would suffice for 25 subsequent wafers giving a higher overall wafer yield.)

We have conducted two series of experiments: one (wafers A–C) recalibrating the layer thicknesses only and the second (wafers D–F) recalibrating the doping as well. Our initial results have given doping profiles that are systematically too high (for which we can correct subsequently), but they are very reproducible.

We have used TEM and SIMS to check our recalibration procedure, although these would not be used in a production environment. Indeed no metrology is accurate enough, and the device results are the real and only test.

5. Metrology results

We present our results on detailed metrology of our wafers in Tables 1 and 2 that give (i) the value of the doping in the top and bottom contact as measured by

Wafer	Top (10^{18}cm^{-3})	Top contact	Base (10^{18}cm^{-3})	Base contact
	SIMS	C–V	SIMS	C–V
Spec.	3.0 ± 0.06	3.0 ± 0.6	3.0 ± 0.6	3.0 ± 0.6
A	3.6 ± 0.1	2.76 ± 0.02	3.72 ± 0.09	3.02 ± 0.04
B	4.0 ± 0.1	3.14 ± 0.02	4.00 ± 0.09	3.18 ± 0.04
C	3.7 ± 0.1	3.10 ± 0.05	3.8 ± 0.1	3.20 ± 0.05
D	4.8 ± 0.2	4.24 ± 0.02	4.9 ± 0.2	4.06 ± 0.03
E	4.7 ± 0.2	4.24 ± 0.02	4.8 ± 0.2	4.12 ± 0.03
F	4.8 ± 0.3	4.04 ± 0.03	4.9 ± 0.2	3.95 ± 0.04

Table 1. Comparison of SIMS and C–V profile values for the doping levels.

Wafer	Depth (nm)		Uniformity (nm)	
	SIMS	TEM	SIMS	TEM
Spec.	796 ± 80	796 ± 80	± 8	± 8
A	748 ± 10	786 ± 1	± 10	± 1
B	771 ± 6	782 ± 5	± 6	± 5
C	777 ± 10	782 ± 5	± 9	± 5
D	740 ± 1	751 ± 5	± 5	± 2
E	745 ± 1	738 ± 5	± 4	± 15
F	731 ± 1	754 ± 5	± 5	± 2

Table 2. Depth of the center of the AlAs layer from the top semiconductor surface, measured by SIMS and TEM, and its uniformity across a 3" wafer.

Wafer	Area (10^3 μm^2)	Current density (A/cm^2)	
		Average	Standard deviation
A	10	237	8
A	27.5	160	11
A	both	201	21
B	10	275	9
B	27.5	222	10
B	both	251	14
C	27.5	199	6
D	10	160	9
D[a]	10	167	2
E	10	123	8
F	10	129	17

[a] Excluding the row of devices nearest the edge of the processed wafer piece.

Table 3. Variability in the current across each wafer at 0.4 V.

SIMS and by $C-V$ profiling and (ii) the depth (and local uniformity of that depth) of the AlAs layer from the wafer surface as measured by SIMS and TEM. These results show the high degree of uniformity and reproducibility between the wafers A–C and D–F. The differences that remain between these two sets of data and indeed between the different measurements of doping can both be explained in terms of the conditions of growth and the difference between atoms measured by SIMS and ions measured by $C-V$. In all cases the TEM data show that the barriers are within 10% of that specified (2.85 nm), but the accuracy of the TEM is not sufficient to predict, or correlate with, the device results.

5.   Electrical results

In Table 3 we collect the electrical data in the form of the average and standard deviation of the 0.4-V forward-bias current density of diodes across the wafer, and in Fig. 1 we also plot the data using a spatial index of row of devices measured from the nearest point to the centre of a wafer. Of course, where different diode sizes are used the data is not strictly comparable as the parasitic resistances have to be de-embedded. Where we can achieve a standard deviation of <6%, we can meet the systems designers' requirements. Where (as in wafer F) there seems to be a radial trend in the data, the standard deviation is greater than might otherwise be expected.

6. Conclusions

We have made a 10-fold improvement in uniformity and reproducibility over the previous results by using 3" wafers and the *ex situ* recalibration technique. There are clear ways to improve on this, so the prospects for high yields are good. Reprocessing part of wafer C has established that the mesa process routine *per se* is partly responsible for the remaining variability in device performance, and a proton isolation process will improve on mesa etching. We appear to have accommodated the extreme sensitivity of the electrical results to the design and the extreme sensitivity of the devices to atomic-scale barrier fluctuations is not apparent. Once a user-friendly simulator is available for reverse engineering and right-first-time design, the target of 10¢ discrete microwave diodes can be met.

To date there have been some similar spot studies on AlAs/GaAs double barrier diodes,[7] achieving 10% variability of the resonant tunneling peak current density within a wafer and 20% variability between wafers. More recent studies[8] on a consecutive sequence of AlAs/InGaAs double-barrier structures grown on InP achieved 2–10% variability of the resonant tunneling peak current density within a wafer and typically 16% variability between wafers. Our frequent recalibration technique is equally applicable to these different structures and materials systems, and can provide comparable improvements.

7. Implications

There is still scope for improving procedures that might lead to achieving the target ±15% absolute current variation about a specified value for tunnel devices.

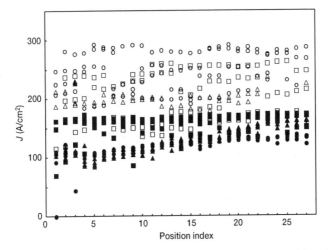

Figure 1. Current densities at 0.4 V forward bias, as a function of position on the wafer for devices from wafers A (□), B (o), C (∆) (open symbols), and D (■), E(●), and F(▲) (filled symbols).

This achievement would enable moderately large-area devices to be made. Once the device diameter shrinks below 0.2 μm for a mesa structure, then new quantum size effects appear in the $I–V$ characteristics.[9] A whole range of new phenomena have been observed, including Coulomb blockade and single-electron effects in resonant tunneling pillars.[10] Coupled quantum dots can also be defined by patterned gates in a two-dimensional electron gas and similar effects observed there.[11] There has been no attempt to reproduce any of these effects systematically as would be a precondition for any practical application. All existing device ideas based on mesoscopic physics must remain firmly in the research laboratory until the challenges of reproducibility and uniformity (far more difficult than for the tunnel devices described here) are tackled successfully.

Acknowledgments

Dr. R. K. Hayden (now of Nortel Networks) and Dr. R. Khan (now at Imperial College, London) worked at the University of Surrey on the contract supported by EPSRC and the EU (under project LOCOM). Dr. M. Missous (UMIST) grew the semiconductor layers, developing the frequent recalibration procedure. The devices were processed by Mr. G. Winiecki at Cambridge University. The TEM was performed by Dr. A. Gunnaes at the University of Surrey, and the SIMS Analysis was performed by Loughborough Surface Analysis.

References

1. T. L. Tansley "Heterojunctions properties," in: R. K. Willardson and A. C. Beer, eds., *Semiconductors and Semimetals*, Vol. 7A, Chapter 6, New York: Academic Press, 1971, pp. 293-368.
2. R. T. Syme, M. J. Kelly, R. E. Smith, A. Condie, and I. Dale, "A tunnel diode with asymmetric spacer-layers for use as a microwave detector," *Electronics Lett.* **27**, 92 (1991).
3. R. T. Syme, "Microwave detection using GaAs/AlAs tunnel Structures," *GEC J. Res.* **11**, 12 (1993).
4. V. A. Wilkinson, M. J. Kelly, and M. Carr, "Tunnel devices are not yet manufacturable," *Semicond. Sci. Technol.* **12** 91 (1997).
5. M. J. Kelly, "New statistical analysis of tunnel diode barriers," *Semicond. Sci. Technol.* **15**, 79 (2000).
6. K. Billen, V. A. Wilkinson, and M. J. Kelly, "Manufacturability of hetero-junction tunnel diodes: Further progress," *Semicond. Sci. Technol.* **12**, 894 (1997).
7. D. E. Mars, L. Yang, M. R. T. Tan, and S. J. Rosner, "Reproducible growth and applications of AlAs/GaAs double-barrier resonant tunneling diodes," *J. Vac. Sci. Technol. B* **11**, 965 (1993).

8. W. Prost, U. Auer, F.-J. Tegude, C. Pacha, K. F. Goser, G. Janssen, and T. van der Roer, "Manufacturability and robust design of nanoelectronic logic circuits based on resonant tunneling diodes," *Int. J. Circuit Theory Applications* **28**, 537 (2000).

9. M. A. Reed, J. N. Randall, R. J. Aggarewal, R. J. Matyi, T. M. Moore, and A. E. Wetsel, "Observation of discrete electronic states in a zero-dimensional semiconductor nanostructure," *Phys. Rev. Lett.* **60**, 535 (1988).

10. M. Tewordt, L. Martin-Morino, J. T. Nicholls, M. Pepper, M. J. Kelly, V. J. Law, D. A. Ritchie, J. E. F. Frost, and G. A. C. Jones, "Single-electron tunneling and Coulomb charging effects in asymmetric double-barrier resonant tunneling diodes," *Phys. Rev. B* **45** 14407 (1992).

11. K. Ishibashi, T. H. Oosterkamp, R. V. Hijman, and L. P. Kouvenhoven, "Coupling characteristics of semiconductor quantum dots in Coulomb blockade regime," *Jpn. J. Appl. Phys. Part 1* **37**, 7161 (1998).

Switching Device Based on a
First-Order Metal-Insulator Transition
Induced by an External Electric Field

Feliks Chudnovskiy
NY State Center for Advanced Sensor Technology, Stony Brook, NY 11794, U.S.A.

Serge Luryi
Dept. of Electrical and Computer Engineering
State University of New York at Stony Brook, Stony Brook, NY 11794, U.S.A.

Boris Spivak
Dept. of Physics, University of Washington, Seattle, WA 98195, U.S.A.

1. Introduction

First-order metal-insulator phase transitions (MITs) in crystalline materials have been known for many years[1] and correspond to a transformation between states with dielectric (semiconductor) and metallic types of conductivity. These transitions occur under the influence of certain external parameters, such as temperature and pressure, as well as with varying material composition. Materials exhibiting these phenomena include many transition metal oxides, of which over forty are known to possess MITs.[2,3]

Among the best-known MIT materials are vanadium oxides. Being able to combine with oxygen in 2-, 3-, 4-, and 5-valent states, vanadium forms a series of oxides of which at least 8 exhibit MIT.[4] The phase transition occurs at a critical temperature T_C = 150 K in V_2O_3 and at T_C = 340 K in VO_2, with the electrical conductivity changing by up to 10 and 5 orders of magnitude, respectively. The electrical changes in MITs in vanadium oxides are also accompanied by discontinuous variations of other properties, such as optical, magnetic, *etc.* Furthermore, vanadium dioxide, VO_2, is of particular interest for technology because its transition occurs near room temperature and furthermore its T_C is tunable over a wide range by doping with impurities such as tungsten.

The nature of the MITs in VO_2 and V_2O_3 is not well understood. Possible mechanisms under discussion are of the Mott-Hubbard and Peierls types.[1,5,6] A profound difficulty of the theory is the lack of a good explanation for the remarkable fact that the electronic bandgap in the dielectric phase of these materials is much larger than kT_C, making the low-temperature phase a good electrical insulator. On the other hand, the electron concentration in the metallic phase of VO_2 and V_2O_3 is typical for a metal ($\sim 10^{21}$ cm^{-3}).

The dramatic difference in the electronic spectrum on the metallic and the dielectric side of the transition opens the possibility for the switching applications discussed in this work. Indeed, since an external electric field penetrates very differently into the dielectric and the metallic phases, the field energy in the two phases is different. As a result, application of an external electric field shifts the critical temperature of the transition.[7-12]

2.   Electric-field induced metal-insulator transition in a thin film

Consider a thin vanadium oxide film sandwiched between two metallic plates insulated from the film by a dielectric layer, Fig. 1. Let us assume that both the bottom metallic plane and the vanadium oxide film are grounded and that an external voltage V is applied to the top metallic plate. Let us compare the free energies F_M and F_D per unit area of the film in its metallic and dielectric states, respectively:

$$F_M(T) = F_{M0}(T) + \tfrac{1}{2} C_M V^2 , \tag{1a}$$

$$F_D(T) = F_{D0}(T) + \tfrac{1}{2} C_D V^2 , \tag{1b}$$

where F_{M0} and F_{D0} are the free energies in the absence of the applied field, while C_M and C_D are the electrostatic capacitances of the system per unit area corresponding to the metallic and the dielectric states of the vanadium oxide film,

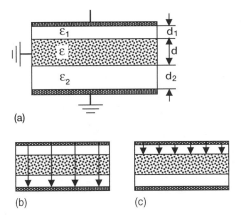

(a)

(b)   (c)

Figure 1. Schematic diagram of an experimental arrangement. A grounded vanadium oxide film (thickness d and permittivity ε) is sandwiched between two insulated metallic gates (a). When the film is in the dielectric state (b), the electric field lines penetrate through the film and terminate on the bottom metallic plate. When the film is metallic (c), the electric field lines terminate on the film surface.

respectively. If the Debye screening length in the dielectric phase is much larger than the film thickness, we can neglect the screening effect and hence the capacitance C_D is given by

$$C_D^{-1} = \left(\frac{d_1}{\varepsilon_1} + \frac{d_2}{\varepsilon_2} + \frac{d}{\varepsilon} \right). \tag{2}$$

In the metallic state, obviously, $C_M = \varepsilon_1/d_1$. The thermodynamic transition temperature T_C is determined by the equation $F_M(T_C) = F_D(T_C)$ — see Fig. 2.

Expanding the functions F_{M0} and F_{D0} in powers of $\delta T_C = (T_C - T_{C0})$, where T_{C0} is the thermodynamic transition temperature in the absence of an applied field, we find to first order that

$$\delta T_C = \frac{(C_M - C_D)V^2}{2(S_M - S_D)}, \tag{3}$$

where S_M and S_D denote the entropy density per unit area of the oxide film in the metallic and the dielectric phase, respectively. Note that we assume that the entire film undergoes the transition, which is true for a thin film if the formation of an in-plane transition boundary is energetically unfavorable. This assumption means we ignore possible phase-separation effects.

The entropy difference can be estimated from the data in Ref. 1, which cites the transition latent heat $\Delta H = 1020$ cal/mol for VO_2 at $T_C = 340$ K. This heat corresponds to the transition entropy $\Delta S = 3$ cal/mol·K. For a VO_2 film of thickness $d = 100$ Å we find, approximately,

$$S_M - S_D \approx 5 \times 10^{-3} \text{ J/K·m}^2 \tag{4}$$

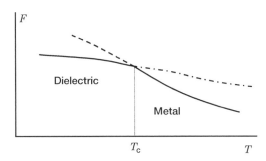

Figure 2. Schematic temperature dependence of the free energy near the metal-insulator transition. Dashed line indicates the metastable "overcooled" metal phase and dash-dotted line the "overheated" dielectric. The slope $\partial F/\partial T$ at the transition point suffers a discontinuity ΔS which is interpreted as the entropy of the transition. In terms of the latent heat ΔH of the transition, $\Delta S = \Delta H/T_C$.

where we have used a mass density of 4.34 g/cm^3 for VO$_2$. Taking $C_M \approx 10^{-2}$ F/m^2 and assuming a conventional SiO$_2$ gate dielectric ($\varepsilon_1 = 3.9\varepsilon_0$, $d_1 = 100$ Å), Eq. (3) gives $\delta T_C \approx 1$ K for $V = 1$ V, which is a substantial shift. Note that δT_C is a positive quantity, which means that application of an external voltage drives the system toward the dielectric state. The effect is stronger for thinner films, since the capacitance increases with decreasing d_1 and the transition entropy decreases with decreasing d. Our freedom to decrease d_1 is limited only by the dielectric breakdown. The thickness d of the vanadium oxide film can, in principle, be made as thin as 10–20 Å, limited only by the electron correlation length near δT_C.

3.   Field-controlled switch

The simplest VO$_2$ switch is shown in Fig. 3. It contains a gated VO$_2$ film with source and drain contacts implemented as VO$_2$ regions that are doped to be metallic in the entire operating temperature regime.   We assume that the temperature is above but close to T_C, so that in the absence of an applied voltage the entire film is in the metallic state. At some gate voltage $\pm V_G$, such that $|V_G| = U_C$, the film goes into the dielectric state. Conversely, starting from a dielectric state at some $|V_G| > U_C$, we bring the film into the metallic state by lowering the magnitude of the gate voltage.

It is important to realize that in our device the switching speed is *not* limited by the transit time of carriers under the gate, as it would be in a field-effect transistor. Indeed, in a FET the transit time delay arises as an *RC* time constant. Let us retrace the argument that leads to this conclusion.[13]   Suppose the total amount of charge Q in the channel is traveling with a velocity v between the source and the drain, separated by a distance L. This flow of charge corresponds to a current $I = Qv/L$. In a field-effect transistor, the magnitude of charge that has to be placed on the gate to induce the channel charge Q must be at least Q, plus any charges on parallel parasitic capacitances. Hence, the gate delay, which is the time required for the current I to charge the next gate to the same value of Q, is at least L/v.

In contrast, the surface density of mobile charge released in the VO$_2$ film upon a field-induced MIT exceeds the gate charge density by orders of magnitude. Therefore, there is no *RC* limitation analogous to transit time in a FET. Instead, the switching delay is limited only by the kinetics of the inhomogeneous phase transition. While this subject is not well understood theoretically at this time, it appears reasonable that the phase boundary should propagate at a rate determined by the Fermi velocity of electrons in the metallic phase. Due to the natural inhomogeneity of T_C over the film dimensions, one expects that near T_C both the metallic and the dielectric phases will contain multiple nucleation centers of the opposite phase, so that the typical distances of phase boundary propagation in switching will be substantially shorter than L. In this limit, the switching speed should be independent of L.

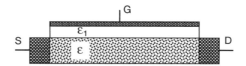

Figure 3. Three-terminal switch with vanadium oxide thin-film channel and a dielectric gate structure of length L. At the operating temperature the channel is in the metallic phase in the absence of a voltage applied to the gate. The source/drain contacts are assumed to be implemented in the same material as the channel but brought into the metallic phase by doping. Application of a gate field at some voltage $|V_G| = U_C$ brings the channel into the dielectric state.

4. Discussion

- *"Weak" and "strong" mechanisms of a field-induced transition*

The electrostatic effect discussed in Section 2 is based on rigorous thermodynamic considerations and is independent of the physical origin of the metal-insulator transition. This mechanism is rather weak, as it allows shifting T_C by at most several degrees K.

Let us now discuss the possibility of a stronger effect, which is based on the assumption that the transition is controlled by the concentration of mobile carriers. Here we assume a model of the transition triggered by the total mobile carrier concentration in the dielectric phase, irrespective of whether it results from doping, thermal excitation, photo excitation — or is induced by the electrostatic field effect. We are interested, of course, in the latter case. In an intrinsic insulator in the absence of an applied field, the chemical potential μ is approximately in the middle of the dielectric gap Δ. As the field E is applied, μ moves toward one of the allowed bands and the concentration of electrons (or holes) in the vicinity of the surface increases. At some critical concentration we can expect a transition to the metallic state. Note that in the induced concentration effect, the electric field drives the system in the *opposite* direction compared to the electrostatic effect. The application of a large enough gate voltage now drives the system into its metallic state.

Unfortunately, we have no quantitative theory to describe such a transition. Nevertheless, the very existence of an induced concentration effect should be regarded as likely. The critical carrier concentration may be estimated to be of the order of the thermal carrier concentration that arises in the dielectric phase at the transition temperature T_C due to thermal excitation across the gap. That concentration is about 10^{18} cm^{-3} and should be easily achievable by the field effect. The induced-carrier effect is much "stronger" than the electrostatic effect and should work in a wide range of temperatures. The expected "strong" effect is very attractive for potential applications not only because it relaxes the requirements for temperature control but, more importantly, because it may enable CMOS-like switching circuits where no significant current flows in either state.

Consider the circuit shown in Fig. 4 where two MIT switches are designated by FET symbols. The two are connected in series between a high voltage level (V_{DD}) and a low level (ground). The gates of both switches are connected together and an input voltage V_{IN} is applied to both. When $V_{IN} = V_{DD}$, the lower switch is in the metallic state. At the same time, the upper switch contains a dielectric region near its drain where the gate-to-channel voltage remains low. A Schottky barrier thus appears in the channel of the upper switch. The output node in this state is low, $V_{OUT} = 0$. However, when $V_{IN} = 0$, the Schottky barrier emerges in the lower switch while the upper switch is fully in the metallic state. In this case, the output node goes high, $V_{OUT} = V_{DD}$. In both configurations the circuit represents a voltage divider with a huge ratio of on/off conductances and dissipates little power in either of its two stable configurations. The circuit thus acts like an inverter similar to that employed in complementary silicon circuits.

We note that if the operating effect were electrostatic, then the circuit of Fig. 4 would still be a switching logic circuit but it would *not* be an inverter. Instead, it would be a voltage amplifier in which V_{OUT} would logically follow V_{IN}.

Either way, the circuit operation relies on the symmetry of the field-induced phase transition with respect to the sign of the field. For the electrostatic effect this symmetry is exact, based on the fact that the electric field energy is quadratic in the field. The exact symmetry is not likely in the induced concentration effect but this misbalance can be readily compensated by doping.

- *Transition speed*

As discussed in Section 2, there is no transit-time limitation for a current switch based on a field-controlled metal-insulator transition. The question arises, what is the real limitation on the switching speed?

We believe this limitation is related to the kinetics of the first-order phase transition. In a truly homogeneous sample, the transition speed would be controlled by the nucleation of critical droplets.[14] In a real situation, however, we can expect a much faster transition, because in a large enough sample, nucleation sites should be present in both phases, due to various inhomogeneities, either natural or built-in by design. In this case, the characteristic delay time should be associated with the propagation of the phase boundary over distances between the phase nuclei and is independent of the device dimension.

We estimate the phase boundary velocity to be of the order of the Fermi velocity of electrons in the metallic phase, which is of the order of 10^8 cm/s. If the typical distance between the nucleation sites is less 1000 Å, the switching time is below 1 ps. Experimentally, there have been studies of the semiconductor-metal transition in VO_2 under femtosecond laser illumination.[15] These studies have demonstrated a transition time of less than 5 ps.

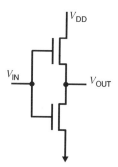

Figure 4. Logic circuit comprising two gate-controlled VO_2 switches. If the mechanism that drives the MIT is such that high $|V_G|$ corresponds to the dielectric phase, then the amplified V_{OUT} voltage follows V_{IN}. If high $|V_G|$ drives the film into its metallic state, then the circuit is a logic inverter.

• *Possible applications*

An ultra-fast switch can find use in electronic as well as optical circuits. Optical applications may be based on the dramatic change in either transmittance or reflectance of light accompanying the metal-insulator transition. Whether or not the range of possible electronic applications may include large-scale integrated circuit applications depends on the transition energetics. The latent heat of transition of 10^3 cal/mol in VO_2 corresponds to about 0.1 eV per electron released. This figure compares favorably with the power dissipation in a single CMOS switching event.

5. Conclusion

First-order metal insulator phase transitions in thin films can be controlled by an applied electric field. The effect can be employed for the implementation of useful devices such as a three terminal ultra-fast switch. Unlike field-effect transistors, the speed of such a switch is not limited by the carrier drift time under the gate. Two field-controlled phase transition switches can be arranged in a CMOS-like inverter circuit in which no significant current flows in either of its steady states.

References

1. N. Mott, *Metal-Insulator Transitions*, London: Taylor & Francis, 1997.
2. P. A. Cox, *Transition Metal Oxides*, Oxford: Clarendon Press, 1995.
3. J. M. Hong, in: P. P. Edwards and C. R. N. Rao, eds., *The Metallic and the Non-Metallic States of Matter*, London: Taylor & Francis, 1985, p. 261.
4. F. A. Chudnovskiy, "Metal-semiconductor phase transition in vanadium oxides and its technical applications," *Sov. Phys. Tech. Phys.* **20**, 999 (1976).

5. T. M. Rice, H. Launois, and J. P. Pouget, "Comment on 'VO$_2$: Peierls or Mott-Hubbard? A view from band theory'," *Phys. Rev. Lett.* **73**, 3042 (1994).

6. H. Nakatsugawa and E. Iguchi, "Electronic structures in VO$_2$ using the periodic polarizable point-ion shell model and DV-X alpha method," *Phys Rev. B* **55**, 2157 (1997).

7. K. A. Valiev, Yu. V. Kopaev, V. G. Mokerov, and A. V. Rakov, "Electron structure and phase transitions in lower vanadium oxides in an electric field," *Sov. Phys. JETP* **33**, 1168 (1971).

8. V. V. Mokrousov and V. N. Kornetov, "Field effects in vanadium dioxide films," *Fiz. Tverd. Tela* **16**, 3106 (1974)

9. G. P. Vasil'ev, I. A. Serbinov, and L. A. Ryabova, "Switching in the VO$_2$-dielectric-semiconductor system," *Pis'ma Zh. Tekhn. Fiz.* **3**, 342 (1977).

10. E. V. Babkin, G. A. Petrakovskii, and A. A. Charyev, "Anomalous features of the conductivity and of the galvanomagnetic properties of vanadium dioxide in strong electric fields," *JETP Lett.* **43**, 697 (1986).

11. C. Zhou, D. M. Newns, J. A. Misewich, and P. C. Pattnaik, "A field-effect transistor based on the Mott transition in a molecular layer," *Appl. Phys. Lett.* **70**, 598 (1997).

12. D. M. Newns, J. A. Misewich, C. C. Tsuei, A. Gupta, B. A. Scott, and A. Schrott, "Mott transition field effect transistor," *Appl. Phys. Lett.* **73**, 780 (1998).

13. S. Luryi, "Field-effect and potential-effect transistors", Chapter 7 in: S. M. Sze, ed., *High-Speed Semiconductor Devices*, New York: Wiley, 1990.

14. E. M. Lifshitz and L. P. Pitaevsky, *Physical Kinetics*, London: Pergamon, 1981.

15. M. F. Becker, A. B. Buckman, R. M. Walser, T. Lepine, P. Georges, and A. Brun, "Femtosecond laser excitation of the semiconductor-metal phase transition in VO$_2$," *Appl. Phys. Lett.* **65**, 1507 (1994).

DNA Conduction Mechanisms and Engineering

Rashid Zia, Andrei Rakitin, and J. M. Xu
Div. of Engineering, Brown University, Providence, RI 02912, U.S.A.

1. Introduction

DNA conductivities reported in the literature vary widely — from metallic to semiconducting to insulating. Given that DNA is a well understood double-helical stacking of periodically-spaced nucleotide base pairs, it is natural to think that not all the accounts of DNA conductivity can be right. However, when one takes into account the fact the periodically-spaced bases can differ in their redox potentials, the situation looks more reasonable. Each sequence of the four bases corresponding to a biologically-different DNA is also a different series of electrical potentials for electron transport. Although the spacing between the stacked base pairs is constant, the location and frequency with which any one of the four bases appears in the sequences are not, and are what could give rise to very different conductivities.

In the extreme limit of all base pairs being identical (*e.g.* the poly-G oligomer), one has an ideal periodic potential. Depending on the coupling strength between the adjacent potential wells, electron transport could in principle take place via band-like conduction or hopping. In the opposite limit, namely complete sequence disorder, no band of the conventional kind can exist because of the absence of any translational symmetry (periodic potential). Here, hopping offers a natural alternative for electron transport. However, if the coupling between adjacent bases is strong, the density of states in a given range of wavevector values can be sufficiently high, and as in amorphous semiconductors, delocalized (band-like) transport can take place.

In this work, we attempt to reconcile the vastly differing accounts of DNA conductivity in light of redox potential (sequence) disorder, and in the frameworks of hopping conduction and delocalized transport mechanisms. With the sequence being variable by biochemical means, we would also like to point out that a better understanding of possible conduction mechanisms could potentially allow one to alter the conductivity of DNA over a wide range. Thus, by combining a sequence-independent technique to incorporate divalent metal ions into the DNA double helix with the selection of a suitable base sequence to introduce energy states or quasi-continuous bands, it seems possible to engineer both the density of states and the relative position of the Fermi level, thereby determining both the conductivity level and type. This procedure is illustrated schematically in Fig. 1.

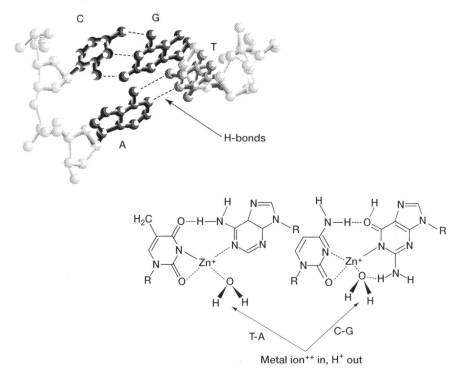

Figure 1. Schematic illustration of the C-G and A-T base pairs in DNA and the hydrogen bonds binding the base pairs together. One of the hydrogen bonds, the imino proton, is replaced by a divalent metal ion in the event of B-DNA dismutation to M-DNA, as shown in the lower part of the figure.

2. Conduction by hopping

In DNA, charge transport along the stacked base pairs has long been assumed to be mediated by π-electron coupling (see the following section). The question of how strong the coupling is between different base pairs is of the utmost importance to understanding the possible conduction mechanisms underlying the vastly differing measured results. It is also a question to which no clear answer has been found despite forty-some years of fascination with the subject and serious efforts by scientists from various communities. In the case of low coupling, electronic transport in a disordered sequence of DNA base pairs is most likely by sequential hopping.[1] If the coupling is large and the π-electron states overlap strongly, the states delocalize. In such a regime, a quasi-continuous band can form and band-like conduction may take place. In doped semiconductors, a similar situation occurs when the impurity concentration is high enough such that the impurity states overlap each other to form a so-called impurity band. In amorphous semiconductors, band tails form for similar reasons.

However, in DNA it is possible that regions of strongly-coupled base-pair stacks or regions of highly periodic sequence are segmented and linked by zones of aperiodic or weakly coupled stacks. In such a situation, hopping once again enters the picture. In fact, hopping likely will dominate the overall conductivity, as is well known in the case of a random network of resistors studied by Miller and Abrahams in the 1960's.[2] This conclusion is a direct consequence of the fact that charge transport in DNA is sequential along the stacking direction, and the total conductance is limited by the segments with the largest resistances.

Even for the case of a low degree of disorder or perfect periodicity (thus in the presence of some translation symmetry and hence a normal conduction band) if the coupling is weak the band will be narrow. Under sufficiently large bias comparable to its bandwidth, such a narrow band will break into a ladder of segmented mini-bands cascading down in energy along the DNA. Hopping of electrons along this ladder is then again what determines the conductivity.

For DNA with sufficiently large sequence disorder, band conduction does not exist and electronic transport is limited by hopping in regions where the redox potential difference between different base pairs is much higher than the room-temperature thermal energy.

Experimentally, it is in the highly aperiodic DNA systems where the DNA conductance measurements reported to date differ the most, often by orders of magnitude. These very large differences that seem to be contradictory can now be reconciled in the framework of the one-dimensional random network of resistors of Miller and Abrahams.[2] The electronic transport between the potential wells (base pairs) can be expressed in terms of the integral rate of the inter-base-pair hopping transition. In the absence of long-range correlation, or in the nearest-neighbor approximation, the rate equation of the standard hopping theory can be applied. The transition rate can be computed from the characteristic disorder energy Δ, defined as the mean-square deviation for the distribution of the energy levels of the potential wells. For a long stacking series, the average inter-base-pair conductivity would have the form $\ln(\sigma) \sim -\eta(\Delta/kT)^2$. For a short series, the conductivity is limited by that of the highest resistance links, as expected intuitively and derived quantitatively by Miller and Abrahams. In such a situation, $R_{ij} \cong (kT/e^2) \exp(2\alpha S_{ij})\exp([E_{ij} + h\nu - \mu]/kT)$, where S_{ij} and E_{ij} are respectively the spatial and energy separations between adjacent base pairs i and j, μ is the Fermi level, and $h\nu$ is the phonon energy. Glossing over the details of the model, given the same degree of unknown details in the reported experiments, we can already see from these general results that the conductance of aperiodic DNA stacking series can differ by orders of magnitude even at room temperature, because of the exponential dependences on the potential difference and the disorder energy, both of which can be significantly larger than kT in DNA with large sequence disorder. Thus, a very large spread in measured conductivities can be expected after all, even if all experiments are carefully carried out.

However, the high conductivity and band-like conduction characteristics reported in some DNA experiments, including very recent ones,[3,4] cannot be accounted for by hopping conduction and call for considerations of delocalized transport, which will be expounded below.

3. Delocalized electronic transport and band-like conduction

Over the past decade or so, an increasing body of evidence seems to indicate that certain molecular materials could demonstrate the characteristic electronic property of metals — the existence of partially filled bands of delocalized electronic energy states.   Responding to these findings, Tobin J. Marks outlined two criteria for such electronic behavior that could "convert an unorganized collection of molecules into an electrically conductive array:"[5]

- the molecules would have to be arranged in close energetically coupled stacks; and

- the molecules would have to be in a partially oxidized (or reduced) state, in a state of incomplete charge transfer.

In terms of band theory, the first requirement provides the basis for the formation of energy bands if the coupling is strong, while the second requirement provides for the partially unfilled states just above the Fermi level. However, Marks did not equate such molecular conduction to crystalline metallic conduction. As reasoned in the proceeding section, simple spatial periodicity of a molecular chain and wavefunction overlap are necessary to create energy bands. Rather, by hinting that such molecular conductors are "highly sensitive to interactions between the electronic systems and various lattice vibrations (phonons)," Marks suggested that this molecular band can be affected by lattice defects and imperfections, resembling the effects commonly attributed to amorphous semiconductors.[6] In amorphous materials, band formation is associated with the energy differences between short-range bonding and anti-bonding configurations. Regardless of the specific origin of band formation, though, the above criteria set forth by Marks have since served as the foundation for the π-band conduction model, named after the readily polarized π-bond formed when two p-orbitals couple.

Since DNA appears to satisfy both criteria, it has long been hypothesized that the internal π-stacks of electrically coupled nitrogenous bases (aromatic heterocycles) could provide for metal-like conduction. In 1993, Barton and co-workers[7] attempted to monitor long-range electron transfer between two metallointercalators that had been incorporated in the DNA double-helix with 15 base pair separation. One metallointercalator was an electron donor complex while the other was an electron acceptor, and electron transfer was monitored by luminescence from the excited donor state. DNA synthesized with only the electron donor would radiate light indefinitely. However, whenever DNA with both acceptor and donor was synthesized, the luminescence was undetectable. Thus, Barton's group concluded that electron transfer must have occurred on a time scale faster than the experimental set-up could detect. They concluded that the electron transfer occurred at a rate indicative of long-range transport, consistent with the delocalized states of the π-band conduction model, as illustrated in Fig. 2. Although these initial results were questioned, particularly for the lack of an observed luminescence quenching, Barton's group as well as others have used

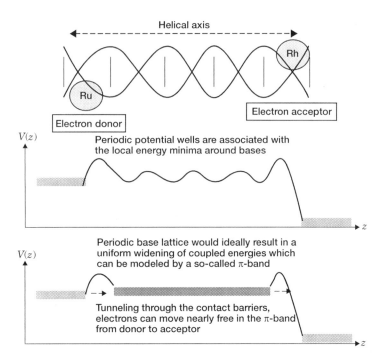

Figure 2. Illustrations of conduction mediated by the π electrons, the periodic potential coupled wells corresponding to the base pairs, and the hypothetical formation of an energy band in the event of strong coupling.

metallointercalators over the years to monitor the rate of electron and hole transfer in DNA.[8] In their following studies, Barton and colleagues found that under certain conditions charge transport was occurring with little dependence on transport distance, again indicative of delocalized energy states.[9–12]

However, fluorescence studies demonstrated not only that such long-range delocalized transport exists, but also the limits of such transport. Even within studies that confirmed delocalized charge transport, conditions were described that prevented or reduced long-range transport. Using the same electron acceptor, Barton's group documented large variations for different donor complexes.[9] Using the same hole acceptor, this group also documented large variations in charge transfer for various base defects in the DNA double-helix.[11] In terms of the π-band, though, these two limits were to be expected. The energy levels associated with certain metallointercalators would correspond to forbidden energy bandgap regions, and base defects would result in a breakdown of the coupling responsible for conduction bands. Hence, band transport in these cases would be impossible, and fluorescence-quenching rates would be drastically reduced. Accordingly, the fluorescence experiments of Barton provided evidence both for the delocalized electronic states associated with conduction bands and for the presence of forbidden energy bandgaps that the π-band model predicted.

4.   Charge localization: Accounting for distance-dependent transport

Yet, some fluorescence experiments demonstrated results that are difficult to reconcile with the π-band model.   In 1997, Lewis *et al.* published results demonstrating distance-dependent charge transport that were inconsistent with delocalized charge transport,[13] but rather pointed to localization of charge along the DNA helix. Giese and colleagues published further findings concerning sequence-dependent transport in DNA due to the potential variations of bases.[14-16] These researchers presented a model of hole localization at guanine sites and hopping dominated transport between guanine sites, a situation described in the proceeding section on hopping conduction.

At first glance, these distance and sequence dependences are not necessarily in conflict with the π-band model.   As in conventional semiconductors, such localized states could be located energetically inside the forbidden energy bandgap, and charge transport between such localized states could be of a hopping nature or could result when charge carriers require thermal excitation from a localized state to the band above.   However, there is a critical difference here — conduction through the π-orbital coupling in DNA is essentially one-dimensional via a single path, and the coupling between two adjacent base pair stacks is much weaker than that of atomic couplings in normal semiconductors.   The simultaneous existence of a band over the localized states in such a one-dimensional conduction system would seem to require a strong coupling between the base pair stacks that is not evidently available, judging by the wide gap and large effective mass obtained by Rinaldi *et al.*[3] on an ideal periodic DNA system.

Although the distance-dependent fluorescence studies can be accounted for by the π-band model, the work and alternative theory of Giese illustrate the limits of such experiments in analyzing the π-band model.   Fluorescence studies cannot by themselves support the existence of bands in DNA, because such experiments can hardly probe the band structure of DNA.   Since the metallointercalators used in fluorescence occupy some finite energy levels, the distance-independent transport of intercalator-associated charge carriers down the DNA double-helix can only prove that energy bands aligned with intercalator states exist.   In other words, the distance-independent transport observed in fluorescence experiments suggests that energy bands are present in DNA, without providing information about the actual Fermi-level position (*i.e.* without discrimination between metal-, semiconductor-, or insulator-like band filling).

Nevertheless, the more recent experiments on a periodic DNA structure (a deoxyguanosine derivative) by Rinaldi *et al.*[3] mentioned earlier indicate that it does behave like a wide-gap semiconductor with an energy gap in the range of 3–3.5 eV and an electron effective mass $m_e \sim 2m_0$.   This result suggests the existence of both band-like conduction, as expected for a periodic stacking series, and strong bonding of electrons to the base pair stacks with weak coupling between the base pairs.

5. Direct current–voltage measurements: Investigating band structure

In order to classify the possible band structure of a particular DNA, experiments must provide for the analysis of long-range transport over a broad continuous range of electronic energies. Ideally, one would use direct current–voltage measurements, but given the size and one-dimensional nature of DNA, making contacts and taking measurements are non-trivial problems.

A problem with direct measurement that should not be overlooked is the possible doping or depletion of charge carriers within DNA by contacts (ohmic or non-ohmic alike). In three-dimensional cases, the decay of the screening effects is exponentially dependent on distance, decaying quickly over short ranges. However, in one-dimensional cases, this decay is logarithmically dependent on distance, decaying slowly over long ranges.[17] In one-dimensional samples, the length of the depletion region varies over a wide range depending on the doping strength. Therefore, given the proposed one-dimensional nature of DNA electronic transport and the small size of measured samples, the possibility exists for injected carriers to bridge two contacts measuring a DNA sample. Such a bridge would demonstrate metal-like behavior regardless of the material being measured.

In order to prevent bridging of DNA by injected current, it is better to conduct STM-like measurements on sufficiently long samples. Although tunneling measurements cannot provide the precise conductivity of a sample, such measurements can still provide a qualitative picture of the band structure of a sample by assessing the bandgap's location relative to the Fermi level and its width. Hence, tunneling contacts can be used to classify a one-dimensionally conductive sample as metallic, semiconducting, or insulating.

Already, such direct current–voltage measurements have been performed. In 2000, Porath *et al.* published measurements of single poly(G)-poly(C) DNA molecules in both air and vacuum down to cryogenic temperatures. These samples exhibited an insulating bandgap at low applied bias, indicating that the band structure of DNA resembled a wide-bandgap semiconductor.[4] However, in 2001, our group published findings including one where λ-DNA molecules in vacuum at room temperature exhibited a small plateau in the measured $I–V$ (~200 mV), which on the surface seems to suggest a narrow bandgap semiconductor.[18] In reality, the plateau in the $I–V$ reflects only the existence of a barrier height measured from the Fermi-level in the electrode to the energy band of the DNA (if a band exists) or to the energy level of the quasi-delocalized states receiving the injected electrons (if a band does not exist).

To add to these contradictory findings, alternative methods of investigating DNA's conductivity have also presented different band structures. Using superconducting contacts, Kasumov *et al.* published findings of proximity-induced superconductivity in DNA samples, implying that DNA's band structure is metallic.[19] Meanwhile, researchers at Gruner's laboratory noninvasively monitored λ-DNA conductivity by measuring resonant cavity losses at microwave frequences.[20] The measured loss spectra were consistent with a bandgap of about 300 meV, which is in line with our direct current–voltage measurements.[18]

6. Bands, band tails, and disorder

Using the analogy of amorphous semiconductors, it is possible to understand the seemingly contradictory direct measurements that characterize different DNA samples that exhibit band-like conduction. Beginning with the uniform poly(G)-poly(C) DNA measured by Porath *et al.*, we see that DNA with minimal sequence disorder has a wide bandgap structure (accompanied by the occurrence of an *I–V* plateau at around zero bias).[4] However, this wide-bandgap structure can effectively shrink as the degree of disordering increases. For the more disordered λ-DNA like the ones we measured, one would expect that varying bonding and anti-bonding configurations in the π-stack would give rise to disorder-induced "band tails" inside the bandgap, effectively narrowing the bandgap.

Here, we again wish to stress the distinction between bandgap and the measured *I–V* plateau. First, as stated before, the measured *I–V* plateau reflects the existence of a barrier height between the Fermi level of the electron or hole reservoir and the energy level of the states receiving the injected electrons or holes, not the bandgap between the conduction band and valence band of the material. Second, the plateau includes the voltage drop across the finite series resistance between the contact and the DNA. Moreover, the existence of an *I–V* plateau is no indication of the existence or absence of a conduction or valence band. Therefore, the *I–V* plateau itself is not a measure of the bandgap, and is specific to both the DNA and the contact conditions.[4,18] There are several classical techniques to probe the conduction mechanisms, including temperature dependent-conductivity measurements, which are however yet to be proven in the case of DNA because of the difficulty in achieving precise control over the contact and sample conditions.

Note that previously discussed fluorescence measurements would seem to suggest not only delocalized states but also localized electronic states (for hopping). Recent works from both Giese and Barton (in collaboration with Zewail's group) move away from single-method transport in DNA, suggesting that charge transport in aperiodic DNA could occur through both coherent delocalized pathways (bands) and discontinuous localized paths (hops).[21,22]

Therefore, the seemingly scattered measurements of band-like conduction in DNA, both direct measurements and fluorescence measurements, can be explained by an amorphous π-band model that takes into account both charge localization induced by disordering and band conduction. However, we are quick to admit that such explanations are by no means foolproof. They are in fact only as good as the assumption of the simultaneous existence of localized states and a continuous band of states, which, as explained in the proceeding section, is questionable in a one-dimensional and weakly coupled system.

7. Metallic DNA: Disorder-independent conductivity change

While sequence selection and gap narrowing provide opportunities for engineering different levels of conductivity, the extent of possible change is rather limited. That is, disorder introduction can to a great extent provide for manipulation of the bandgap, but to a much lesser extent for manipulation of the actual Fermi-level position. According to experimental results, for uniform poly(G)-poly(C) and random B-DNA alike, the Fermi-level is located inside a bandgap region.[4,18] Therefore, all DNA sequences, regardless of effective bandgap, still maintain the band structure of a semiconductor. However, this situation can now be changed, thanks to the discovery of a method of engineering metallic conductivity into DNA by incorporating metal ions between base pairs.[23,24,18]

In 1992, our biochemist collaborators J. Lee and colleagues published their findings of a possible transformation of DNA in metal binding experiments, and they named this possible form of DNA dismutation-metal-DNA (M-DNA).[23] Together with these biochemists, we conducted a set of direct experiments and measurements over a period of two years on the transformation of DNA under various controlled conditions. We obtained direct evidence of metallic conduction arising from the precise substitution of all imino protons between standard beta-form DNA (B-DNA) base pairs with divalent metal cations (either Zn^{2+}, Ni^{2+}, or Co^{2+}).[18] Despite the precision of the transformation, M-DNA is synthesized easily from B-DNA, and M-DNA can be restored easily to B-DNA.

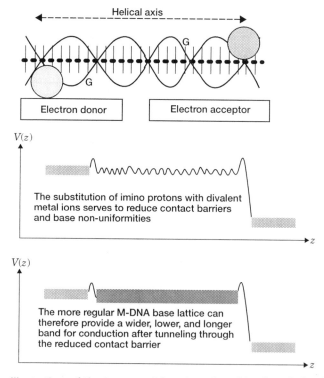

Figure 3. Illustration of the two possible roles played by the chain of metal ions that are inserted in the place of the imino proton in each base pair: reducing contact barriers and the degree of sequence disorder, and providing a highly conductive second pathway for electron transport.

In addition, through fluorescence experiments, Aich *et al.* demonstrated that fast long-range electron transfer occurred in M-DNA under conditions in which such transfer would not occur in B-DNA.[24] Thus, similar to the first fluorescence experiments in biological DNA, these results indicate that delocalized electronic energy states exist in M-DNA, and more precisely, that such delocalized states exist for energy levels where delocalized transport is not possible in B-DNA.

By substituting the imino protons in each base-pair regardless of sequence, the metallic ions of M-DNA appear to form a one-dimensional periodic chain introducing a metal-associated electronic band aligned with the DNA Fermi-level, as illustrated in Fig. 3. Thus, by doping DNA with metallic ions, M-DNA offers a method to engineer the position of the Fermi-level and to add an additional conduction pathway (through the periodic and strongly coupled metal ion chain), and thereby to produce metallic conductivity in DNA.

In closing, combining the sequence-dependent bandgap engineering of band tails and disorder with the sequence-independent Fermi-level engineering and the addition of a second conduction pathway via the metal ion chain of M-DNA, it appears that the conditions are ready for us to engineer both levels and types of DNA conductivity. Clearly, there are many experiments to be performed to further investigate the conduction mechanisms and the validity of the models as presented here. However, regardless of the precise techniques that will finally provide for the controlled manipulation of DNA electronic properties, it seems that the movement toward engineering DNA conductivity is inevitable.

Acknowledgments

This work was supported by CIAR, NSF, and ONR. We are thankful to Jeremy Lee and his group with whom we obtained the first direct proof of the metallic conductivity in M-DNA and from whom we learned the basic biochemistry for forming and manipulating the M-DNA.

References

1. Y.-J. Ye , R.-S. Chen, A. Martinez, *et al,* "Calculation of hopping conductivity in aperiodic nucleotide base stacks," *Solid State Commun.* **112**, 139 (1999).
2. A. Miller and E. Abrahams, "Impurity conduction at low concentrations," *Phys. Rev.* **120**, 745 (1969).
3. R. Rinaldi, E. Branca, R. Cingolani, *et al*, "Photodetectors fabricated from a self-assembly of a deoxyguanosine derivative," *Appl. Phys. Lett.* **78**, 3541 (2001).
4. D. Porath, A Berzryadin, S. deVries, *et al*, "Direct measurements of electrical transport through DNA molecules," *Nature* **403**, 635 (2000).
5. T. J. Marks, "Electrically conductive metallomacrocyclic assemblies," *Science* **227**, 881 (1985).

6. K. Morigaki, *Physics of Amorphous Semiconductors*, London: Imperial College Press, 1999.

7. C. H. Murphy, M. R. Arkin, Y. Jenkins, *et al*, "Long-range photoinduced electron transfer through a DNA helix," *Science* **262**, 1025 (1993).

8. For a thorough review, see M. E. Nunex and J. K. Barton, "Probing DNA charge transport with metallointercalators," *Curr. Op. Chem. Biol.* **4**, 199 (2000).

9. M. R. Arkin, E. D. A. Stemp, R. E. Holmlin, *et al*, "Rates of DNA-mediate electron transfer between metallointercalators," *Science* **273**, 475 (1996).

10. D. B. Hall, R. E. Holmlin, and J. K. Barton, "Oxidative DNA damage through long-range electron transfer," *Nature* **382**, 731 (1996).

11. P. J. Dandlike, R. E. Holmlin, and J. K. Barton, "Oxidative thymine dimer repair in the DNA helix," *Science* **275**, 1465 (1997).

12. S. O. Kelley and J. K. Barton, "Electron transfer between bases in double helical DNA," *Science* **283**, 375 (1999).

13. F. D. Lewis, T.Wu, Y. Zhang, *et al*, "Distance-dependent electron transfer in DNA hairpins," *Science* **277**, 673 (1997).

14. E. Meggers, M. E. Michel-Beyerle, and B. Giese, "Sequence dependent long range transport in DNA," *J. Am. Chem. Soc.* **120**, 12950 (1998).

15. B. Giese, S. Wessely, M. Spormann, *et al*, "On the mechanism of long-range electron transfer through DNA," *Agnew Chem. Int. Ed.* **38**, 996 (1998).

16. M. Bixon, B. Giese, S. Wessely, *et al*, "Long-range charge hopping in DNA," *Proc. Natl. Acad. Sci. USA* **96**, 11713 (1999).

17. A. A. Odintsov, "Schottky barriers in carbon nanotube heterojunctions," *Phys. Rev. Lett.* **85**, 150 (2000).

18. A. Rakitin, P. Aich, C. Papaopoulos, Y. Kobzar, A. S. Vedeneev, J. S. Lee and J. M. Xu, "Metallic conduction through engineered DNA: DNA nanoelectronic building blocks," *Phys. Rev. Lett.* **86**, 3670 (2001).

19. A. Yu. Kasumov, M. Kociak, S. Gueron, *et al*, "Proximity-induced superconductivity in DNA," *Science* **291**, 280 (2001).

20. P. Tran, B. Alavi, and G. Gruner, "Charge transport along the l-DNA double helix," *Phys. Rev. Lett.* **85**, 1564 (2000).

21. C. Wan, T. Fiebig, O. Schiemann, *et al*, "Femtosecond direct observation of charge transfer between bases in DNA," *Proc. Natl. Acad. Sci. USA* **97**, 14052 (2000).

22. B. Giese, J. Amaudrut, A. K. Kohler, *et al*, "Direct observation of hole transfer through DNA by hopping between adenine bases and by tunneling," *Nature* **412**, 318 (2001).

23. J. S. Lee, L. J. P. Latimer, and R. S. Reid, "A cooperative conformational change in duplex DNA induced by Zn^{2+} and other divalent metal ions," *Biochem. Cell Biol.* **71**, 162 (1993).

24. P. Aich, S. L. Labiuk, L. W. Tari, *et al*, "M-DNA: A complex between divalent metal ions and DNA which behaves as a molecular wire," *J. Mol Biol.* **294**, 477 (1999).

New Cold Cathode Paradigms for Vacuum Microelectronics Applications

M. Cahay, Y. Modukuru, J. Thachery,
A. Malhotra, H. Tang, and P. Boolchand
Dept. of Electrical and Computer Engineering and Computer Science
University of Cincinnati, Cincinnati, OH 45221, U.S.A.

1. Introduction

Recently, there has been renewed interest in cold cathode emitters for application to a variety of electronic devices, including, among others, microwave vacuum transistors and tubes, pressure sensors, flat-panel displays, and high-temperature and radiation-tolerant sensors.[1]

Many cold cathodes based on the concept of field emission have been proposed. The idea is to produce a sufficiently high external electric field at the surface of the cathode so as to decrease the width of the potential barrier sufficiently for reasonable tunneling currents to flow. Among the most widely investigated field emitters are the Spindt-type cathodes, which utilize the cone geometry field enhancement to increase locally the electric field near the tip and enhance electron field emission.[1] Spindt-type cathodes have received considerable attention since their invention in the early 1970's. Several prototype field emission displays (FEDs) based on metallic and semiconductor tips have been demonstrated at several display industry events over the last few years and are in the process of being commercialized.[2] However, these field emission cathodes still require too large a voltage for operation (several hundred volts). In practical field-emitter applications, where many field-emitter tips are to be used simultaneously, a critical requirement is that the tip arrays be highly uniform in order to maintain a uniformly high emission-current density. For arrays of Spindt cathodes, this uniformity in the distribution of the tips is a challenge to the most advanced lithographic techniques and has led to malfunction of field arrays. Indeed, there are many experimental reports showing that the malfunction of one tip out of millions can lead to the destruction of the entire array.[1,2]

For that reason, several other approaches have been proposed to realize field emission from various metallic and semiconducting materials (porous and poly-crystalline silicon, ferroelectrics, and carbon nanotubes, among others). A class of emitters based on the concept of negative electron affinity (NEA) and intended to achieve cold cathode emission has been proposed. Negative electron affinity can be achieved through bandgap engineering of semiconductor materials by covering the surface of a wide bandgap semiconductor with a low work function (WF) material, such that the surface vacuum barrier is brought below the bulk conduction band edge.[3] The most conventional approach to NEA has been the use of cesiated semiconductor surfaces. However, cesium suffers from poor stability due to its willingness to release bonding electrons. Cesium melts at 28.5 °C, has a high vapor pressure $(10^{-3}$ torr at 100 °C), and eventually tends to migrate everywhere. It also spreads to inter-electrode spaces where it will provide a relatively easy path for gaseous conduction and electrical arcs. Recent demonstrations of NEA using diamond and other wide bandgap materials are welcome alternatives to cesiated surfaces, but these investigations are still in their infancy and far from providing a technological breakthrough in the near future.[4]

Several excellent reviews of the electrical properties of the cold-cathode approaches mentioned above have been published recently.[1-7] Herein, we present a few unconventional approaches to realize cold cathodes that constitute paradigm shifts to the approaches listed above in a rapidly booming field of vacuum microelectronics. These alternatives could alleviate some of the difficulties linked with other approaches, especially in terms of reliability and reproducibility of the electronic properties of the cathodes.

2.   Paradigm shifts

- *The solid-state field-controlled electron emitter*

This device was recently proposed by Vu Thien Binh et al.[8,9] It consists of an ultra thin (few nm) wide-bandgap n-type semiconductor deposited on a metallic substrate (Fig.1, left). Explanation for the observed current density vs. applied external electric field (Fig.1, right) has been given in terms of a two-step mechanism consisting of injection of electrons across the metal-semiconductor Schottky junction followed by emission of electrons from the wide bandgap semiconductor/vacuum interface. During this two-step process, the charge pile-up

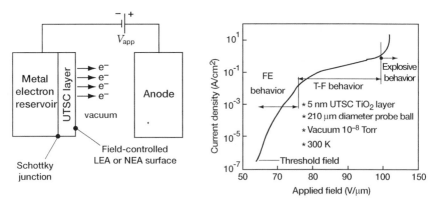

Figure 1. Schematic representation of a solid-state field-controlled electron emitter, SSE (left). Experimental current density *J vs.* applied field F_{app} characteristics of a SSE (right). [Reprinted with permission from *J. Vac. Sci. Technol. B* **19**, 1044 (2001). Copyright 2001, American Institute of Physics.]

of electrons in the wide bandgap semiconductor eventually leads to low or even negative electron affinity at the semiconductor/vacuum interface.[8,9] One advantage of this cold cathode is that the threshold field for substantial emitted current into vacuum was observed to be around 50 V/μm for a prototype Pt/TiO$_2$ cold cathode, about two orders of magnitude below the threshold electric field required to observe substantial emission from field emitter tips or diamond films.

- *InGaN/GaN field emitter arrays with piezoelectric surface barrier lowering*[10]

To increase the emitted current density of cold cathodes, Mishra *et al.* have proposed the use of wide bandgap semiconductors to lower the effective electron affinity at the surface via the piezoelectric effect in these materials.[10] For instance, the piezoelectric effect in a thin InGaN surface layer grown on top of a GaN substrate is due to strain which, in certain crystal directions, can lead to a separation of core valence electrons. The latter generates an internal electric field in the thin InGaN layer as a result of the pseudomorphic growth of this layer on the GaN substrate. Mishra *et al.* have shown that, for a structure grown in the (0001) direction, the internal electric field in the InGaN layer points towards the GaN substrate, leaving the positive end of the dipole in the InGaN layer at the InGaN/vacuum interface. This field leads to a potential drop across the InGaN layer and reduces the effective electron affinity χ_{eff} at the surface of the cathode, as illustrated on the left in Fig. 2. As a result, the turn-on voltage is reduced and the emitted current density is increased. For field emitter arrays with similar cathode to anode spacing, Mishra *et al.* observed a reduction of the turn-on voltage from 450 V to 70 V going from GaN to InGaN/GaN devices (Fig. 2, right).

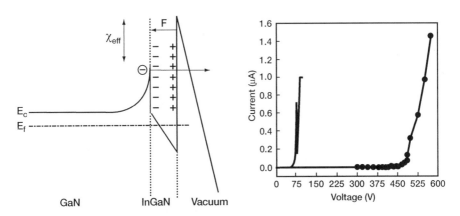

Figure 2. Schematic conduction band diagram of an InGaN/GaN field emitter. Electrons travel ballistically across the InGaN layer and, thus, effectively tunnel from the maximum of the GaN conduction band at the GaN/InGaN interface (left). *I–V* characteristics of InGaN/GaN (solid line) and GaN (line with circles) field emitter array (right). [Reprinted with permission from *Appl. Phys. Lett.* **73**, 405 (1998). Copyright 1998, American Physical Society.]

- *Rare-earth monosulfides as a means to achieve negative electron affinity*[11,12,13]

Typically, the operating lifetime of cathodes using cesiated surfaces is good as long as the tube is not subjected to excessive cathode currents from high-level usage. Otherwise, the emitted electrons from the cathode are accelerated and, depending on the accelerating voltage and cathode current density, there can be an electron scrubbing effect of the anode resulting in positive ions being liberated and accelerated back onto the NEA surface, causing a continuous increase in the cathode WF. An accelerated cathode degradation that occurs under high-current operating conditions is electron stimulated desorption of the activation layer. In general, III-V NEA surfaces are not stable in ultra-high vacuum.

At 300 K, Cs will slowly desorb from the surface in UHV at a rate dependent, in part, on the partial pressure of Cs within the vacuum system. Additional Cs will restore a fraction of the original surface escape probability. A small quantum yield can still give rise to a large current if the excitation intensity is sufficiently strong, such as light from a laser. However, excessive radiation may lead to the destruction of the activation layer by light-induced desorption. This effect must be counteracted by continuous cesiation during operation of the source.

Recently, we have proposed the use of sulfides of rare-earth elements as more stable alternatives to reach NEA at various III-V semiconductor surfaces.[12] These compounds do not suffer from all the limitations of cesiated surfaces. Table 1 lists a summary of some of the material properties of sulfides of rare-earth elements in their cubic form. Of particular interest is the fact that the WFs at room temperature of these compounds, when extrapolated from high-temperature measurements,[14] are quite small. It is therefore expected that these materials can be used to reach NEA when deposited on *p*-type doped semiconductors. For instance, LaS has a lattice constant (5.854 Å) very close to the lattice constant of InP (5.8688 Å) and NdS has a lattice constant (5.69 Å) very close to the lattice constant of GaAs (5.6533 Å). Since the room-temperature WFs of LaS (1.14 eV) and NdS (1.36 eV) are, respectively, below the band gap of InP (1.35 eV) and GaAs (1.41 eV), NEA can therefore be reached at InP/LaS and GaAs/NdS interfaces using heavily *p*-type doped semiconductors. Recently, we confirmed this result by means of a first-principles electronic-structure method based on a local-density approximation to density-functional theory.[11] This analysis predicted a WF of 0.9 eV and 1.1 eV for LaS and NdS, respectively, at low temperature.

Two other important features of the fcc cubic rare-earth compounds listed in Table 1 are the high melting temperature (about 2000 °C) and the fairly low resistivity. Hence thin films of these compounds used to promote NEA at semiconductor surfaces should be stable and should not suffer from current crowding effects which would lead to non-uniformity in the emitted current into vacuum.

Recently, we have proposed a new InP/CdS/LaS cold cathode emitter concept making use of lanthanum sulfide (LaS) to reach NEA. The architecture of the structure is shown in Fig. 3. A schematic energy band diagram throughout the proposed structure is shown in Fig. 4. The main elements in the design and functioning of such an emitter are: first, a wide bandgap semiconductor slab equipped on one side with a heavily doped *n*-type InP substrate that supplies electrons at a sufficient rate into the conduction band and, second, a thin semimetallic film (LaS) on the opposite side to facilitate the tunneling of electrons from the semiconductor conduction band into vacuum.

As shown in Fig. 3, an array of Au contacts is defined on the surface of the LaS thin film to bias the structure. The bias is applied between the InP substrate and the metal grid with emission occurring from the exposed LaS surface. For the InP/CdS/LaS cold cathode, the choice of a LaS semimetallic thin film grown on nominally undoped CdS is quite appropriate since the lattice constant of CdS (5.83 Å) is very close to the lattice constant of LaS (5.85 Å) in its cubic crystalline form.

Sulfide	Lattice constant (Å)	Work function (eV)	Melting temp. (°C)	Resistivity ($\mu\Omega$-cm)
ErS	5.424			
YS	5.466		2060	
NdS	5.69	1.36	2200	242
GdS	5.74			
PrS	5.747	1.26	2230	240
CeS	5.778	1.05	2450	170
LaS	5.854	1.14	2200	25
EuS	5.968			
SmS	5.97		1870	

Table 1. Materials parameters of some sulfides of rare-earth metals (cubic form).[14]

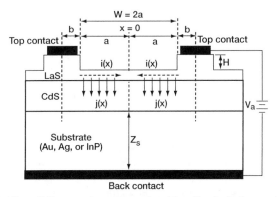

Figure 3. Cross-section of the newly proposed cold cathode between two emitter fingers. Trapping of electrons by the LaS semimetallic thin film leads to a lateral current flow and current crowding in the structure.

Additionally, LaS is expected to have a quite low work function (1.14 eV),[14] a feature which combined with the large energy gap of CdS (2.5 eV) leads to NEA of the semiconductor material.[12] The electronic structure of the CdS/LaS interface was recently investigated by means of a first-principles electronic-structure method based on a local-density approximation to density-functional theory.[11] The extrapolated 1.14-eV low work function of LaS was reproduced by that theory. It was found that NaCl structured layers of LaS should grow in an epitaxial way on a CdS substrate with a ZnS structure.[11]

We have also successfully deposited CdS on InP substrates using RF magnetron sputtering.[15] Recently, we also reported the successful growth of bulk cubic LaS and NdS samples.[13] Powder XRD scans of the samples show the rocksalt phase with a lattice constant $a_0 = 5.857(2)$ Å and 5.694(2) Å for LaS and NdS, respectively. In Raman scattering, we have observed the vibrational density of states (cubic symmetry) and identified the longitudinal optical (LO) and transverse acoustic (TA) phonons at 261 (284) cm^{-1} and 100 (92) cm^{-1} with LaS (NdS), respectively. Steps are now underway in our lab to fabricate targets of LaS and NdS to be used in RF magnetron sputtering of thin films on various substrates, as discussed above. To insure reproducibility of deposition, a fixed

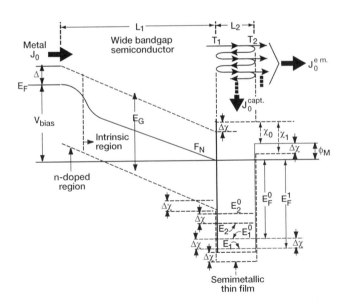

Figure 4. Schematic representation of the conduction band profile throughout the cold cathode emitter described in the text. Under forward bias, a fraction of the emitted current is captured in the LaS semimetallic slab. The subsequent excess sheet carrier concentration in the quantum well formed by the semimetallic slab leads to a shift of the Fermi level in the thin film which is similar to a lowering of the work function of the thin film.[12] [Reprinted with permission from *J. Appl. Phys.* **19**, 1044 (2001). Copyright 2001, American Institute of Physics.]

starting stoichiometry is desirable. Given the stability of LaS and NdS, the bulk material would appear to be a useful cathode material. Our experience with a stoichiometric CdS target shows that stoichiometric CdS thin films can be grown by RF sputtering.[15] It is therefore quite likely that the same procedure would work for deposition of LaS and NdS thin films. In particular, the experimental conditions (argon pressure, RF power, substrate temperature) established for CdS thin films[15] could serve as a guide in LaS and NdS thin-film deposition by RF sputtering. Large area InP/CdS/LaS cold cathodes should therefore be realizable using the cost-efficient RF magnetron sputtering technique.

3. Conclusions

In this contribution, we have described three recent proposals for the design of new planar cold cathodes that present paradigm shifts from the widely investigated Spindt field emitter arrays and diamond-based field emitters.[1-5] The first cathode proposed by Vu Thien Binh *et al.*[8,9] is based on a metal/wide-bandgap n-type semiconductor/vacuum structure whose surface barrier can be lowered in the regime of negative electron affinity following the creation of space charge in the wide-bandgap semiconductor region. Prototypes of this solid-state field-controlled electron emitter have been realized recently using a Pt/TiO_2 configuration. Another way for lowering the surface barrier for field emission based on the piezoelectric effect has been proposed and demonstrated recently by Mishra and coworkers[10] using InGaN/GaN field emitter arrays. Finally, we have proposed to use sulfides of rare-earth elements as a way to achieve negative electron affinity at the surface of various III-V and II-VI compounds. Rare-earth sulfides offer several advantages over the commonly used cesiated surface. Not only do they possess high chemical stability (melting temperatures above 2000 °C), but they also display metallic conduction. In the past, we have proposed and analyzed a new InP/CdS/LaS cold cathode based on this concept[12] that can provide emission current densities of several tens of A/cm^2. Following our successful growth of the cubic phase of rare-earth sulfides (LaS and NdS),[13] we are currently working on the fabrication of prototypes of the InP/CdS/LaS emitter based on RF magnetron sputtering of the CdS and LaS layers. This conventional deposition technique could lead to a cost-effective technology to produce large-area cold cathodes for flat-panel display applications, among others.

Since the market for cold cathodes also includes microwave vacuum transistors and tubes, pressure sensors, and high-temperature and radiation-tolerant sensors, among others,[1,2] the quest for new paradigm shifts in cold cathode design has just begun and should lead to exciting developments in the field of vacuum microelectronics within the next decades.

Acknowledgments

This work is supported by the National Science Foundation under award ECS-9906053 and by the Air Force Research Laboratory, Sensors Directorate, at Wright-Patterson Air Force Base under Contract No. F33615-98-C-1204.

References

1. I. Brodie and C. A. Spindt, "Vacuum microelectronic devices," *Proc. IEEE* **82**, 1006 (1994).
2. D. Temple, "Recent progress in field emitter array development for high performance applications," *Mater. Sci. Eng. Reports* **24**, 185 (1999).
3. P. R. Bell, *Negative Electron Affinity Devices*, Oxford: Clarendon Press, 1973.
4. J. E. Jaskie, "Diamond-based field emission displays," in: P. Boolchand, ed., *Series on Directions in Condensed Matter Physics*, Vol. 17, Singapore: World Scientific, 2000.
5. See *Proc. First Int. Symp. Cold Cathodes*, 198th Meeting Electrochemical Soc., Phoenix, AZ, 20-27 October 2000.
6. J. Ristein, "Electronic properties of diamond surfaces – blessing or curse for devices?" *Diamond Related Mater.* **9**, 1129 (2000).
7. G. Rosenman, D. Shur, Ya. E. Krasik, and A. Dunaevsky, "Electron emission from ferroelectrics," *J. Appl. Phys.* **88**, 6110 (2000).
8. Vu Thien Binh, "Electron spectroscopy from solid-state field-controlled emission cathodes," *Appl. Phys. Lett.* **78**, 2799 (2001).
9. Vu Thien Binh, V. Semet, J. P. Dupin, and D. Guillot, "Recent progress in the characterization of electron emission from solid-state field-controlled emitters," *J. Vac. Sci. Technol. B* **19**, 1044 (2001).
10. R. D. Underwood, P. Kozodoy, S. Keller, S. P. DenBaars, and U. K. Mishra, "Piezoelectric surface barrier lowering applied to InGaN/GaN field emitter arrays," *Appl. Phys. Lett.* **73**, 405 (1998).
11. O. Eriksson, J. Willis, P. D. Mumford, M. Cahay, and W. Friz, "Electronic structure of the LaS surface and LaS/CdS interface in a new cold cathode configuration," *Phys. Rev. B* **57**, 4067 (1998).
12. P. D. Mumford and M. Cahay, "Dynamic work function shift in cold cathode emitters using current-carrying thin films," *J. Appl. Phys.* **79**, 2176 (1996).
13. Y. Modukuru, J. Thachery, H. Tang, A. Malhotra, M. Cahay, and P. Boolchand, "Growth and characterization of rare-earth monosulfides for cold cathode applications," *J. Vac. Sci. Technol. B* **19**, 1958 (2001).
14. G. V. Samsonov, *High Temperature Compounds of Rare-Earth Metals with Nonmetals*, New York: Consultants Bureau Enterprises, Inc., 1965.
15. Hai Tang, M.Sc. Thesis, University of Cincinnati (1999).

3 THE FUTURE ALONGSIDE SILICON: OPTICAL

Contributors

3.1 H. van Houten

3.2 N. N. Ledentsov

3.3 L. V. Asryan and S. Luryi

3.4 G. Belenky, L. Shterengas, C. W. Trussell, *et al.*

3.5 Q. Hu, B. S. Williams, M. R. Melloch, and J. L. Reno

3.6 M. A. Green

3.7 M. Razeghi, Y. Wei, and G. Brown

3.8 A. Zukauskas, M. S. Shur, and R. Gaska

3.9 A. Zaslavsky, C. Aydin, G. J. Sonek, and J. Goldstein

3.10 M. A. Sánchez-García, E. Calleja, E. Muñoz, *et al.*

The Evolution of Optical Data Storage

Henk van Houten
Philips Research Laboratories, 5656 AA Eindhoven, The Netherlands

1. Some thoughts on technical evolution

Gradual step-by-step technical progress is commonly called "evolutionary," to distinguish it from more radical innovation. In this sense, Moore's law or the areal density improvement in hard disc drives is evolutionary, even though the speed of progress is astounding. Equating gradual rather than breakthrough technical progress with the term evolution reflects a 19th century grasp of biology. Whereas Darwin[1] and his contemporaries envisioned a process of gradual environmental change and survival of the fittest, Gould and Eldridge[2] have more recently introduced the idea of punctuated equilibrium: "We expect no slow and steady transition, but a break with essentially sudden replacement of ancestors by descendants ... this break may record the extinction or emigration of a parental species, and the immigration of a successful descendant rapidly evolved elsewhere in a small peripherally isolated population."

Also in the realm of technology, "evolutionary progress" may imply radical innovation as well as gradual improvement. Breakthroughs like the steam engine replacing manual labour, the incandescent lamp replacing candle light, the motor car replacing the horse-drawn carriage, or the transistor replacing the vacuum tube, are all part of technical evolution. Although these inventions may initially have been conceived as replacement technologies, their impact has been so large that the very fabric of society has been affected.

In examining future trends, and in particular in thinking about possible successors for well established technologies, it is important to identify the key factors that made the established technology successful in the first place. But one should also consider carefully whether the environment has altered to a sufficient degree that the original success factors might be on the verge of becoming obsolete. The term "environment" should be taken broadly: it may include order-of-magnitude changes in the performance of competing technologies, but also changes in patterns of behavior of end users (which in turn may be the result of technical or social innovation in totally different areas).

2. Origin of the species: The compact disc

The Edison gramophone pioneered the use of a needle for the dual purpose of tracking a continuous groove and picking up the high frequency signal stored in the groove modulation. The original recordable cylindrical rolls were rapidly replaced by a vinyl disc suitable for mass replication, and thus for music distribution. The gradual improvement of technology lead to the replacement of the 78 rpm disc, and to the eventual widespread acceptance of the long playing record as the *de facto* standard for music playback.

- *The optical needle*

    In optical storage, a "needle" is formed by a focused beam of light, as illustrated in Fig. 1. The optical pick-up is never in contact with the disc, yet the focused spot can keep on track and pick up the HF signal on the disc, just like a needle in a gramophone. This innovation had to await the invention and miniaturisation of the HeNe gas laser, because of the need for a sufficiently bright light source. Because the diffraction-limited spot has a sub-micron diameter, the track pitch can be made very small. Also the optical needle can easily pick up very high frequency signals from the disc. The question therefore was what to do with the large (analog) storage capacity. The answer was thought to be the playback of video information. Thus the "Video Long Play" (VLP) optical disc system was developed by Philips in 1972.[3] The VLP was given the same diameter as a long-playing record, because this size was thought to facilitate consumer acceptance. It used analog signal modulation, like the gramophone. The VLP system was introduced in 1976 on the US market as the "Laser Vision" system. Optical disc mastering and replication was invented to enable low-cost disc replication, which was known to be a key success factor in low-cost music distribution.

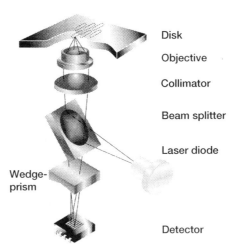

Disk

Objective

Collimator

Beam splitter

Laser diode

Wedge-prism

Detector

Figure 1. The light path of an optical disc drive. The detector yields the high-frequency data signal, in addition to the servo signals for closed-loop control of the 2D actuator used to adjust the position of the objective lens. The wedge is used as part of the focus error-signal generation (Foucault method).

- *Digital encoding and error correction*

The non-contact playback by an optical needle and the resulting absence of wear suggested the idea of a high-quality audio disc. But the analog storage capacity of the VLP was much too large for audio applications. The real breakthrough came with the application of digital channel encoding and error correction of the information on the disc, which allowed a flawless and permanent sound quality. The Compact Disc Digital Audio system was designed around the AlGaAs double heterojunction laser, emitting infrared light (780-nm wavelength). This source enabled the construction of a very compact light path, which fitted naturally with the choice for a compact 12-cm diameter disc. The CD was designed for robustness and low cost, as required for a consumer product. It was introduced on the market in 1982.[4]

The rest is history: as we write this article (summer of 2001), over 750 million CD audio players have been installed world wide, supplemented with 380 million CD-ROM players, and 50 million video-CD players attached to TVs (mainly in China). A more recent development adding to the CD's success is the invention of the write-once or recordable CD (CD-R), which after recording plays back on any standard CD player. In 2000 alone, the astounding number of 3 billion CD-R discs were sold.

3. Survival of the fittest: success factors of optical storage

While in hard discs magnetic fields are used to read and write data, optical disc systems use a focused beam of light. Magnetic fields can be concentrated within soft magnetic materials, serving as a flux guide. But a free-space "magnetic lens" does not exist. Consequently, magnetic recording is intrinsically a near-field technology. The minimum bit length depends strongly on the head-disc separation. In contrast, light can be focused to an extremely small spot at a distance far away from the light source or the objective lens.

- *Removability*

In a CD system, a semiconductor laser generates a divergent light beam. The beam is focused by a collimator and an objective lens, and it enters a 1.2 mm thick transparent plastic substrate on the surface opposite to the one carrying the information layer (see Fig. 2). The substrate protects the information layer from scratches or wear. Since the laser beam is out of focus at the entrance surface of the disc, the system is insensitive to dust or fingerprints on the disc. This is one of the crucial advantages of optical disc storage over hard disc, because it enables media removability (and thus exchange of media, and distribution of pre-recorded content on a disc). In addition, the large lens-disc separation (more than 1 mm in CD) guarantees that crashes between the optical pick-up and the disc can be avoided at all times. These advantages were of key importance when the Compact Disc Digital Audio system was conceived by Philips and Sony.[4]

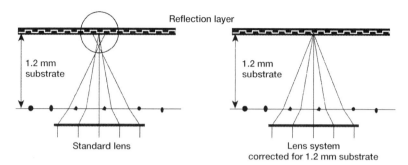

Figure 2. Substrate-incident recording (information layer away from the entrance surface) has the advantage that particles or fingerprints are out of focus, but it requires the use of a lens system corrected for the spherical aberration introduced by the propagation of the convergent beam through the plastic substrate.

The price to pay for "substrate incident" optical recording is that the propagation of the focused laser beam through the 1.2-mm-thick plastic substrate introduces aberrations of the optical wave front. As a result, the read-out spot is distorted, leading to a loss of resolution. Spherical aberration can be corrected using a special lens system, designed for the 1.2-mm-thick substrate. To reduce drive complexity, preferably a single aspherical objective lens is used.

- *Standardization*

Additional aberrations are introduced if the disc is slightly tilted with respect to the optical axis. This tilt cannot be avoided, because the removable plastic disc may be slightly warped, and the disc is not always clamped in the same way in the drive. This problem is addressed by assigning a tolerance budget to both the disc and the drive. The definition of such tolerances makes sense only if all disc and drive manufacturers adhere to the same specifications. Optical storage therefore needs stable well-protected worldwide standards. These standards deal with the basic physical parameters, such as disc capacity and data rate, and with specifications such as reflectivity and modulation level of the light. Also the disc format (including topics such as the channel modulation code, the error correction method, the addressing format) is fully specified.

- *Low-cost replication*

Besides substrate incident read-out, low-cost mass-manufacturing media replication technology is a key factor. The pits representing the digital information on the disc are embossed. Replication is done using an injection molding technique and a stamper carrying the negative of the desired pit pattern.

As a result of the unique advantages explained above, the CD has become the preferred medium for the distribution and exchange of pre-recorded digital content (both for digital audio and data).

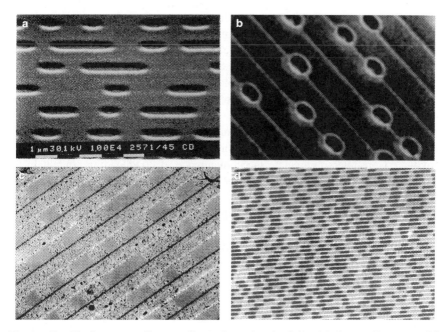

Figure 3. Marks on a disc: replicated read-only (a), ablative write once (b), rewritable phase change (c), and rewritable MO (d).

4. Recordable and rewritable CDs

Many ways exist to define marks on a disc. They can be replicated pits as in CD-ROM (Fig. 3(a)), pits written by laser ablation of material on the disc as in early write-once media (Fig. 3(b)), amorphous marks in a polycrystalline matrix as used in CD-RW (Fig. 3(c)), or oriented magnetic domains characteristic of MO recording (Fig. 3(d)).[5,6]

- *Compatibility*

    An early magneto-optic version of the CD failed because of incompatibility with the CD. A write-once 12-cm disc with 650 MB capacity, compatible with CD, was defined subsequently. Known as *recordable compact disc,* or CD-R, it has recently conquered the world. The CD-R can be used for applications such as photo archiving, back-up, or the copying of read-only discs for personal use. The compatibility of the CD-R discs was made possible by the use of an organic dye medium that has the required high reflectivity and contrast, so that the same signal levels are achieved as in a read-only CD.

- *Rewritable: CD-RW*

    A rewritable CD based on phase change media was introduced in 1996 under the name of CD-RW. CD-RW is the first example of a really successful disc format based on phase-change recording, where the information is written reversibly in the form of amorphous marks in a crystalline background.[5,6] After recording is finalized, a CD-RW disc meets all specifications of the CD-ROM (or CD-Digital Audio) standard, with the exception of the reflection level, which is lower. Full read-compatibility with play-back equipment therefore required a small modification to the pre-amplifier settings in the drives, which has now been standardized as the "multi-read" function. Its usage is widespread in new drives.

CD-RW is well positioned as the successor to the floppy disc. Both CD-R and CD-RW have further enhanced the role of CD as a digital convergence product. Consumers appreciate the convenience of downloading content in the PC, and subsequently using the disc for playback in the car or the living room. The breakthrough of CD-R and CD-RW has also proven the point that compatibility with the existing read-only format used for distribution of digital content is a key factor of success in introducing a recordable format for optical storage. This statement is corroborated by the insignificant market penetration of various incompatible magneto-optical disc formats that have been developed over the past decades.

- *System aspects of the blank disc format*

Writing, erasing, or reading of the small marks in CD-R and CD-RW imposes a rather tight tolerance on the precision with which the laser beam is focused onto the dye or phase-change layer. The same holds for the accuracy with which the focused beam is kept on the center of the track on which the amorphous marks are positioned. The focusing and tracking error signals are derived from the reflected laser beam. In the case of CD-ROM or CD-Audio, the tracking signal is derived from the embossed pits. In rewritable and recordable discs, tracking is also necessary while writing on a blank disc. In this case, the tracking error signal is generated from a pre-embossed groove on the surface of the plastic substrate.

The pre-embossed groove can also be used to generate timing signals by providing a small-amplitude modulation on its radius of curvature. This is called a "wobble." If the modulation has a sufficiently high spatial frequency, the focused beam is not able to follow the wobble, because of the finite bandwidth of the electromechanical tracking actuator on which the objective lens is mounted. But the detector that generates the tracking error signal is still capable of detecting the wobble. In this way, the system clock can be locked to the wobble period, ensuring appropriate synchronization with the disc velocity of the laser pulses used to write or erase amorphous marks on the disc. The groove on the disc may be interrupted by short intervals of pre-embossed pits carrying addressing information. This scheme is employed in disc formats with headers (a recent example is the DVD-RAM format). Alternatively, the addressing information may be coded in the wobble (for example by phase or frequency modulation). This method is the mainstream choice, used in CD-RW, DVD+RW, and DVR (see below).

- *Land and groove recording*

It is possible to write the amorphous marks only in the groove (as in CD-RW or DVD+RW). Alternatively, the lands between the grooves may be used as well to write additional data (this is called land-groove recording, and it is employed for example in DVD-RAM). Unfortunately, the gain in density is much less than a factor of two, because in land-groove recording the track pitch has to be chosen considerably wider to avoid cross-talk between adjacent tracks. The net density gain is typically only 15%. In addition, compatibility with write-once and read-only discs is hampered (because of the different track pitch, and the need to jump from land to groove after each disc revolution).

4.   The x-game

The first CD-ROM drives had a user data rate of 1.2 Mbit/s, which equaled that of CD-Audio. That data rate, in turn, was a consequence of the required audio quality (no data compression was used). For computer applications even higher data rates are desired, and in recent years the CD-ROM industry has been involved in a fierce competition, called the "x-game" or the speed race. One speaks of a "40×" drive, to denote a CD drive that can achieve a user data rate 40 times higher than the original 1.2 Mbit/s.

High speed CD "burners" need more powerful lasers and optics with an improved light efficiency. Also, with increasing writing speeds, the precision

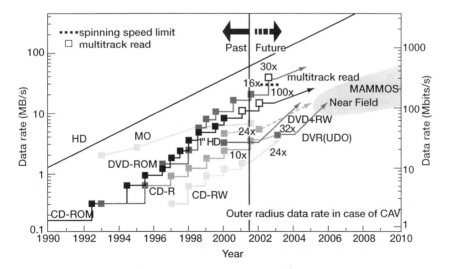

Figure 4. Speed race for hard disc drives (HD), magneto-optical drives (MO), and optical disc systems (CD, DVD, DVR). The horizontal dashed lines indicate the speed limit for single-spot read-out in CD (64×) and DVD (20×), imposed by the mechanical strength of the substrate strained by centrifugal forces.

with which transitions of amorphous to crystalline areas are written has to be improved, in order to keep the timing jitter in the play-back signal sufficiently low. This requirement can be achieved by a more elaborate write strategy that calls for high-precision and high-speed laser driver electronics (with timing resolutions of better than 1 ns).

The write data rate for CD-RW is primarily limited by media properties. The transformation from amorphous to crystalline states takes a minimum amount of time, the crystallization time, corresponding to the time to first form nuclei on which crystallites can grow. This time depends on the phase change material used. For a 1× CD-RW disc, this time is on the order of 0.5 ms. Faster crystallization can be achieved by using different materials, and by clever design of the stack of thin films surrounding the phase-change layer. Crystallization times of less than 100 ns represent the state of the art, and 10-12× CD-RW (12–15 Mbit/s) are being introduced on the market now. The write-once CD-R media allow higher recording speeds already, with 24× drives being introduced on the market now. We expect this trend to saturate around 50–60× (close to the mechanical strength limit of the disc also encountered in ROM systems, see Fig. 4).

6. A step in capacity: From CD to DVD

* *Spot-size scaling*

The storage capacity in optical disc systems is determined by the diffraction-limited focused spot diameter, which is about $\lambda/(2NA)$. Here λ is the wavelength of the laser light in air. The numerical aperture, $NA = n \sin\theta$, is determined by the opening angle θ of the objective lens (formed by the optical axis and the ray with the largest opening angle), and by the refractive index n of the medium. In air $n = 1$, and necessarily $NA < 1$. The storage capacity of 650 Mbytes on a single-sided 12-cm diameter CD corresponds to a density of 1 bit per μm^2. This density is close to the fundamental limit of diffraction imposed by the use of an infrared laser with a wavelength of 780 nm of the focused beam and an $NA = 0.45$. In the early 1980's, this density surpassed that of magnetic hard discs by two orders of magnitude. At the introduction of the CD in 1982 it was difficult to imagine that as early as the year 2000 this storage capacity would no longer suffice.

Since that time, a software explosion has been triggered by the rapid increase of the storage capacity of hard discs and solid-state memory chips, and of the processing power of microprocessors. Today, the world's most popular operating system and children's computer games easily fill one or even a few CD-ROMs, and particularly the latter require a high data rate in order to satisfy the users. It is thus natural that research has been carried out worldwide to find ways to store higher and higher capacities on a 12-cm disc. But the optical storage business has certain characteristics that impose stringent boundary conditions on innovation. One must ensure that the intrinsic benefits of optical disc storage are maintained: cheap replication of pre-recorded ROM discs, robustness or "playability," removability, and a long data life exceeding 30 years.

Removability also implies interchangeability of discs and players made by different manufacturers. Recordable and rewritable discs have to play back on read-only players as well. This requirement explains why worldwide standardization — with its associated company politics related to intellectual property rights — is important in optical disc storage, in contrast to the hard disc industry. It also explains the historical failure of many proposals for alternative optical storage technologies or systems. And it explains why the areal bit density of optical discs progresses step-wise rather than continuously, each step taking a considerable number of years.

- *DVD*

In 1996, a new read-only optical system, Digital Versatile Disk, or DVD, was introduced with enough data storage capacity to hold 135 min of MPEG-2 compressed video, sufficient for more than 90% of all feature movies. A smaller focused spot diameter is achieved by using a shorter wavelength (red) laser and higher numerical aperture optics (an objective lens with $NA = 0.60$, while CD uses 0.45). In combination with a tightening of all system margins, this change leads to a storage capacity of 4.7 Gbyte. Because of the stronger optics, a 1.2 mm thick plastic substrate through which the data are accessed would introduce too strong optical aberrations under conditions of tilt. A thinner substrate, 0.6 mm, was therefore chosen for DVD. Two such thin substrates are glued back to back to get a disk that is as rigid as CD.

Like CD, DVD was introduced as a read-only format, but soon after the introduction of DVD, it became clear that rewritable versions would again be needed. What followed was the emergence of a range of competing rewritable system proposals, each with its own characteristics. Initially, systems with less storage capacity than DVD-ROM were introduced: DVD-RAM, holding 2.6 GB, and DVD+RW at 3.0 GB. Recently, upgraded versions of DVD-RAM and DVD+RW able to store the full 4.7 GB have been standardized. To make matters more confusing, a 4.7 GB format called DVD-RW has been introduced. The way the data is arranged on DVD-RAM makes it primarily suitable for PC use, whereas the format of DVD-RW is optimized for consumer electronics (CE) use. Neither disk will play back in standard DVD-video or ROM players, so both formats should be considered to be incompatible with DVD. To fulfil the need for a truly compatible rewritable DVD format for both CE and PC use, a consortium including Philips Electronics has developed a 4.7 GB DVD+RW.[7] The DVD+RW disk uses the same phase-change recording materials as CD-RW. It also employs a wobbled groove arranged in a continuous spiral. The compatibility between DVD+RW and DVD-Video and -ROM has been achieved by using the same physical parameters as in the existing read-only DVD media. During writing on a DVD+RW disc, tracking is done by following the wobbled groove. During playback, tracking is done using the written marks, as in DVD-ROM and DVD-Video.

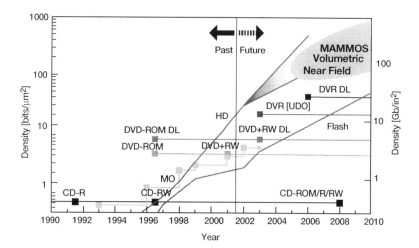

Figure 5. Trend in user density for hard disc drives (HD), magneto-optical storage (MO), nonvolatile solid state memory (flash), and optical disc systems (CD, DVD, DVR). The fourth-generation technology is still quite open. Candidates are, for example, advanced MO (MAMMOS), volumetric, or near field recording. Dual-layer extensions nearly double the capacity (DL).

7. The next step: DVR

In 1995, Shuji Nakamura invented the blue-violet diode laser.[8] Sony and Philips have worked out the technology for an optical disc system based on the use of this laser (405 nm wavelength) and an objective lens with a numerical aperture of 0.85. This powerful lens has to be built up of two separate lenses. In DVR, the information layer is addressed through a 0.1-mm thin cover layer. The substrate is just a mechanical carrier, the transparent cover now plays the role of the substrate that was illustrated in Fig. 2 (0.1 mm still being sufficient for adequate protection). The reason for this change is the need for sufficient disc tilt margin, which otherwise would be reduced to unacceptably low values due to the use of a high numerical aperture lens. With this system, 23 GB can be recorded on a single layer of a 12-cm disc — the storage capacity of about 35 CDs!

- *HD-video*
  The first application for this system will be the recording of high-definition digital video, hence the technical name DVR (for digital video recording).[9-12] Broadcasting of HDTV is starting now in Japan and the U.S. The data rate of this

HD compressed digital video stream is about 22 Mb/s, so that the 23-GB storage capacity of the new system guarantees a 2-hour recording time. Of course, the system may also be used for downloading standard definition (SD) movies at a lower data rate, with correspondingly larger recording time. Data applications are expected to emerge as well. Also read-only versions of the DVR system will be possible, and ultimately, DVR may be able to replace DVD, just as DVD can be seen as the successor of the CD. The basic physical concept of the three disc families is shown in Fig. 6.

Apart from the changes in the physical parameters (wavelength, *NA*, cover layer thickness), also the channel modulation code, the error correction scheme, and the address format have been newly designed for the DVR system, to ensure that the system has at least the same robustness as its predecessors. DVR is starting out as a video recorder, with rewritable media based on phase change recording (like CD-RW and DVD+RW). Compatible read-only and write-once versions will follow later.

8. Options for a fourth generation

In this section we give a brief review of the most promising options to increase the capacity of optical storage systems beyond that of DVR.

- *Spot size reduction*
  A natural approach to take — given the path followed in the case of DVD and DVR — would be to reduce the spot size further, by using an even shorter wave-

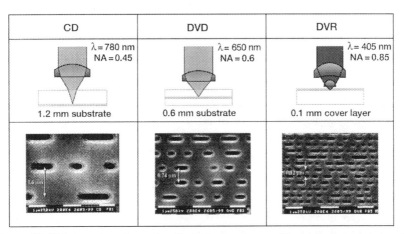

Figure 6. Three generations of optical disc systems. Progress in areal density is made in big steps (CD has a capacity of 650 MB, DVD 4.7 GB, and DVR 22 GB), by reducing the spot size through a shorter wavelength λ, and a stronger objective lens (with a higher numerical aperture *NA*). Read-only discs with replicated pit patterns are shown in the electron micrographs.

length and an even higher numerical aperture. Recent reports on efficient free-exciton recombination in thin-film diamond *pn* junctions[13] suggest that a deep UV solid state diode laser based on diamond (emitting at 235 nm) might be feasible. The use of deep UV light would imply more expensive optics, and probably one would have to resort to air-incident optical recording (no cover layer), because glass and plastic are not transparent at 235 nm.

Since $NA = n \sin\theta$, the numerical aperture for an imaging system in air ($n = 1$) cannot be increased beyond unity. A higher NA can be achieved, however, when the focus is formed in a medium with a higher refractive index. The optical principle of solid or liquid immersion lenses has been known in microscopy for a long time. The use of a solid immersion lens (SIL) in optical storage was proposed in 1995.[14] This approach is sometimes referred to as near-field optical storage, because it relies on evanescent wave coupling between the solid immersion lens and the disc (for those rays that would otherwise be totally internally reflected). An enhancement of NA from 0.85 as in DVR to 1.4 or even 2 might be possible by using the appropriate high refractive index materials.[15,16]

The combination of a SIL and a (future) deep-UV diamond laser would in principle allow a rather spectacular further increase of the optical storage density by a factor of up to 15. Unfortunately, one may have to pay a heavy penalty for this approach. The reason is that efficient evanescent wave coupling requires a very small lens-medium gap (on the order of $< \lambda/10$, which would be 20–30 nm in the case of deep UV light). Such small gaps may require the use of slider technology as in a hard disc, and this raises the question whether media removability could still be guaranteed.

- *Magnetic superresolution and domain expansion*

Alternatively, tricks can be used to enhance the resolution of optical storage beyond the limit imposed by the spot size (superresolution). In magneto-optical recording, it is possible to write crescent-shaped marks with a length much shorter than the focused spot by using pulsed magnetic field writing and erasing, a technique known as magnetic field modulation (MFM). Selective copying of a single mark in the storage layer to a read-out layer is possible using thermally enhanced magnetic coupling. This technique, known as magnetic superresolution (MSR), suppresses the interference of signals from adjacent marks on the disc, thereby allowing a higher storage density.[17,18] An additional gain in resolution is possible by expanding the copied marks temporarily. This expansion can be achieved by heating the read out layer by the laser beam in the presence of an external magnetic field. In combination with DVR-type optics ($NA = 0.85$ and a 405 nm laser) 20 GB/in^2 has been demonstrated, and up to 100 Gb/in^2 may be feasible[19-22] (see also Fig. 6).

This approach is quite interesting scientifically, but rather challenging from an engineering point of view. The media tend to become complicated, with many magnetic layers that have to be manufactured to tight specifications. The detection channel is nonlinear (although free of intersymbol interference) and is faced with synchronization difficulties, because the domain expansion is a dynamic process.

Also, the tolerances on the read power are very narrow, a magnetic coil has to be maintained in close proximity to the information layer to allow high data rates, and multilayer extensions seem to be impractical or impossible. More significantly, there is no simple replicated read-only or write-once counterpart for this approach.

- *Using the third dimension*

  Far-field optical storage would seem to lend itself naturally to an extension to the third dimension. This extension would seem to be more attractive than near-field storage using a SIL, because media removability would be safeguarded. The most straightforward approach to the third dimension is to define a medium containing two or more "standard" information layers, separated by a spacer layer. The DVD-9 standard refers to a dual layer read-only disc, with a capacity slightly below two single layer DVDs. The second layer is addressed through the first one, which is semi-transparent. Robust focusing, small interlayer cross-talk, and small additional spherical aberration are the main criteria that determine the disc design (in particular the spacer layer thickness). More recently, reports have announced the feasibility of dual-layer phase-change media, for DVD as well as for DVR conditions.[23] It seems difficult to go beyond two layers for such rewritable media.

  More ambitious 3D storage schemes have been proposed as well, including holographic discs, 2-photon recording, and multilayer fluorescent storage. Breakthroughs are still needed in these fields of research, in particular to allow sufficient signal-to-noise ratio. In addition, these approaches typically are hampered by issues such as materials stability and the complexity of media manufacturing.

- *Coding and detection*

  A quite different approach resorts to advanced signal-processing techniques. In a CD player, the zero crossings of the signal are used to detect the bits on the disc. As in hard disc drives, one may use more advanced algorithms, such as adaptive equalization, partial response maximum likelihood, and Viterbi bit detection. These algorithms allow smaller marks to be detected reliably (using information about permitted bit sequences for the relevant channel modulation code, and knowledge about the temporal status of the optical channel through which the binary bit stream is generated). In addition, the track pitch can be reduced if electronic cross-talk cancellation is used.[24] These techniques are likely to be applied already in conjunction with the DVR technology, allowing a further increase in capacity and enhancing the system robustness.

  Yet another approach is grey-scale or multi-level coding, which allows storing more than one bit of information in a single mark by making a more efficient use of the available signal-to-noise ratio in the optical channel.[25,26]

  Taking the effect on all system margins into account, such techniques typically can be used to increase the density by 50–100 %, so that they are useful in particular in combination with physics-based approaches to higher capacity, such as superresolution, shorter wavelength, or near field recording.

9.   Changing environment: bio-diversity or extinction?

A meaningful assessment of future opportunities for optical storage requires a broader perspective, because the world we live in is rapidly changing. As in biology, we have to ask ourselves the question whether a variety of storage media will survive (bio-diversity), or whether some will become extinct (as is probably the case for magnetic tape storage, which is not compatible with the rapid random access requirements of digital applications). Let us take a look at some of the relevant factors.

- *Improved data compression*
  CD digital audio is based on the recording of uncompressed signals. Digital video signals always require data compression, and DVD video is based on the MPEG-2 standard. Rapid developments in data compression technology might slow down the need for more storage capacity. Downloading MP3 audio files on portable appliances with solid-state storage is already becoming fashionable for portable audio. On the other hand, CD-R is also an ideal medium for the storage of a large number of MP3 files, and MP3 CD players are already on the market.

- *Increasing bandwidth of digital services*
  The emergence of wireless or wired networks and of the Internet is leading to different business models for storage and distribution of digital content. Network-attached storage may replace storage in devices in the home (as it is already doing in the office). Distributing digital content over the net could be a threat to optical data storage, if one thinks about business models like video-on-demand. But downloading on a recordable optical disc is a viable alternative. It is instructive to look at Table 1, which lists the time required to download the equivalent of a CD or a DVD for various network tranmission protocols (assuming that the transmission bandwidth is the limiting factor).
  On the one hand, this table tells us that downloading of a movie will become feasible. On the other hand, it is clear that this activity requires optical disc drives to keep up with the transmission data rates, unless downloading is done on a hard disc first (see also Fig. 4).

- *Decreasing cost per bit for various media*
  In the early 1980's, it was believed that magnetic hard discs would never be able to compete with optical storage in terms of cost per bit. Here is a representative quote from a 1985 book, *Principles of Optical Disc Systems* (Ref. 4,

Download	Phone 30 kb/s	ISDN 100 kb/s	ISDN+ 2 MB/s	Dig. cable 100 Mb/s	ATM 500 Mb/s
CD 650 MB	2 days	15 hr	40 min	48 s	10 s
DVD 4.7 GB	12 days	4 days	5 hr	5 min	1 min

Table 1. Time required to download the equivalent of a CD or DVD.

p. 266): "Magnetic disc systems ... offer the same accessibility as optical discs. However, high capacity Winchester-type magnetic disc memories are very expensive and are of the fixed spindle type. Both aspects are unacceptable when considering storage costs per bit, the essential parameter in electronic filing." In fact, the cost for HDD-type storage has dropped today (summer 2001) below 0.01 $/MB, with a further 2-orders-of-magnitude price reduction still expected before the end of the decade. This price point should be compared with the price of a blank 650 MB CD-R disc, which retails at about $1 right now (the drive cost is a one-time expense). It should be expected that blank recordable DVD or DVR discs will remain somewhat more expensive, so that their cost per bit will go down less than proportionally to their larger capacity. The conclusion is that hard discs will eventually outperform optical storage in terms of cost per bit, and would thus be the most attractive medium for temporary storage.

The tremendous progress in solid-state storage and hard-disc storage can be illustrated by the fact that the amount of embedded memory in a PC or in CE-type equipment has increased by a factor of 1000 in 15 years — from kBs to MBs, while optical storage in the same time frame (1982–1997) went only from CD (650 MB) to DVD (4.7 GB). Looking another fifteen years ahead, to 2012, we expect DVR to have become a success (23 GB or a bit more), and a fourth-generation optical storage system to have been introduced. This evolution should be put in the perspective of Moore's law, which tells us that by that time, solid-state memories will have progressed by another factor of 1000. Hard discs may be able to accomplish the same (see Fig. 5).

A very interesting emerging application of hard disc recording is automated personalized video recording in the home (using adaptive user profiles). This application may lead to obsolescence of the traditional VCR, and it could be seen as a threat to an optical-disc-based video recorder. On the other hand, optical storage is preferred if movies are to be preserved for future viewing, and combinations of HDD and optical disc drives in a single box are to be expected.

Solid-state storage (nonvolatile flash) typically is two orders of magnitude more expensive than HDD storage in terms of cost per bit. So in this case optical media are cheaper. Yet, one has to be careful: for a given application it is not the cost per bit that is most significant, but the cost for a given function, which is why for a digital still camera solid-state storage is most commonly used (although some cameras feature an optical disc drive). Other applications may follow. By 2012, a camcorder based on solid-state storage is a possibility.

- *A mobile world*

Now that cellular phones have become commonplace, it is clear that more and more applications will be ported to mobile appliances. For portable memories, a key consideration will be the power consumption. A useful measure is the cost in energy for the storage of a single bit.

The reference is the energy required to transmit a single bit over a wireless network. Whereas today the estimated energy is 15,000 nJ/bit for a handheld terminal in a mobile network, this figure will go down to 30 nJ/bit in wireless local area networks such as "HiperLAN/2" (operating at 32 MB/s).

For a small HDD and for flash memory, it is estimated that the energy required to store a bit is also about 30 nJ/bit (although this amount depends heavily on the usage model). Future solid-state storage technologies such as ferroelectric random access memories offer the potential for an improvement by an order of magnitude. A portable CD-RW drive still consumes much more power (our estimate is 2000 nJ/bit), but can be optimized significantly. A coin-sized optical disc drive is conceivable with an acceptable power consumption.

- *End-user behavior*

The toughest question of all is: how will end-user behavior evolve? Will people still give each other digital content in a tangible form, like a CD, as a birthday present by 2012? Or will they prefer "pay per usage" models? We refrain from making predictions, but note the recently announced decision to reissue the Encyclopedia Britannica in book form rather than only through the Internet.

10. Conclusions

Based on the remarks made in the previous sections, one might develop the extreme view that in the future, storage will be distributed on the Internet, end users will only view content on demand, and local storage needs will eventually be met with nonvolatile solid-state memory (portable) or HDD (home server, set-top box). In this view, there would be little or no need for removable discs and optical storage would eventually become obsolete, just as vinyl records did before, and as magnetic tape storage will soon.

An opposite view is possible as well. It is based on the realization that optical storage uses inexpensive, removable, standardized media. This medium is a unique feature and a key benefit, as it allows for very cheap publishing (which is of interest for the content industry), offering low-cost capacity increments and exchangeability between drives (of interest to the end user).

Personally, I expect that packaged optical media will remain attractive and that optical storage may be able to survive in a networked world, in peaceful coexistence with hard disc and solid state storage. Optical storage will be used where the required update rate is low; or for long-term storage, because of its intrinsic portability and exchangeability; and for personal collections. Depending on developments in digital rights-management policies, optical storage will remain attractive as a copy system, and as a semi-integral part of network services for downloading of digital content. Prerequisites are that recordable optical media keep up with bandwidths of networks in terms of write data rate, and that the cost per bit is kept quite low.

The last sentence of Charles Darwin's famous book on evolution, entitled "The Origin of Species by Means of Natural Selection, or The Preservation of Favoured Races in the Struggle for Life" provides us a nice quote with which to end this paper: "There is grandeur in this view of life, with its several powers, having been originally breathed into a few forms or into one; and that, whilst this planet has gone cycling on according to the fixed law of gravity, from so simple a beginning endless forms most beautiful and most wonderful have been, and are being, evolved."

Acknowledgments

I would like to thank Dr. C. Busch for preparing Figs. 4 and 5, and Dr. J. Schleipen for preparing Fig. 6.

References

1. Charles Darwin, *The Origin of Species*, New York: Bantam Classics, 1999 (also http://www.literature.org/authors/darwin-charles/the-origin-of-species/).
2. S. J. Gould and N. Eldredge, "Punctuated equilibria: the tempo and mode of evolution reconsidered," *Paleobiology* **3**, 115 (1977).
3. K. Compaan and P. Kramer, "The Philips 'VLP' system," *Philips Tech. Review* **33**, 178 (1973).
4. G. Brouwhuis, J. Braat, A. Huijser, J. Pasman, G. van Rosmalen, and K. Schouhamer Immink, *Principles of Optical Disc Systems*, Bristol: Adam Hilger, 1985.
5. George H. Johnson and Gary E. Thomas, eds., *Information Storage Materials*, Ullmann's Encyclopedia of Industrial Chemistry, 6th ed., New York: Wiley, 1999.
6. H. van Houten and W. Leibbrandt, "Phase change recording," *Commun. ACM* **43**, 64 (2000).
7. Chris Buma, Robert Brondijk, and Sorin G. Stan, "DVD+RW: 2-way compatibility for video and data applications," *Digest Tech. Papers IEEE Int. Conf. Consumer Electronics*, Los Angeles, 2000.
8. Shuji Nakamura and Gerhard Fasol, *The Blue Laser Diode*, Berlin: Springer, Berlin, 1997.
8. Tatsuya Narahara, Shoei Kobayashi, Masayuki Hattori, *et al.*, "Optical disc system for digital video recording," *Jpn. J. Appl. Phys.* **39**, 912 (2000).
9. Kees Schep, Bert Stek, Roel van Woudenberg, Martijn Blum, Shoei Kobayashi, Tatsuya Narahara, Tamotsu Yamagami, and Hiroshi Ogawa, "Format description and evaluation of the 22.5 GB DVR disc," *Jpn. J. Appl. Phys.* **40**, 1813 (2001).
10. Herman Borg, Paul M. Blom, Ben Jacobs, Benno Tieke, Sandy Wilson, Igolt Ubbens, and Guo-Fu Zhou, "AgInSbTe materials for high-speed phase change recording," *Jpn. J. Appl. Phys.* **40**, 1592 (2001).
11. Benno Tieke, Martijn Dekker, Nicola Pfeffer, Roel van Woudenberg, Guo-Fu Zhou, and Igolt P. D. Ubbens, "High data-rate phase-change media for the

digital video recording system" *Jpn. J. Appl. Phys.* **39**, 762 (2000). See also: Maarten Kuijper, Igolt Ubbens, Louis Spruijt, Jan Matthijs ter Meulen, and Kees Schep, "Groove only recording under DVR conditions," *Tech. Digest ODS* (2001).

12. K. Horiuchi, A. Kawamra, T. Ide, T. Ishikura, K. Nakamura, and S. Yamashita, "Efficient free-exciton recombination emission from diamond diode at room temperature," *Jpn. J. Appl. Phys.* **40**, L275 (2001).

13. B. D. Terris, H. J. Mamin, D. Rugar, W. R. Studenmund, and G. S. Kino, "Near-field optical data storage using a solid immersion lens," *Appl. Phys. Lett.* **65**, 388 (1994).

14. Kiyoshi Osato, Kenji Yamamoto, Yuji Kuroda, and Kimihiro Saito, "Near-field phase-change optical recording of 1.36 numerical aperture," *Jpn. J. Appl. Phys.* **39** (Part 1), (2000), 962 .

15. F. Zijp and Y. V. Martynov, "Static tester for characterization of optical near-field coupling phenomena," *Proc. SPIE* **4081**, 21 (2000).

16. A. Fukumoto and S. Kubota, "Superresolution of optical disks using a small aperture. (Magneto-optical disks)," *Jpn. J. Appl. Phys.* **31**, 529 (1992).

17. M. Kaneko, F. Aratani, A. Fukumoto, and S. Miyaoka, "IRISTER – magneto-optical disk for magnetically induced superresolution," *Proc. IEEE* **82**, 544 (1994).

18. H. Awano and N. Ohta, "Magnetooptical recording technology toward 100 Gb/in^2," *IEEE J. Selected Topics Quantum Electron.* **4**, 815 (1998).

19. T. Miki, A. Nakaoki, and M. Yamamoto, "20 Gbit/in^2 recording on magneto-optical disk using NA = 0.85 and 405 nm optics," *J. Magn. Soc. Jpn.* **25**, 343 (2001).

20. H. W. van Kesteren, Yu. V. Martynov, F. C. Penning, R. J. M. Vullers, M. van der Aa, and C. A. Verschuren, "Towards 100 Gb/in^2 MO storage using a blue laser, high-NA far-field optics and domain-expansion media," *J. Magn. Soc. Jpn.* **25** 334 (2001).

21. S. Yoshimura and A. Fukumoto, "Challenge to several-tens-nm/bit recording using domain wall displacement detection medium," *Jpn. J. Appl. Phys.* **39**, 986 (2000).

22. T. Akiyama, M. Uno, H. Kitaura, K. Narumi, R. Kojima, K. Nishiuchi, and N. Yamada, "Rewritable dual-layer phase-change optical disk utilizing a blue-violet laser," *Jpn. J. Appl. Phys.* **40**, 1598 (2001).

23. Y. Tomita, H. Nishiwaki, S. Miyanabe, H. Kuribayashi, K. Yamamoto, and F. Yokogawa, "25 GB read-only disk system using the two-dimensional equalizer," *Jpn. J. Appl. Phys.* **40**, 1716 (2001).

24. J. P. de Kock, S. Kobayashi, T. Ishimoto, H. Yamatsu, and H. Ooki, "Sampled servo read only memory system using single carrier independent pit edge recording," *Jpn. J. Appl. Phys.* **35**, 437 (1996).

25. G. Langereis, W. Coene, and L. Spruijt, "An implementation of limited multi-level (LML) optical recording," *Jpn. J. Appl. Phys.* **40**, 1711 (2001).

Long Wavelength Quantum Dot Lasers: From Promising to Unbeatable

N. N. Ledentsov

Institut für Festkörperphysik, TU Berlin,, D-10623 Berlin, Germany and
A. F. Ioffe Physico-Technical Institute, 194021, St. Petersburg, Russia

1. Introduction

Edge-emitting GaAs-based quantum dot (QD) lasers operating at 1.3 μm are realized with properties close to those for the best quantum well devices on InP substrates, giving the advantage of better cost efficiency. Maximum continuous wave (CW) output powers of 3 W (multimode) and 330 mW (single transverse mode) are realized at a heat sink temperature of 20 °C. GaAs-based QD vertical-cavity surface-emitting lasers (VCSELs) operating at 1.3 μm also are fabricated. Continuous wave output power of 0.8 mW (25 °C) has been realized. Extended CW operation lifetimes at 35-°C heat-sink temperature (>700 hours) are demonstrated for laboratory VCSEL technology. Maximum differential efficiency is above 60% and the maximum wall-plug efficiency is about 20%. Good prospects exist for shifting the wavelength of QD lasers towards the 1.5–1.6 μm range.

2. Edge-emitting and surface-emitting lasers

Semiconductor lasers dramatically changed many areas of human life. The first electrically-driven semiconductor laser was proposed in 1959 by Basov, Vul, and Popov.[1] Population inversion between ionized impurities and free carriers was thought to be achieved via impurity ionization by the application of a pulsed electric field. The boundaries of the sample were to provide the reflection of light for a laser feedback mechanism. In 1961, the idea of a current-injection laser using a degenerately doped p^+-n^+ junction was introduced.[2] The authors additionally mentioned that "It should be noted that lower current densities can be used if the semiconductors forming the p-n junction have forbidden gaps of different widths."[2] Basov and co-workers also proposed the possibility of lasing in the direction perpendicular to the p-n junction plane. The authors wrote that the boundaries of the current injection region may serve as reflectors due to the difference in nonequilibrium carrier concentration and a "resonant cavity" is formed. This paper may be considered the first proposal for a vertical-cavity surface-emitting laser (VCSEL). In fact, the first stimulated emission in p^+-n^+ junctions was realized, however, in an edge geometry.[3–5]

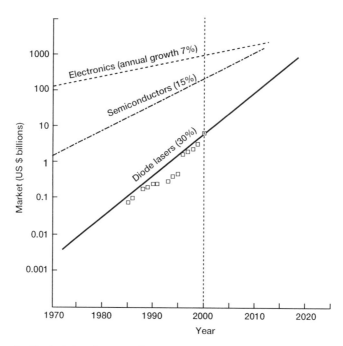

Figure 1. Market development for the electronics, semiconductor, and diode laser industries (source: *Laser Focus World*).

The p^+-n^+ devices operated at very high current densities and continuous-wave (cw) operation at room temperature, necessary for practical applications, was not achieved. The breakthrough became possible after the concept of the double-heterostructure laser, proposed independently in 1963 by Alferov and Kazarinov[6] and by Kroemer[7] was experimentally proven in 1968[8] to result in lasers having low threshold current densities at room temperature. In 1970, cw operation was realized and the history of commercial laser applications began.[9,10]

During the last two decades, the laser diode market has demonstrated a steady growth with an impressive averaged annual growth rate of slightly above 30% (see Fig. 1), approaching now the same level as the market for the entire semiconductor industry in the late 1970's or early 1980's.

Most of the diode laser market is linked to two major applications: telecommunications and optical storage. Lasers used in telecommunications have demonstrated explosive growth in recent years due to the enormous increase in Internet data traffic (see Fig. 2). In telecommunications, most of the laser diode applications are linked to the spectral range beyond 1.25 µm. The range near

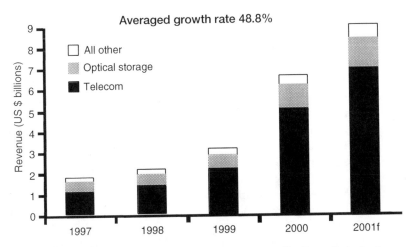

Figure 2. Breakdown of the laser diode market by application. Note faster growth of the telecommunications sector (source: Laser and Optoelectronics Marketplace Seminar, San Jose, January 22, 2001). Forecast for 2001 is also given.

1.28–1.32 μm is particularly important for datacom applications at distances up to 2–10 km (single or multimode fiber), with the advantage of ultrafast data links based on time division multiplexing (TDM) and, more recently, code wavelength division multiplexing (CWDM). High-power single-transverse-mode long-wavelength laser diode pumps are required for the ~1.5-μm wavelength range, where Raman fiber amplifies for long-haul telecom applications are used for dense wavelength division multiplexing (DWDM). Wavelength-tunable lasers are needed for the 1.5–1.65 μm range used in DWDM. This classification became to some extent obsolete after the appearance of Lucent "Allwave" fiber, which does not contain an OH^+ absorption hump between the 1.3-μm and 1.55-μm optical fiber transparency windows. In the near future, all applications between 1.2 and 1.7 μm may be merged in unified CWDM/DWDM systems (transmitters, Raman amplifiers).[11] In addition to telecommunications, eye-safe lidars for terrestrial and free-space applications need long-wavelength lasers.

Currently, both edge-emitting and vertical-cavity devices are used, as shown in Fig. 3. Edge-emitting devices are usually long (0.3–3 mm depending on application). In this situation, the loss of light at the crystal boundary due to the finite facet reflectivity may be compensated easily by sufficiently increasing the length of the device and/or the population inversion current. For short cavity devices, however, the role of the facet losses increases, and the gain may not be increased due to the gain saturation effect in the active medium. For short cavities, multilayer dielectric mirrors may be deposited on one or both facets of the device to increase the facet reflectivity. The advantage of edge-emitting devices is the possibility of fabricating long- and narrow-stripe devices. Narrow-stripe devices are necessary to provide the single-lateral-mode operation important for pumping of optical fibers, *e.g.* in Er-doped fiber amplifiers or in Raman fiber amplifiers. In the case of protected (coated) facets and an extended waveguide design, the total output power of the laser is limited by overheating of the active region, and thus by the efficiency of heat dissipation in cw operation. Heat dissipation takes place via the sample surface (the upper *p*-contact closest to the active region is mounted on a heat sink in "*p*-side down mounting"). For the same current density, the device having a larger surface will provide higher power, assuming the same conversion efficiency. Thus, edge-emitting single-lateral-mode devices can provide much higher powers as compared to single-lateral-mode VCSELs because of the larger surface area available for heat dissipation for edge emitters. On the other side, a disadvantage of the currently existing edge-emitting lasers is the wide divergence of the laser beam in the vertical direction caused by self-diffraction at the thin waveguide layer output aperture. Typically, this value is close to 50–60° (full width at half maximum) and in advanced designs may go down to 20–30°, still several times larger than for single-lateral-mode VCSELs. This divergence makes coupling to fiber less efficient and more costly due to the need for special optics. Careful waveguide design potentially allows reduction of

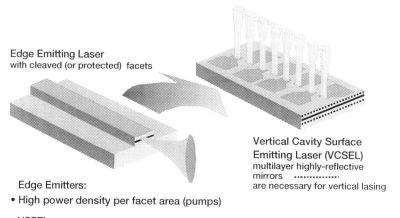

Edge Emitting Laser
with cleaved (or protected) facets

Vertical Cavity Surface
Emitting Laser (VCSEL)
multilayer highly-reflective
mirrors ················
are necessary for vertical lasing

Edge Emitters:
• High power density per facet area (pumps)

VCSELs:
• Ultralow threshold currents
• High beam quality, single-mode, coupling to fiber
• Temperature stable wavelength
• Monolithically-integrated mirrors
• Planar technology, dense arrays, on-chip integration
• On-wafer testing, small size, cost-effective
• Tunable (e.g. membrane) VCSELs

Figure 3. Schematic comparison of edge- and surface-emitting lasers.

the beam divergence down to 3–7°, but edge-emitting lasers of this kind are still to be proven for commercial applications.

For signal transmission in optical fibers, high-power laser beams cannot be used due to nonlinear effects. Here a cheap and compact device with narrow beam divergence and vertical light output is desirable. In VCSELs,[12] the active region is embedded in the resonant cavity (size being a multiple of the half-wavelength in the crystal), confined by two multilayer (usually all-epitaxial) distributed Bragg reflectors (see Fig. 3) having high reflectivities (typically above 99.5%). These reflectivity values are sufficient to reduce external mirror losses to the level where they become less than the saturated modal gain in the vertical cavity and thus, lasing becomes possible. VCSELs also provide single-longitudinal-mode operation and thus demonstrate temperature stability of the wavelength, limited by the temperature dependence of the refractive index. A small current aperture also means a small operating current. As compared to edge-emitting single-mode wavelength-tunable lasers, which require complicated post-growth technology for wavelength tuning,[13] advanced approaches for wavelength tuning exist for VCSELs (*e.g.* when the top DBR is placed on a membrane, with the membrane positioning tunable by an electric field).[12] Planar technology, on-chip integration, and on-wafer testing result in a cost-efficient technology. All these advantages make the VCSEL sector of the market the most quickly growing.

The main field where VCSELs can replace edge emitters is related to the 2–10 km optical fiber range where cost-efficient 1.3-µm VCSELs are required. However, they may also replace short-wavelength 0.85-µm VCSELs dominating the datacom market for short-reach and very-short-reach interconnections. For one reason, the necessity of 0.85-µm to 1.3-µm signal conversion disappears. For another, 1.3-µm single-mode silica fiber is better suited for TDM[11] than plastic fiber and solves the problem of the "end user" (or the "first access"). Additionally, the 1.3-µm wavelength provides higher eye-safe power levels, making possible higher total light powers, further lifting the problem of the maximum connection

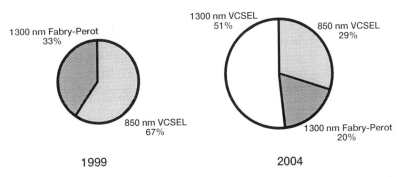

Figure 4. Forecast for 1.3-µm VCSEL applications (source: Laser Marketplace Seminar, San Jose, January 26, 2000[14]).

distance and multiwavelength CWDM operation. Rapid expansion of VCSEL technology into datacom and telecom applications will drive the 1.3-μm VCSEL market growth in the near future,[14] with 1.3-μm VCSELs expected to take 51% of the total datacom market in 2004[14] — see Fig. 4. Until recently, InP-lattice-matched materials (InGaAsP, InGaAlAs) have been utilized exclusively to fabricate 1.3–1.55 μm VCSELs, because the InP-lattice-matched alloy compositions provide forbidden bandgaps that fit the wavelength ranges requested in telecommunications. Development of cost-efficient InP-based VCSELs, however, is limited by several fundamental drawbacks of this technology. Cost-effective VCSEL applications are commercially available now only for GaAs-based structures, where high-reflectivity GaAs-AlAs multilayer distributed Bragg reflectors can be realized easily.

In contrast, InP-based structures suffer from a low refractive index difference in the III-V alloys lattice-matched to InP, important for the fabrication of sufficiently thin DBRs with a reasonable stop-band reflectivity, and the lack of a developed current-aperture oxidation technology. The low thermal conductivity of the substrate, and very low thermal conductivity of the alloy DBRs make VCSEL applications on InP much less advantageous as compared to GaAs-based VCSELs. This situation motivates attempts to develop new types of laser active media on GaAs to bring GaAs-based VCSEL technology to the long-wavelength (1.25–1.7 μm) spectral range. Currently, two techniques are competing: InGaAsN alloys and InAs quantum dots (QDs). Despite its late start in 1998, long-wavelength InAs-GaAs QD laser technology has gained some significant advances recently.

3. Generations of semiconductor diode lasers

Development of advanced active media for semiconductor diode lasers was the mainstream direction that enabled enormous progress in applications. Each time a new approach came, the properties of the devices were greatly improved and the application niche broadened. The first step was the proposal of current-injection lasers[1,2] and soon after, their realization.[3-5] The decisive step for commercial applications was related to the concept of double heterostructures,[6,7] giving devices with the low threshold current density[8] permitting cw operation.[9,10] Further progress relates to the use of quantum size effects in semiconductor heterostructures.[15] The advances due to different device structures are illustrated in Fig. 5 using the threshold current density as a figure of merit.

In 1975, Dingle and Henry proposed the idea to "exploit quantum effects in heterostructure semiconductor lasers to produce wavelength tunability" and achieve "lower lasing thresholds" via "the change in the density of states which results from reducing the number of translational degrees of freedom of the carriers."[15] They wrote: "As the number of degrees of freedom is reduced the density of states decreases at high energies. This causes the electrons to occupy the lower energy states and thereby to contribute more effectively to population inversion and laser gain."

For structures with size quantization in more than one direction, a singularity in the density of states occurs[15] and the above positive effects are further enhanced over the quantum well (QW) case. The ultimate example of size quantization in solids is realized in a quantum dot, which represents a semiconductor crystal only several nanometers in size, coherently inserted in a larger bandgap semiconductor matrix. Quantum dots mimic the basic properties of atoms, providing a geometrical size that allows the practical application of atomic physics to the field of semiconductor devices.

The proposal of Dingle and Henry concerning structures with dimensionality lower than two did not draw much attention, as the QW concept was widely doubted at the time. However, in the early 1980's, progress in QW lasers and experimental advances in crystal growth on misoriented[17] or patterned[18] substrates revived interest in heterostructures with size quantization in more than one dimension. In 1982, Arakawa and Sakaki[19] theoretically considered some effects in lasers based on heterostructures with size quantization in one, two, and three dimensions and concluded: "Most important, the threshold current of such a laser is reported to be far less sensitive than that of conventional laser reflecting the reduced dimensionality of electronic state." After the initial observation of QD

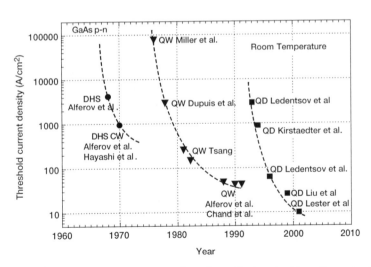

Figure 5. Evolution of semiconductor lasers from GaAs p^+-n^+ junctions, to double heterostructure (DHS), quantum well (QW), and quantum dot (QD) structures. Minimum threshold current density is the value required to overcome internal and external losses, such as free-carrier absorption, optical scattering loss, finite facet reflectivity, substrate leakage loss, *etc.*, and achieve modal gain. This value is *not* directly linked with the population inversion current density and represents a key figure of merit. The last result for a QD laser structure is from the team at the University of New Mexico, Albuquerque (in print).

lasing in the early 1990's, progress in QD-based structures has been rapid — see Fig. 5.

4. Self-organized quantum dots

All discussions of QD lasers had a merely academic character until 1993, when the first experimental demonstration of lasing in self-organized QDs[20] was reported. The real application of QDs to lasers became possible when the approach of self-organized growth (SOG) was developed to a level permitting fabrication of dense arrays of uniform QDs and, simultaneously, significant reduction of all types of defects. Self-organized QDs represent an example of a more general field of growth effects related to spontaneous formation of macroscopic order from initially random distributions. In 1980 spontaneous periodic nano-faceting at crystal surfaces[21] was predicted (experimentally reported for GaAs surfaces in 1991,[22] and confirmed for overgrown GaAs/AlAs structures in 2001[23]). In 1980 also spontaneous formation of ordered ("parquet") structures was proposed in the case when two phases coexist on the crystal surface[24] (*e.g.* surface reconstruction domains, submonolayer deposits[25]).

The most widely used practical solution to fabricate QDs with properties sufficient for laser applications came from an effect that was traditionally considered undesirable by crystal growers. It was found that the growth of a strained material on a lattice-mismatched substrate first proceeds in a planar mode and a so-called "wetting" layer is formed. However, at some critical thickness this planar growth is interrupted and *three-dimensional* (3D) nanoscale islands on top of the thin wetting layer are formed on the surface, as was demonstrated for the case of the growth of indium arsenide on gallium arsenide in 1985 by the group at CNET, France.[26] When these islands are covered with GaAs, a dense array of coherent nanoscale InAs insertions in a GaAs matrix may be formed. Since InAs has a much smaller band-gap energy than GaAs, an array of InAs QDs is formed. The authors demonstrated that for multi-stack deposition InAs QDs tend to grow in a vertically correlated way. They studied x-ray and transmission electron microscopy, and demonstrated a broad luminescence band at 1.15 eV at 77 K; *i.e.* at surprisingly long wavelengths given the small average amount of InAs deposited (~2 monolayers). The authors attributed this luminescence to InAs nanodomains and concluded, "these kind of structures are thus proved to be of interest to study low-dimensional (<2) objects showing good optical properties." In 1987, the same group reported on detailed transmission electron microscopy studies of the InAs nanoislands, determined their lateral size, observed local ordering of nanoislands in chains along the [001]-like directions, and demonstrated that InAs deposition in excess of about 2 monolayers results in the formation of a high concentration of mesoscopic dislocated clusters.[27] This approach gained particularly strong attention in the beginning of the 1990's.[28-31]

In 1993, the first photopumped lasing in InGaAs-GaAs QDs was realized.[20] In 1994, low-threshold injection lasing from self-organized InGaAs-GaAs QDs was demonstrated.[32] The lasing was observed at low and at room temperature. At 77 K it occurred at 1.24 eV and at 1.31 eV for two different structures, being in both cases clearly in the range of the QD luminescence. The threshold current density was remarkably low (<100 A/cm^2) and remained practically unaffected by temperature up to about 150 K in agreement with the basic expectation for a QD laser.[19] At higher temperatures the threshold current increased due to the thermally activated escape of carriers from quantum dots to the surrounding GaAs matrix.

Several promising ways to fabricate QDs using self-organization phenomena at crystal surfaces and in the bulk have been demonstrated:[33]

- spontaneous quasiperiodic faceting of crystal surfaces and heteroepitaxial growth on faceted surfaces;

- spontaneous phase separation in semiconductor alloys during growth or slow cooling;

- spontaneous alloy decomposition upon high-temperature annealing;

- formation of two-dimensional islands during submonolayer hetero-epitaxial deposition.

Quantum dots also can be fabricated in a combined approach involving two or more processes. Here we will concentrate on the approaches to fabricate defect-free self-organized QDs suitable for applications in long wavelength (1.25–1.7 μm) GaAs-based light-emitting devices.

5.   Long-wavelength InGaAs-GaAs QDs

- *Large three dimensional (3D) quantum dots*

These quantum dots can be formed in highly lattice-mismatched systems,[34] such as InAs-(Al)GaAs, GaSb-GaAs, InAs-Si, *etc*. Both Stranski-Krastanow (with wetting layer, WL) and Volmer-Weber (without it) growth mechanisms may be realized for the same system, depending on the growth sequence and the deposition parameters. Ultrathin-layer (<1 nm) AlAs overgrowth of small InAs QDs[35] results in a replacement of In atoms of the WL with Al atoms, increasing the height and the volume of the QDs (see Fig. 6).

An increase in the substrate temperature increases the density of adatoms on the surface and decreases the total number of atoms arranged in QDs. Reversible and irreversible phenomena in Stranski-Krastanow (SK) growth of strained three-dimensional (3D) InAs islands on the GaAs(001) surface were studied in Refs. 35 and 36. Transmission electron microscopy combined with photoluminescence (PL) spectroscopy has revealed that an increase in the substrate temperature during the InAs deposition results in a *decrease* in the island density, an *increase* in their lateral size, and a *decrease* in the island volume. If the substrate temperature is reduced *after* the InAs deposition, the adatom condensation results in an increase

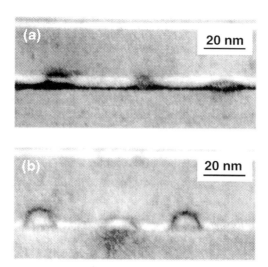

Figure 6. Cross-section transmission electron microscopy (TEM) images of InAs islands ($d_{InAs} \approx 6.5$ Å) covered by a 40 Å-thick $Al_{0.2}Ga_{0.8}As$ layer (a) and a 9-Å thick AlAs layer (b) with further GaAs overgrowth.

in the density of the islands, an increase in the average volume of a single island, while the lateral sizes undergo shrinkage compensated by the strong increase of the island height. The temperature ramping allows the realization of high-density large-volume InAs-GaAs QDs emitting at 1.3 μm via the In adatom condensation process.[36]

- *Vertically coupled QDs*

   An opportunity to approach even longer wavelengths is to grow vertically-coupled InAs-AlGaAs QDs.[37,38] Electronic coupling causes a significant shift of PL emission towards longer wavelengths with an increase in the number of stacks, even in the case where very small islands are used as the stacking objects. Strongly electronically coupled quantum dots, forming QD superlattices in the vertical direction, have been demonstrated both in MBE[37] and in MOCVD.[39] Using this approach it is possible to fabricate QD structures emitting with a photoluminescence (PL) intensity maximum in the range up to 1.42 μm at 300 K[38] and a PL half-maximum on the low-energy side at a wavelength of 1.46 μm. Dramatic modification of the cigar-like heavy-hole wavefunctions in vertically-coupled QDs is manifested by the polarization reversal (TE to TM) of the QD photoluminescence in structures with a large number of QD stacks.[40]

- *Activated phase separation*

   This effect allows a controlled increase of the volume of the strained island by alloy overgrowth and was demonstrated for InAs QDs on GaAs substrates covered

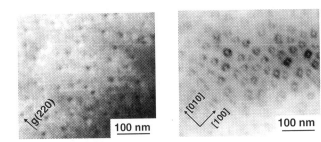

Figure 7. Plan-view TEM images of 2.5-ML InAs QDs covered by GaAs (left) and a 4-nm-thick $In_{0.15}Ga_{0.85}As$ layer (right).

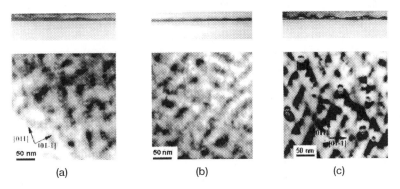

Figure 8. Cross-section (top) and plan-view TEM images of 6-nm-thick InGaAsN insertions in GaAs. The nitrogen concentration is 1%, In concentration is 25% (a), 30% (b), and 35% (c).

by an (In,Ga)As alloy layer.[41,42] Initially, an array of small coherent InAs stressors is formed. If these InAs islands formed by ~2 ML InAs deposition are covered by GaAs, the resulting QDs have a high density, but a rather small lateral size <10 nm. Once the dots are covered by InGaAs alloy, it is energetically favorable for InAs molecules to nucleate at the elastically-relaxed islands, where the lattice parameter is close to that in InAs, as illustrated in Fig. 7. This effect makes it possible to create a dense array of large coherent QDs with a lateral size up to 20–25 nm, emitting up to 1.32 μm at room temperature.

One should note that phase-separation effects were found to be very important also in InGaAsN-based structures, resulting in the formation of nanodomain structures (see Fig. 8), whereas InGaAs layers without nitrogen doping remain stable towards morphological transformation.[43]

- *Defect reduction techniques*

    Realization of bright luminescence at 1.34 μm from InGaAs insertions in a GaAs matrix was reported already in 1991 by the team at the University of Illinois.[44] Alternate InAs and GaAs deposition cycles were used to form the nominally 7-nm-thick InGaAs layer. On the other hand, no lasing at 1.3 μm was reported for GaAs-based structures until 1998–1999.[41,45] Different kinds of defects formed during the highly lattice-mismatched growth prevent population inversion and laser gain. To succeed with growth of laser-quality material one usually needs to apply various defect-reduction techniques.[46–48] An example of selective elimination of defects has been demonstrated for QDs emitting at 1.3 μm in Ref. 47. In Fig. 9 we show plan-view (only the upper sheet of QDs is trapped by the TEM foil) and cross-sectional TEM images of 3-fold-stacked InGaAs/GaAs QDs formed by InGaAs deposition using MOCVD. In this example of a defect-reduction technique, a thin GaAs layer covered coherent QDs but did not cover dislocated InAs islands, which impose higher lattice mismatch with respect to the GaAs cap layer and remain open. A properly executed annealing step left the coherent islands, while the dislocated islands were selectively evaporated. The structure grown without the annealing step was completely dislocated — see Fig. 9(a). Coherent dots persisted only in the lower sheet in this case, while in the upper sheet all the InGaAs exceeding the thickness of the wetting layer was absorbed by the dislocations.

- *Defect engineering*

    An extension of the defect-reduction technique can be used for direct fabrication of nanostructures using dislocation networks as templates. An elastic interaction of the deposited material with existing defects and dislocations in the lattice-mismatched layer may result in either decoration of the dislocations with the deposit material or just the opposite—repulsion of the deposit from the dislocation regions, depending on the lattice mismatch between the deposit, the

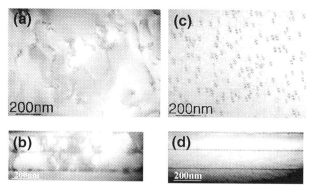

Figure 9. Plan-view (a,c) and cross-section (b,d) TEM images of stacked MOCVD QDs grown without (a,b) and with (c,d) the defect-reduction step.

substrate, and the dislocated layer. If the deposit undergoes a repulsive interaction with respect to the plastically-relaxed dislocated region (*e.g.* AlAs deposition on top of a plastically-relaxed InGaAs layer grown on a GaAs substrate), the defect-containing region remains uncovered by the deposit. If the deposit has higher temperature stability than the defect-rich layer, an *in situ* evaporation technique can be applied.[46,48] We note that structures with dense and ordered arrays of dislocations can be used both for nanostructure fabrication and for producing coherent pedestals for further nanoepitaxy, such as epitaxial lateral overgrowth.

InGaAs domains having a size of 30–50 nm were successfully fabricated from plastically relaxed 20-nm-thick $In_{0.3}Ga_{0.7}As$ layers by 20-nm-thick AlAs cap layer deposition and annealing at 750 °C. After the annealing step the structure was capped with AlGaAs material.[46,48]

- *Low-temperature QDs*

Long-wavelength (1.5–1.7 μm) emission on GaAs substrates can be realized using lateral agglomerates of InAs quantum dots (LAQDs). These LAQDs are formed on the GaAs(100) surface at relatively low substrate temperatures (350–400 °C).[49] After overgrowth these 1D and 2D chains of laterally-coupled InAs-GaAs QDs may emit up to 1.7 μm at room temperature. However, TEM studies demonstrated that the structural quality of the LAQDs is low. Dislocated clusters in high concentrations are formed and only a very limited share of LAQDs is defect free. Integrated PL intensity of the structures is also weak, being four orders of magnitude lower than those of 1.3-μm QD structures.

LAQDs represent chains of individual QDs emitting at 1.5–1.7 μm, which appear only after some critical InAs layer thickness and critical low deposition temperature is reached. LAQDs do not appear at temperatures above 400 °C at any InAs average layer thickness. They do not appear at temperatures 320–350 °C for average layer thickness below 2.5 ML InAs (individual QDs form at 1.9–2 MLs at 320 °C as compared to the 1.7 ML necessary for the 2D-3D growth transition at 450–500 °C). LAQDs do not form at temperatures below 300 °C, nor do individual QDs form at this temperature for InAs deposits up to 4 ML thick. Increase of the InAs layer thickness to 4 MLs results in an increase of the density of LAQDs (to above 10^9 cm^{-2}). At the same time, the defect density is also increasing and most of the LAQDs become dislocated. At thickness below 2.7 MLs the density of LAQDs is too small and irreproducible, even though the defect density is also small. Our defect-reduction approach[46–48] helps to restore the quality also in this case.

The key points of the defect-reduction technique in this case are:[50]

- Deposition of a thin GaAs cap layer which covers completely coherent InAs nanoobjects, while plastically relaxed objects giving higher strain remain uncovered due to GaAs/AlAs repulsion from regions with high lattice mismatch with respect to GaAs;

- Annealing of the InAs QD structures with thin GaAs caps, to evaporate defects;

- Application of multistage annealing using AlAs sealing to evaporate the remaining defects trapped in GaAs without losing the coherent regions.

6.   Edge-emitting long-wavelength QD lasers

The general requirements of semiconductor lasers are linked to power, differential and conversion efficiency, temperature stability of the threshold current and lasing wavelength, and sufficient broadening of the gain peak. The latter requirement of at least a 50-meV-wide gain peak originates from the need for wavelength-stable operation in vertical-cavity surface-emitting lasers (VCSELs) and in wavelength-stabilized (*e.g.* fiber-grating wavelength-stabilized) edge emitters. As the bandgap shrinks with temperature, the need to keep high enough gain at the chosen cavity mode or at the grating reflection peak wavelength, respectively, requires that the gain broadening be larger than the bandgap shift in the range of temperatures used (*e.g.* –20 to +70 °C). The typical width of the ground-state gain peak in QD lasers is about 30–40 meV and is therefore at the low boundary of the range of values needed.

Operation of a QD laser depends crucially on the operating temperture and the depth of the QD localization potential, as non-quasiequilibrium carrier populations in QDs can be realized due to the strongly suppressed lateral transport in QDs.[51] The use of defect-free large InAs/GaAs QDs obtained by activated spinodal decomposition allows cw operation at room temperature up to 2.8 W (see Fig. 10) at 1.3 μm wavelength in a conventional GaAs-AlGaAs separate confinement heterostructure design.[41] For a stripe width of 100 μm and cavity lengths (*L*) exceeding 1200–1600 μm (uncoated) it was found that the threshold current

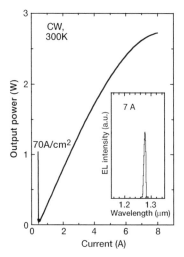

Figure 10. High-power operation of a QD laser. Cavity length is 2 mm.

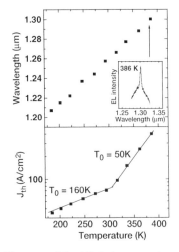

Figure 11. High-temperature operation of a QD laser.

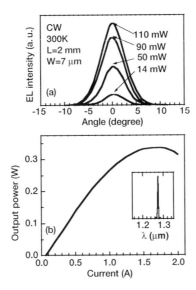

Figure 12. High-power kink-free single-mode operation of a long-wavelength QD laser. Emission spectrum in the insert is taken at 0.5 A. Cavity length is 1.9 mm, uncoated facets.

density was about 70–90 A/cm² and only weakly dependent on L. The lasing wavelength is near 1.3 μm and corresponds to the QD ground-state emission. The estimated maximum modal gain for the QD ground-state lasing is about 14 cm^{-1}, and can be increased up to 35 cm^{-1} for 10-fold-stacked QD active regions.[15] The internal losses, derived from the slope of the inverse differential efficiency as a function of L, are as low as 1.5 cm^{-1}. Quantum dot lasers demonstrate a $T_0 \sim 160$ K (see Fig. 11) in the pulsed regime and ~50 K for cw operation. The differential efficiency was 56% for $L = 1.9$ mm (uncoated) and was weakly temperature-dependent. Narrow 7-μm stripes were also fabricated from similar wafers. Low-threshold single-transverse-mode kink-free operation up to 330 mW was demonstrated for uncoated facets (see Fig. 12).

7. Operating lifetime test

We have found high resistance to cw degradation in edge-emitting long-wavelength QD lasers, superior to that of conventional 0.98-μm QW devices grown in the same molecular beam epitaxy machine and tested in similar conditions and environment. The laser structures are grown by solid-source molecular beam epitaxy[52] on n^+-GaAs (001) substrates using a conventional

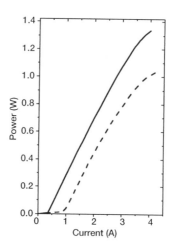

Figure 13. Continuous wave L–I characteristics of the ridge stripe quantum dot laser with a cavity length of 2 mm and a stripe width of 100 µm at 15 °C (solid line) and 60°C (dashed line).

double-heterostructure separate-confinement design.[53] Sheets of InGaAs QDs are formed by 2.5-ML InAs deposition followed by overgrowth of a 5-nm-thick $In_{0.13}Ga_{0.87}As$ layer at 480 °C. The QD density, estimated from transmission electron microscopy images, is about 4×10^{10} cm^{-2}. Three sheets of QDs stacked with 28.5-nm-wide GaAs spacers are used. The QDs are symmetrically confined by 30-nm-thick GaAs layers on both sides. Shallow-mesa ridge structures were fabricated and tested. The samples were mounted on copper heat sinks using indium solder. The heat-sink temperature was fixed at 15 °C or at 60 °C.

The resulting cw light-current (L–I) characteristics of the devices are shown in Fig. 13. The differential efficiency was about 42% and was weakly affected by temperature. Quantum efficiency and internal losses estimated from the dependence of the inverse differential efficiency on cavity length were 95±5% and 7 cm^{-1}, respectively. The cw threshold current density was 170 A/cm^2 at 15 °C and 390 A/cm^2 at 60 °C. Five devices were mounted and tested at 60 °C with a cw output power of ~0.3 W. No hermetic sealing or facet protection was used. No inert gas flow was applied either and the devices were tested in a room environment. After 100 hours of operation, no noticeable degradation of the performance of the devices was observed, and four devices were dismounted and replaced by conventional quantum well (QW) AlGaAs-InGaAs lasers for the 0.98-µm range grown in the same MBE machine. One QD laser device was tested for 450 hours and then was dismounted. The output power versus time for this device is shown in Fig. 14. Continuous-wave electroluminescence spectra at 60 °C at two currents are given in the inset. No noticeable degradation was detected.

On the other hand, QW devices demonstrated a 10% drop in the output power

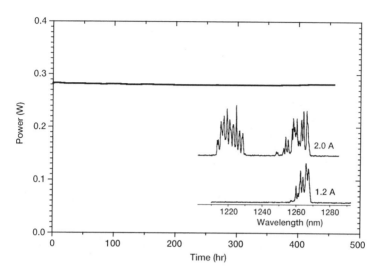

Figure 14. Light output power in the continuous wave (cw) regime *vs.* time measured at 60 °C. The inset shows cw electroluminescence spectra at 60 °C for two drive currents.

after 100 hours of cw operation. This drop may be due to facet degradation caused by accumulating facet contamination under cw operation. A higher stability of unprotected facets in the case of QD lasers may be explained by reduced carrier diffusion towards the facets. The effect of carrier confinement in QDs reduces both overheating of the facets and free-carrier-stimulated chemical reactions.

Further tests are to be performed for QD lasers with protected facets, or operating in an inert-gas environment. At the same time, the results already presented indicate the high stability of QD lasers with respect to accelerated degradation. The observed behavior is in agreement with the recently reported enhanced radiation hardness of QD lasers to high-energy proton irradiation.[54]

8. Long-wavelength vertical-cavity surface-emitting lasers

As discussed above, GaAs-based VCSELs currently are designed for plastic fiber connections and operate at distances below 300 m. The drawbacks of existing devices will encourage 1.3-μm VCSELs to replace both InP-based FP and DFB lasers and also 0.85-μm GaAs VCSELs.[55] An additional advantage of VCSELs is the natural possibility of vertical integration of the device with wavelength modulators (*e.g.* for chirp compensation), intensity modulators, and photodetectors, important for advanced applications in DWDM. The VCSEL has an important counterpart — the cavity-enhanced selective photodetector — allowing DWDM data transmission links based on laser frequency modulation. Wavelength tunable VCSELs also can be used to simplify the DWDM protocol. Electronic (electrooptic) tuning of the wavelength is preferable for ultrahigh-speed frequency modulation, while membrane tuning may be used in DWDM applications. Thus, it is today of the utmost importance to develop a technology that fits the requirements of modern VCSELs — high uniformity, high yield, and precise control of fluxes for DBR growth — and at the same time provides defect-free active media enabling long-wavelength gain on GaAs substrates.

Self-organized QDs are particularly advantageous for VCSELs,[56] as nonequilibrium carriers are localized in the QDs and thus spreading of nonequilibrium carriers out of the injection region can be suppressed, allowing also microscale devices. For ultrasmall apertures, ultralow threshold currents (<70 μA) have been demonstrated for QD VCSELs.[57] Furthermore, the use of QDs does not require nitrogen plasma sources and the technology can be transferred easily to large-scale production MBE systems.

The epitaxial structures are prepared by solid-source MBE on n^+-GaAs (100) substrates.[58] The self-organized QDs consist of planar sheets of initially small InAs pyramidal islands formed by a 2.5-ML InAs deposition covered by a 5 nm-thick $In_{0.15}Ga_{0.85}As$ alloy layer. The QDs are followed by a 25-nm-thick GaAs barrier/separation layer. Three-fold-stacked QDs are used. The dot density per stack is 5×10^{10} cm^{-2}. The microcavity is surrounded by p- and n-$Al_{0.98}Ga_{0.02}As$ layers (less than λ/4 thick) followed by λ-thick p- and n-GaAs current spreading/intracavity contact spacer layers doped 10^{18} cm^{-3}. Intracavity contacts are used and the device geometry is similar to that described in Refs. 58 and 59. The spacer layers are clad by DBRs composed of alternating $Al_{0.98}Ga_{0.02}As$ and λ/4-thick GaAs layers.[60] The $Al_{0.98}Ga_{0.02}As$ layers in the DBR, as well as those surrounding the optical cavity, are selectively oxidized to form Al(Ga)O. The QDs are centered in a λ-thick GaAs optical microcavity, whose edges are doped to 10^{17} cm^{-3}. The ends of the microcavity are composed of $Al_xGa_{1-x}As$ linearly graded from $x = 0.02$ up to 0.98.[60]

The cw light power-current-voltage (L–I–V) characteristics of a QD VCSEL at various heat sink temperatures are shown in Fig. 15 on the left. The threshold current is 1.2 mA and remains practically unchanged with increasing temperature.[60] The electroluminescence measurements from QD LED test structures indicate that the lasing proceeds via the QD ground-state transition. The maximum differential efficiency is 64% and the maximum wall-plug efficiency is up to 20%. The emission wavelength is near 1.3 μm (1.28–1.306 μm) depending on the particular position on the wafer. Variation in the threshold current across the wafer is ~10%, as shown on the right in Fig. 15.

We find that the threshold current demonstrates only a weak dependence on the aperture size down to submicrometer cavities, while the photon confinement effect becomes increasingly important — see Figs. 16(a) and 16(b), respectively. During lifetime testing in excess of 700 hours cw at 35 °C, no change in the

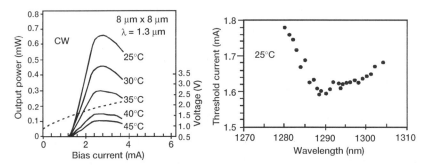

Figure 15. Continuous wave *L–I–V* curves of a GaAs-based QD VCSEL emitting at 1.3 μm (left), together with the threshold current and the lasing wavelength variation across the wafer (right). Note the very high (64%) maximum differential efficiency of the device at 25 °C.

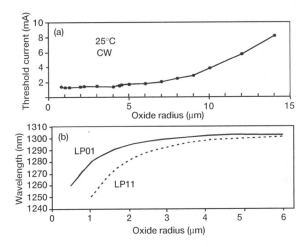

Figure 16. The dependence of the threshold current (a) and the confined photon modes (b) on the oxide aperture size.

performance of the devices was observed.[60] In the near future, work will be concentrated on the practical implementation of 1.3-μm GaAs QD VCSELs in optical interconnects and on further improvement of the device parameters.

8. Conclusions

Recent progress in quantum dot heterostructures[61] has led to practical devices that may revolutionize modern optoelectronics. Long-wavelength GaAs-based QD lasers have been proven to be suitable for advanced applications in both edge- and surface-emitting structures operating at 1.3-μm wavelength. The devices are shown to be reliable and simple to fabricate. Further work is underway to extend QD lasers towards the 1.5–1.7 μm wavelength range.

Acknowledgments

This work is supported by NanOp CC, INTAS, and Volkswagen Foundation. I am grateful to the DAAD Guest Professorship programme and the Alexander von Humboldt Foundation (equipment donation). Helpful discussions with Zh. I. Alferov, A. Yu. Egorov, P. S. Kop'ev, A. R. Kovsh, I. L. Krestnikov, J. A. Lott, M. V. Maximov, N. A. Maleev, N. Grote, C. Ribbat, R. Sellin, V. A. Shchukin, Yu. M. Shernyakov, A. F. Tsatsul'nikov, V. M. Ustinov, B. V. Volovik, and A. E. Zhukov are gratefully acknowledged.

References

1. N. G. Basov, B. M. Vul, and Yu. M. Popov, "Quantum mechanical semi-conductor generators and amplifiers of electromagnetic oscillations," *JETP* **37**, 416 (1959).
2. N. G. Basov, O. N. Krokhin, and Yu. M. Popov, "Production of negative-temperature states in *p-n* junctions of degenerate semiconductors," *JETP* **40**, 1320 (1961).
3. R. N. Hall, G. E. Fenner, J. D. Kingsley, T. J. Soltys, and R. O. Carlson, "Coherent light emission from GaAs junctions," *Phys. Rev. Lett.* **9**, 366 (1962).
4. M. I. Nathan, W. P. Dumke, G. Burns, F. H. Dill, Jr., and G. Lasher, "Stimulated emission of radiation from GaAs *p-n* junctions," *Appl. Phys. Lett.* **1**, 62 (1962).
5   N. Holohyak, Jr. and S. F. Bevacqua, "Coherent (visible) light emission from Ga(As$_{1-x}$P$_x$) junctions," *Appl. Phys. Lett.* **1**, 82 (1962).
6. Zh. I. Alferov and R. F. Kazarinov, "Double heterostructure laser," Author's Certificate No. 27448, Application No. 950840 (priority from March, 30, 1963).
7. H. Kroemer, "A proposed class of heterojunction injection lasers," *Proc. IEEE* **51**, 1782 (1963).
8. Zh. I. Alferov, V. M. Andreev, E. L. Portnoi, and M. K. Trukan, "AlAs-GaAs heterojunction injection lasers with a low room-temperature threshold," *Sov. Phys. Semicond.* **3**, 1107 (1970).
9. Zh. I. Alferov, V. M. Andreev, D. Z. Garbuzov, Yu. V. Zhilyaev, E. P. Morozov, E. L. Portnoi, and V. G. Trofim, "Effect of heterostructure parameters on the laser threshold current and the realization of continuous generation at room temperature," *Sov. Phys. Semicond.* **4**, 1573 (1970).

10. I. Hayashi, M. B. Panish, P. W. Foy, and S. Sumski, "Junction lasers which operate continuously at room temperature," *Appl. Phys. Lett.* **17**, 109 (1970).
11. S. V. Kartalopoulos, *Introduction to DWDM Technology. Data in a Rainbow*, New York: IEEE Press and Wiley-Interscience, 2000.
12. J. Cheng and N. K. Dutta, eds., *Vertical Cavity Surface-Emitting Lasers: Technology and Applications*, Amsterdam: Gordon and Breach, 2000.
13. M. C. Amann and J. Buus, *Tunable Laser Diodes*, Boston: Artech House, 1998.
14. J. F. Day "The future of laser diode applications," in: *Materials of the Laser Marketplace Seminar* (CD), San Jose, January 26, 2000.
15. N. N. Ledentsov, M. Grundmann, F. Heinrichsdorff, D. Bimberg, V. M. Ustinov, A. E. Zhukov, M.V. Maximov, Zh.I. Alferov and J. A. Lott, "Quantum-dot heterostructure lasers," *IEEE J. Selected Topics Quantum Electronics* **6**, 439 (2000).
16. R. Dingle and C. H. Henry, "Quantum effects in heterostructure lasers," U.S. Patent No. 3982207 (issued September 21, 1976).
17. P. M. Petroff, A. C. Gossard, R. A. Logan, and W. Weigman, "Towards quantum well wires: Fabrication and optical properties," *Appl. Phys. Lett.* **41**, 635 (1982).
18. E. Kapon, M. C. Tamargo, and D. M. Hwang, "Molecular beam epitaxy of GaAs/AlGaAs superlattice heterostructures on nonplanar substrates," *Appl. Phys. Lett.* **50**, 347 (1987).
19. Y. Arakawa and H. Sakaki, "Multidimensional quantum well lasers and temperature dependence of its threshold current," *Appl. Phys. Lett.* **40**, 939 (1982).
20. N. N. Ledentsov, V. M. Ustinov, A. Yu. Egorov, A. E. Zhukov, M. V. Maximov, I. G. Tabatadze, P. S. Kop'ev, "Optical properties of heterostructures with InGaAs-GaAs quantum clusters," *Semicond.* **28**, 832 (1994).
21. V. I. Marchenko, "On the theory of the equilibrium shape of crystals," *Sov. Phys. JETP* **54**, 605 (1981).
22. R. Nötzel, N. N. Ledentsov, L. Däweritz, M. Hohenstein, and K. Ploog, "Direct synthesis of corrugated superlattices on non-(100)-oriented surfaces," *Phys. Rev. Lett.* **67**, 3812 (1991).
23. N. N. Ledentsov, D. Litvinov, A. Rosenauer, D. Gerthsen, I. P. Soshnikov, *et al.*, "Interface structure and growth mode of quantum wire and quantum dot GaAs-AlAs structures on corrugated (311)A surface," *J. Electron. Mater.* **30**, 463 (2001).
24. V. I. Marchenko, "Possible structures and phase transitions of crystal surfaces," *JETP Lett.* **33**, 381 (1981).
25. I. L. Krestnikov, N. N. Ledentsov, A. Hoffmann, and D. Bimberg, "Arrays of two-dimensional islands formed by submonolayer insertions: Growth, properties, devices," *Phys. stat. sol. (a)* **183**, 207 (2001).
26. L. Goldstein, F. Glas, J. Y. Marzin, M. N. Charasse, and G. Leroux, "Growth by molecular-beam epitaxy and characterization of InAs/GaAs strained-layer superlattices," *Appl. Phys. Lett.* **47**, 1099 (1985).

27. F. Glas, C. Guille, P. Hénoc, and F. Houzay "TEM study of the molecular beam epitaxy island growth of InAs on GaAs," in: A. G. Cullis and P. D. Augustus, eds., *Microscopy of Semiconducting Materials*, Institute of Physics Conf. Series Vol. 87, 1987, pp. 71-76.

28. J. M. Moison, F. Houzay, F. Barthe, L. Leprice, E. Andre, and O. Vatel, "Self-organized growth of regular nanometer-scale InAs gots on GaAs," *Appl. Phys. Lett.* **64**, 196 (1994).

29. D. Leonard, M. Krishnamurthy, C. M. Reaves, S. P. Denbaars and P. M. Petroff, "Direct formation of quantum-sized dots from uniform coherent islands of InGaAs on GaAs surfaces," *Appl. Phys. Lett.* **63**, 3203 (1993).

30. A. Madhukar, Q. Xie, P. Chen, and A. Koknar, "Nature of strained InAs three-dimensional island formation and distribution on GaAs (100)," *Appl. Phys. Lett.* **64**, 2727 (1994).

31. R. Nötzel, J. Temmyo, H. Kamada, T. Furuta, and T. Tamamura, "Strong photoluminescence emission at room temperature of strained InGaAs quantum discs (200-30 nm diameter) self-organized on GaAs (311)B substrates," *Appl. Phys. Lett.* **65**, 457 (1994).

32. N. Kirstaedter, N. N. Ledentsov, M. Grundmann, D. Bimberg, V. M.Ustinov, *et al.*, "Low-threshold, large T_0 injection laser emission from (InGa)As quantum dots," *Electronics Lett.* **30**, 1416 (1994).

33. N. N. Ledentsov, "Self-organized quantum wires and dots: New opportunities for device applications," *Prog. Crystal Growth Charact.* **35**, 289 (1997).

34. N. N. Ledentsov, M. Grundmann, N. Kirstaedter, O. Schmidt, R. Heitz, *et al.*, "Ordered arrays of quantum dots: Formation, electronic spectra, relaxation phenomena, lasing," *Solid State Electronics* **40**, 785 (1996).

35. A. F. Tsatsul'nikov, A. R. Kovsh, A. E. Zhukov, Yu. G. Musikhin, *et al.*, "Volmer-Weber and Stranski-Krastanow InAs-(Al,Ga)As quantum dots emitting at 1.3 μm," *J. Appl. Phys.* **88**, 6272 (2000).

36. N. N. Ledentsov, V. A. Shchukin, D. Bimberg, V. M. Ustinov, N. A. Cherkashin, *et al.*, "Reversibility of the island shape, volume, and density in Stranski-Krastanow growth," *Semicond. Sci. Technol.* **16**, 502 (2001).

37. N. N. Ledentsov, V. A. Shchukin, M. Grundmann, N. Kirstaedter, J. Böhrer, *et al.*, "Direct formation of vertically coupled quantum dots in Stranski-Krastanow growth," *Phys. Rev. B* **54**, 8743 (1996).

38. B. V. Volovik, D. S. Sizov, A. F. Tsatsul'nikov, Yu. G. Musikhin, N. N. Ledentsov, *et al.*, "The emission from structures with arrays of coupled quantum dots grown by submonolayer epitaxy in the spectral range of 1.3–1.4 μm," *Semicond.* **34**, 1316 (2000).

39. N. N. Ledentsov, J. Böhrer, D. Bimberg, I. V. Kochnev, M. V. Maximov, *et al.*, "Formation of coherent superdots using metal-organic chemical vapor deposition," *Appl. Phys. Lett.* **69**, 1095 (1996)

40. P. Yu, W. Langbein, K. Leosson, J. M. Hvam, N. N. Ledentsov, *et al.*, "Optical anisotropy in vertically coupled quantum dots," *Phys. Rev. B* **60**, 16680 (1999).

41. Yu. M. Shernyakov, D. A. Bedarev, E. Yu. Kondrat'eva, P. S. Kop'ev, *et al.*, "1.3 μm GaAs-based laser using quantum dots obtained by activated spinodal decomposition," *Electronics Lett.* **35**, 898 (1999).

42. M. V. Maximov, A. F. Tsatsul'nikov, B. V. Volovik, D. S. Sizov, Yu. M. Shernyakov, *et al.*, "Tuning quantum dot properties by activated phase separation of an InGa(Al)As alloy on InAs stressors," *Phys. Rev. B* **62**, 16671 (2000).

43. B. V. Volovik, A. R. Kovsh, H. Kuenzel, N. Grote, N. A. Cherkashin, Yu. G. Musikhin, N. N. Ledentsov, D. Bimberg, and V. M. Ustinov, "Optical and structural properties of self-organized InGaAsN/GaAs nanostructures," *Semicond. Sci. Technol.* **16**, 186 (2001).

44. E. J. Roan and K. Y. Cheng, "Long-wavelength (1.3 μm) luminescence in InGaAs strained quantum-well structures grown on GaAs," *Appl. Phys. Lett.* **59**, 2688 (1991).

45. D. L. Huffaker, G. Park, Z. Zou, O. B. Shchekin, and D. G. Deppe, "1.3 μm room-temperature GaAs-based quantum dot laser," *Appl. Phys. Lett.* **73**, 2564 (1998).

46. N. N. Ledentsov, "Self-organization effects in crystal growth: Nanoepitaxy, nanostructures, defect engineering," in: *Proc. IX National Conf. Crystal Growth (NCCG-2000)*, Moscow, 16-20 October 2000.

47. N. N. Ledentsov, M. V. Maximov, D. Bimberg, T. Maka, C. M. Sotomayor Torres, *et al.*, "1.3 μm luminescence and gain from defect-free InGaAs-GaAs quantum dots grown by metal-organic chemical vapour deposition," *Semicond. Sci. Technol.* **15**, 604 (2000).

48. I. L. Krestnikov, N. A. Cherkashin, D. S. Sizov, D. A. Bedarev, I. V. Kochnev, V. M. Lantratov, and N. N. Ledentsov, "InGaAs nanodomains formed *in situ* on the surface of (Al,Ga)As," *Tech. Phys. Lett.* **27**, 233 (2001).

49. M. V. Maximov, A. F. Tsatsul'nikov, B. V. Volovik, D. A. Bedarev, A. Yu. Egorov, *et al.*, "Optical and structural properties of InAs quantum dots in a GaAs matrix for a spectral range up to 1.7 μm," *Appl. Phys. Lett.* **75**, 2347 (1999).

50. N. D. Zakharov, P. Werner, D. Bimberg, G. E. Cirlin, N. A. Maleev, *et al.*, "Structure and luminescence of low-temperature InAs quantum dots (QDs) grown on 001 GaAs by MBE and subjected to thermal flashing," *Proc. MRS* (2001), paper O8.5.

51. M. Grundmann and D. Bimberg, "Gain and threshold of quantum dot lasers: Theory and comparison to experiments," *Jpn. J. Appl. Phys.* **36**, 4181 (1997).

52. N. N. Ledentsov, *Growth Processes and Surface Phase Equilibria in Molecular Beam Epitaxy*, Berlin: Springer, Berlin, 1999, p. 81.

53. E. Yu. Lundina, Yu. M. Shernyakov, M. V. Maximov, I. N. Kaiander, A. Yu. Egorov, *et. al.* "Long operation lifetime of long-wavelength GaAs-based quantum dot lasers," unpublished.

54. C. Ribbat, R. Sellin, M. Grundmann, D. Bimberg, N.A. Sobolev, and M.C. Carmo "Enhanced radiation hardness of quantum dot lasers to high energy proton irradiation," *Electronics Lett.* **37**, 174 (2001).

55. See, for example, *Compound Semicond.* **7**, No. 5 (2001).
56. J. A. Lott, N. N. Ledentsov, V. M. Ustinov, A. Yu. Egorov, A. E. Zhukov, P. S. Kop'ev, Zh. I. Alferov, and D. Bimberg, "Vertical cavity lasers based on vertically coupled quantum dots," *Electronics Lett.* **33**, 1150 (1997).
57. N. N. Ledentsov, D. Bimberg, V. M. Ustinov, M. V. Maximov, Zh. I. Alferov, V. P. Kalosha and J. A. Lott, "Interconnection between gain spectrum and cavity mode in a quantum dot vertical cavity laser," *Semicond. Sci. Technol.* **13**, 99 (1999).
58. J. A. Lott, N. N. Ledentsov, V. M. Ustinov, N. A. Maleev, A. E. Zhukov, *et al.*, "InAs-InGaAs quantum dot VCSELs on GaAs substrates emitting at 1.3 μm," *Electronics Lett.* **36**, 1384 (2000).
59. N. N. Ledentsov, D. Bimberg, V. M. Ustinov, J. A. Lott, and Zh. I. Alferov, in: *Advanced Nanoelectronics: Devices, Materials and Computing*, Memoirs Institute Sci. Industrial Res. Vol. 57, Osaka, Japan, 2000, pp. 80-87.
60. N. N. Ledentsov, D. Bimberg, V. M. Ustinov, Zh. I. Alferov, and J. A. Lott, "Self-oganized InGaAs quantum dots for advanced applications in optoelectronics," *Proc. Int. Conf. Indium Phosphide Related Mater.*, Nara, Japan (2001), paper TuPl2, pp. 5-8.
61. D. Bimberg, M. Grundmann and N. N. Ledentsov, *Quantum Dot Heterostructures*, New York: Wiley, 1999.

A Temperature-Insensitive Semiconductor Laser

Levon V. Asryan
State University of New York at Stony Brook, Stony Brook, NY 11794-2350, U.S.A.
and *A. F. Ioffe Physico-Technical Institute, St Petersburg 194021, Russia*

Serge Luryi
State University of New York at Stony Brook, Stony Brook, NY 11794-2350, U.S.A.

1. Introduction

Highly temperature-stable operation is an essential feature required of long-wavelength semiconductor heterostructure lasers for telecommunications. Commercial lasers with either bulk or quantum-well (QW) active regions suffer from rather poor temperature stability. The all-important parameter T_0, which describes empirically the temperature dependence of the threshold current density j_{th} and is defined as $T_0 = (\partial \ln(j_{th})/\partial T)^{-1}$, does not exceed 100 K in the best commercial lasers. Quantum-dot (QD) lasers, exploiting a zero-dimensional (0D) active medium were proposed years ago[1,2] and one of their main predicted advantages was high temperature stability. Nevertheless, despite significant recent progress in the fabrication of QD lasers,[3–12] their temperature stability has fallen far short of expectations. Although the best results for T_0 in QD lasers are quite respectable, matching and even exceeding the best QW results at room temperature, so far they have been nowhere near the predicted "infinite" values that would justify regarding the laser as temperature insensitive.

In all semiconductor diode lasers, electrons and holes are injected from 3D contact regions where carriers are free, into an active region where lasing transitions take place and where carriers may be dimensionally confined. At relatively high temperatures (300 K and above), the dominant source of the T dependence of j_{th} in all semiconductor lasers originates from carriers that do not contribute to the lasing transition. In lasers with 3D (bulk) or 2D (QW) active regions, there is always a population of carriers distributed according to Fermi-Dirac statistics within some energy range (determined by the temperature and the injection level) around the lasing transition energy (Fig. 1). These carriers reside in the active region itself and their recombination contributes to a T-dependent threshold.

It is the absence of parasitic recombination in the active medium itself that gave rise to the original hopes of ultra-high temperature stability in QD lasers, where optical transitions occur between discrete levels. However, in all conventional QD laser designs the problem of parasitic recombination has not been removed.

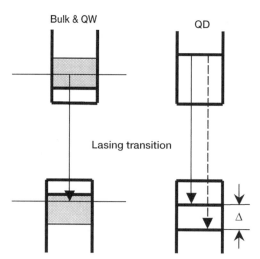

Figure 1. Carrier populations in bulk, QW, and QD. The dashed arrow shows the excited-state transition in a QD.

This recombination arises primarily from carriers residing in layers adjacent to the active medium, primarily in the optical confinement layer (OCL). Consider the conventional QD laser heterostructure (Fig. 2). Carriers are injected from cladding layers into the OCL where an approximate equilibrium with the QDs is established at room temperature. High occupation of QD electron and hole levels embedded in the OCL is therefore accompanied by an appreciable 3D population of both types of carriers in the OCL itself. These carriers give rise to a temperature-dependent recombination current, while not contributing to the lasing. Such a mechanism of T dependence is also present in other semiconductor lasers, but in QD lasers it plays the central role at room temperature.

Another source of T dependence arises from the fact that not all QDs are alike. Owing to the unavoidable inhomogeneous broadening, a sizeable fraction of them does not contribute to the lasing transition, while still adding to the parasitic recombination. As far as T_0 is concerned, this effect is similar to that due to carriers residing in the OCL. Despite the impressive recent progress in controlling QD parameters during epitaxial growth, even the best devices show a significant QD size dispersion, as indicated by the measured gain and spontaneous emission spectra.

The effect of QD size dispersion on j_{th} and its T dependence was first considered in Refs. 13 and 14, respectively. If we define separately the characteristic temperatures T_0^{QD} and T_0^{out} for the threshold current components associated with the recombination inside the QDs, $j_{QD}(T)$, and outside the QDs, we find that T_0^{QD} is much higher than T_0^{out}. Indeed, the value of T_0^{out} is below 100 K

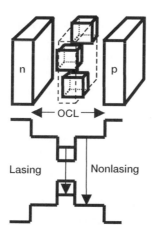

Figure 2. Schematic view and energy band diagram of a conventional QD laser.

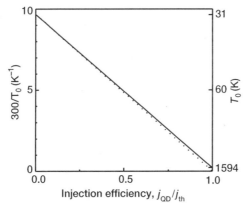

Figure 3. Characteristic temperature *vs.* the injection efficiency at room temperature. The dotted line depicts T_0^{-1} in the absence of inhomogeneous line broadening ($T_0^{QD} = \infty$). The values of T_0 are indicated on the right axis.

at room temperature,[14] whereas the calculated value of T_0^{QD} due to inhomogeneous broadening is over 1500 K.[15] We see that the effect of QD size dispersion is relatively small compared to the OCL recombination. Defining the injection efficiency as the QD fraction of the total injection current, *i.e.* as j_{QD}/j, we can write the reciprocal of the characteristic temperature in the form:

$$\left(T_0\right)^{-1} = \frac{j_{QD}}{j_{th}}\left(T_0^{QD}\right)^{-1} + \left(1 - \frac{j_{QD}}{j_{th}}\right)\left(T_0^{out}\right)^{-1}. \tag{1}$$

The characteristic temperature T_0 increases dramatically with injection efficiency as can be seen from Fig. 3. When the entire injection current is consumed in the QDs, the dominant remaining contribution to temperature dependence is from inhomogeneous broadening and hence $T_0 = T_0^{QD}$ should be very high.

Thus we can expect that suppression of the OCL recombination alone will result in a dramatic improvement of the temperature stability. One way of accomplishing this suppression is based on tunneling injection of carriers into the QDs.[15] This novel design, discussed in Section 2, both suppresses the parasitic components of threshold current and diminishes the effect of the inhomogeneous line width. Another novel approach, discussed in Section 3, is to suppress the parasitic recombination by introducing heterojunction barriers that block the passage of minority carriers into the "wrong" side of the QD layer, while remaining transparent for majority-carrier injection into the QDs themselves.

The relative merits of these two approaches and the remaining sources of temperature dependence are discussed in Section 4 and our conclusions are summarized in Section 5.

2.   Tunneling-injection temperature-insensitive QD laser

A schematic view of the structure and its energy band diagram are shown in Fig. 4. Basically, we have a separate confinement double-heterostructure laser. Electrons and holes are injected from *n*- and *p*-cladding layers, respectively. The QD layer, located in the central part of the OCL, is clad on both sides by QWs separated

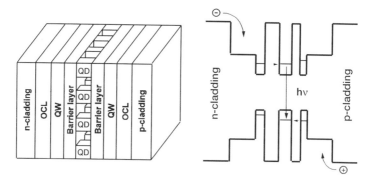

Figure 4. Schematic view (left) and energy band diagram (right) of a tunneling-injection QD laser. The QWs and the QDs are implemented in the same material, although this need not be the case. The electron-injecting QW is wider than the hole-injecting QW and both QWs are narrower than the QD to accomplish resonant alignment of the majority-carrier subbands with the QD energy levels. The tunnel barrier on the electron-injecting side is made thicker to suppress hole leakage from the QD.

from the QDs by thin barrier layers. Injection of carriers into the QDs occurs by tunneling from the QWs.

The key idea of the device is that the QWs themselves are not connected by a current path that bypasses the QDs. Electrons (coming from the left in Fig. 4) approach the right QW only through the confined states in the QDs. Similarly, holes cannot directly approach the left QW. To realize this idea, the following conditions must be met:

1. The material and the thickness of the QWs should be chosen so that the lowest subband edge in the injecting QW matches the quantized energy level for the corresponding type of carrier in the average-sized QD (the QWs may or may not be of the same material as the QDs).

2. The barriers should be reasonably high to suppress thermal emission of carriers from the QWs.

3. The material separating the QDs from each other in the QD layer should have a sufficiently wide bandgap to suppress all tunneling other than that via the QD levels. This material may be the same as that of the barrier layers.

4. The barrier layers should be thin enough to ensure effective tunneling between the QW and QD states. At the same time, the separation between the adjacent QDs in the QD layer should be large enough to prevent any significant tunnel splitting of the energy levels in neighboring QDs (otherwise, such a splitting would effectively play the same role as the inhomogeneous line broadening).

- *Suppression of escape tunneling*

We should note that a fraction of the injected carriers might not recombine in the QD but escape in a second tunneling step into the "foreign" QW and recombine with the majority carriers there. The ratio of the escape tunneling rate to the QD recombination rate is given by

$$\frac{j_{\text{leak}}}{j_{\text{QD}}} = \frac{\tau_{\text{QD}}}{\tau_{\text{leak}}}, \tag{2}$$

where τ_{QD} and τ_{leak} are the spontaneous radiative recombination time in the QDs and the tunneling-mediated leakage time from the QDs. Since neither τ_{QD} nor τ_{leak} depends on T, there should be only a weak temperature dependence of j_{leak}, faithfully following that of j_{QD}. (As discussed below, the only remaining source of T dependence of j_{QD} is due to the inhomogeneous line broadening and the violation of charge neutrality in QDs). Hence, the escape tunneling does not lead to a considerable temperature dependence of the threshold current; nevertheless, in a sensible design this form of leakage should be minimized to lower the value of j_{th}.

A possible way of suppressing the escape tunneling is illustrated in Fig. 4(b). This way takes advantage of the lower electron effective mass compared to the hole mass, but uses this advantage differently on the electron- and hole-injecting sides of the structure. On the *p*-side, when the hole level in the QD is aligned with the hole subband in the hole-injecting QW, the electron subband edge in that QW will be necessarily above the electron level in the QD, thus suppressing the tunneling escape of electrons. On the *n*-side, this trick does not work, since the resonant alignment of the electron subband in the QW and the electron level in the QD does not prevent tunneling of QD holes into the electron-injecting QW. However, due to the effective-mass difference, we can design a wider tunnel barrier on the electron-injecting side, such that it effectively suppresses the tunneling escape of holes while still being relatively transparent for electrons.

Suppression of the leakage on both sides also can be organized in a different way. Depending on the heterostructure composition, the barrier heights for electrons and holes can be varied independently. Using this freedom, one can optimize the structure by increasing the barrier height for out-tunneling particles while keeping it low for in-tunneling particles. An exemplary heterostructure that meets this criterion is illustrated in Fig. 5.

- *Previous tunneling-injection QD laser structures*

Carrier injection by tunneling has been used previously in both QW and QD laser designs.[16-20] Tunneling-injection QD laser designs[19,20] were intended primarily to minimize hot-carrier effects. Consider the band diagram in Fig. 6, which describes the QD lasers reported in Refs. 19 and 20. The tunneling injection barrier on the electron side allows hot electrons to thermalize before entering the QD region. However, a quasi-equilibrium bipolar carrier density and hence substantial parasitic thermo-activated recombination remains on the hole-injecting side of the structure. Therefore, this design does not address the issue of temperature sensitivity associated with recombination in the OCL.

Figure 5. Energy band diagram of a tunneling-injection QD laser, wherein the tunneling leakage is suppressed by independent variation of the barrier heights for electrons and holes.

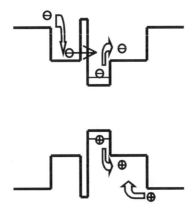

Figure 6. Schematic band diagram of the tunneling-injection QD lasers reported in Refs. 19 and 20. Carriers of one type only (electrons) are tunneling. The tunneling barrier separates the region where electrons are hot from the rest of the structure containing the active layer with QDs. Electrons thermalize before tunneling through the barrier, so that the tunneling electrons — to be captured by the QDs — are already cold.

3.   Temperature-insensitive QD laser by bandgap engineering

Tunneling injection suppresses parasitic recombination by ensuring that the electron density is high where the hole density is negligible and *vice versa*. However, tunneling is not the only way of accomplishing this goal. An alternative approach can be based on the ability to control independently the potential barriers and fields acting upon electrons and holes in the same physical region.

Consider the structures illustrated in Fig. 7. In these structures the QD layer is embedded in the OCL in such a way that there are only low barriers or no barrier at all (see Figs. 7(a) and (b), respectively) for the injection of electrons (from the left) and holes (from the right) into the QDs. On the other hand, the structures are provided with large "impenetrable" escape barriers that are those blocking electron injection into the right-hand side of the structure, where holes are majority carriers, and hole injection into the left-hand side of the structure, where electrons are in abundance. Heterostructure barriers as in Fig. 7(a) can be found readily within the manifold of quaternary III-V heterojunctions, both strained and lattice-matched. The structure in Fig. 7(b) is a "limiting case," which serves to illustrate that no barrier for injected carriers is necessary on the injecting side.

The space within the QD layer between the quantum dots can comprise either of the two barriers or be implemented as a wider gap semiconductor providing blocking of both carrier types.

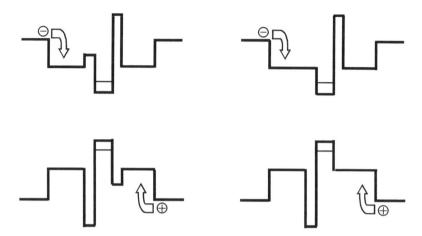

Figure 7. Prevention of thermo-activated parasitic recombination in the OCL by band-gap engineering. Quantum dots (shown with energy levels) are clad by heterostructure barrier layers that block only the minority-carrier transport.

4.   Remaining sources of the temperature sensitivity

In the proposed structures carriers cannot bypass the QDs on their way from one contact to other, so that the QDs play the role of the sole reservoir of electrons (holes) for the OCL regions adjacent to p-cladding (n-cladding). Therefore, the density of minority carriers in these regions will be negligible and outside the QDs there will be no region in the structure where both electron and hole densities are simultaneously high. The spontaneous radiative recombination rate hence is nonvanishing solely in the QDs, strongly suppressing the parasitic components of the threshold current, which would otherwise give the main contribution to the temperature dependence.

With the parasitic recombination channels suppressed, we can expect only a slight temperature dependence of j_{th} caused by the size dispersion, violation of neutrality, and parasitic recombination from higher energy levels in the QDs.

- *Inhomogeneous line broadening*

  Calculations show that the characteristic temperature T_0 as limited by the inhomogeneous line broadening is above 1500 K, which is to say the device is temperature insensitive for most practical purposes.[15] We emphasize that this dramatic improvement results solely from the suppression of recombination channels outside the active region.

It is interesting to note that the tunneling-injection structures (Figs. 4 and 5) benefit from another still "finer" effect that will further enhance the temperature stability. This effect stems from the resonant nature of tunneling injection and leads to an effective narrowing of the inhomogeneous linewidth. Indeed, tunneling injection inherently selects QDs of the "right" size, since it requires the carriers' confined levels to be in resonance with the lowest subband states in the QW. When this condition is met by the QDs of average size, $i.e.$ when QDs with $a = a_{av}$ are resonant, the number of active QDs is maximized. For QDs with $a > a_{av}$, tunneling transitions can be only phonon-mediated. The rates of such transitions are much lower and can be neglected safely. Hence, QDs of sizes larger than the average are effectively cut off. Smaller-size QDs are also cut off, although perhaps less efficiently, because their energy levels would be pumped from higher-momentum states in the injecting QW subband. The higher the in-plane momentum of a 2D carrier in the QW, the lower is the probability of a tunneling transition that results in its capture by the QD.

Selective injection means that non-lasing QDs are not pumped either. Hence, the threshold current will decrease and the temperature stability of j_{th} will be further enhanced.

- *Recombination from higher QD levels*

If a QD can support excited electron and hole states, thermally activated parasitic recombination can serve as another source of T dependence of j_{th}. Denoting the characteristic temperature limited by the presence of excited states in QDs by T_0^{exc}, the ratio T_0^{exc}/T is a universal function of Δ/T (where Δ is the separation of the transition energies in the QDs — see Fig. 1). Determination of this function leads to a problem identical to that considered in the Appendix of Ref. 15 for a bimodal ensemble of QDs. The result is:

$$\frac{T_0^{exc}}{T} = \frac{1}{2}\left(\frac{1}{1-r}\right)\frac{\left[\left(f_n^{-1}-1\right)e^{\Delta/T}+1\right]^3}{\frac{\Delta}{T}\left(f_n^{-1}-1\right)e^{\Delta/T}}\left\{rf_n^2+\frac{(1-r)}{\left[\left(f_n^{-1}-1\right)e^{\Delta/T}+1\right]^2}\right\} \qquad (3)$$

where f_n is the occupancy of the ground state in the QDs, $r = I_1/(I_1 + I_2)$, and I_1 and I_2 are the rates of the ground- and excited-state transitions, respectively.

This function is shown as a solid line in Fig. 8 for a particular value of Δ, chosen so that the minimum of T_0^{exc} would occur at room temperature. From this plot it appears that recombination from higher QD levels may be the dominant remaining source of T dependence at room temperature, because the excited-state transition rate I_2 was chosen to be equal to that of the ground-state transition and hence $r = 0.5$. However, it should be noted that small enough QDs might not support more than one electronic level even though there might be an excited hole level. Recombination to this level from the electron ground state is suppressed, at least partially, by the selection rules. Indeed, in a symmetric (cubic) QD this transition is forbidden and while in a realistic (pyramidal) QD this transition is allowed, its rate is suppressed by at least an order of magnitude ($I_2 < 0.1\ I_1$) and r

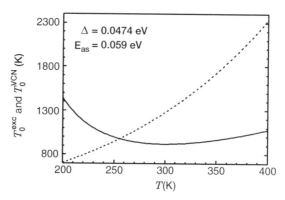

Figure 8. Characteristic temperature limited by the presence of excited states in QDs (the solid curve) and by violation of charge neutrality (VCN) in QDs (the dashed curve) against temperature.

is close to unity ($r > 0.9$). We may conclude safely that T dependence arising from thermally excited higher levels is negligible in small enough QDs.

- *Violation of charge neutrality*

The only remaining contribution to the temperature dependence of j_{th} results from the violation of charge neutrality in QDs. Unconstrained by neutrality, the occupancies of the electron and hole levels in the QD are no longer fixed by the generation condition and become temperature dependent.[14] Violation of charge neutrality is the dominant mechanism of temperature sensitivity at low temperatures but it is unimportant at 300 K. Temperature dependence of the characteristic temperature limited by this effect is shown in Fig. 8 by the dashed line. At room temperature the calculated value of T_0 is well over 1000 K.

Note that the dimensionless ratio T_0^{VCN}/T (where T_0^{VCN} is the characteristic temperature limited by violation of charge neutrality (VCN) in QDs) can also be viewed as a universal function of a dimensionless ratio E_{as}/T. The parameter E_{as} that controls the QD charge depends on the asymmetry of the QD and its environment and is given by $2E_{as} = (F_p + \Delta E_v - \varepsilon_p) - (F_n + \Delta E_c - \varepsilon_n)$ where $\Delta E_{c,v}$ are the conduction and valence band differences between the materials of the cladding layers and the QDs, $F_{n,p}$ are the quasi-Fermi levels in n- and p-claddings respectively, and $\varepsilon_{n,p}$ are the quantized energy levels in the QDs. Since $F_{n,p}$ depend on T, the parameter of asymmetry E_{as} itself is also T-dependent.

It is again interesting to note that tunneling injection structures offer an additional advantage. Indeed, the resonant nature of tunneling injection favors a correlation between the occupancies of any given QD by electrons and holes. In an idealized structure, we can expect that all the active QDs will remain neutral and then T_0 will be literally infinite.

5. Conclusions

A novel approach to the design of temperature-insensitive lasers has been proposed. The approach, based on blocking the parasitic recombination of carriers outside of the QDs, offers the possibility of achieving ultrahigh temperature stability — which has been the key desired advantage of QD lasers. Temperature stability can be accomplished in at least two ways by special tailoring of the heterostructures that surround the QDs. Conceptually, the simplest way is to introduce potential barriers blocking the minority-carrier transport on both sides of the QDs while leaving majority-carrier injection unimpeded. Suppression of the parasitic recombination outside of the QDs alone leads to characteristic temperatures T_0 above 1000 K. A particularly favorable way of accomplishing this suppression is by using resonant-tunneling injection of majority carriers. Tunneling injection offers further enhancement of T_0 owing to an inherently suppressed pumping of the nonlasing QDs and correlated occupancies of any given QD by electrons and holes.

Acknowledgment

This work was partially supported by AFOSR MURI Grant F49620-00-1-0331 and by the Semiconductor Research Corporation Grant 98-SJ-612.

References

1. R. Dingle and C. H. Henry, "Quantum effects in heterostructure lasers," U.S. Patent 3982207 (1976).
2. Y. Arakawa and H. Sakaki, "Multidimensional quantum well laser and temperature dependence of its threshold current," *Appl. Phys. Lett.* **40**, 939 (1982).
3. N. Kirstädter, N. N. Ledentsov, M. Grundmann, D. Bimberg, V. M. Ustinov, *et al.*, "Low threshold, large T_0 injection laser emission from (InGa)As quantum dots," *Electronics Lett.* **30**, 1416 (1994).
4. R. Mirin, A. Gossard, and J. Bowers, "Room temperature lasing from InGaAs quantum dots," *Electronics Lett.* **32**, 1732 (1996).
5. N. N. Ledentsov, M. Grundmann, F. Heinrichsdorff, D. Bimberg, V. M. Ustinov, *et al.*, "Quantum-dot heterostructure lasers," *IEEE J. Selected Topics Quantum Electronics* **6**, 439 (2000).
6. P. Bhattacharya, D. Klotzkin, O. Qasaimeh, W. Zhou, S. Krishna, and D. Zhu, "High-speed modulation and switching characteristics of In(Ga)As-Al(Ga)As self-organized quantum-dot lasers," *IEEE J. Selected Topics Quantum Electronics* **6**, 426 (2000).

7. L. Harris, D. J. Mowbray, M. S. Skolnick, M. Hopkinson, and G. Hill, "Emission spectra and mode structure of InAs/GaAs self-organized quantum dot lasers," *Appl. Phys. Lett.* **73**, 969 (1998).

8. P. M. Smowton, E. J. Johnston, S. V. Dewar, P. J. Hulyer, H. D. Summers, A. Patane, A. Polimeni, and M. Henini, "Spectral analysis of InGaAs/GaAs quantum-dot lasers," *Appl. Phys. Lett.* **75**, 2169 (1999).

9. D. L. Huffaker, G. Park, Z. Zou, O. B. Shchekin, and D. G. Deppe, "Continuous-wave low-threshold performance of 1.3-μm InGaAs-GaAs quantum-dot lasers," *IEEE J. Selected Topics Quantum Electronics* **6**, 452 (2000).

10. M. Sugawara, K. Mukai, Y. Nakata, K. Otsubo, and H. Ishikawa, "Performance and physics of quantum-dot lasers with self-assembled columnar-shaped and 1.3-μm emitting InGaAs quantum dots," *IEEE J. Selected Topics Quantum Electronics* **6**, 462 (2000).

11. L. F. Lester, A. Stintz, H. Li, T. C. Newell, E. A. Pease, B. A. Fuchs, and K. J. Malloy, "Optical characteristics of 1.24-μm InAs quantum-dot laser diodes," *IEEE Photonics Technol. Lett.* **11**, 931 (1999).

12. J. K. Kim, R. L. Naone, and L. A. Coldren, "Lateral carrier confinement in miniature lasers using quantum dots," *IEEE J. Selected Topics Quantum Electronics* **6**, 504 (2000).

13. L. V. Asryan and R. A. Suris, "Inhomogeneous line broadening and the threshold current density of a semiconductor quantum dot laser," *Semicond. Sci. Technol.* **11**, 554 (1996).

14. L. V. Asryan and R. A. Suris, "Temperature dependence of the threshold current density of a quantum dot laser," *IEEE J. Quantum Electronics* **34**, 841 (1998).

15. L. V. Asryan and S. Luryi, "Tunneling-injection quantum-dot laser: ultrahigh temperature stability," *IEEE J. Quantum Electronics* **37**, 905 (2001).

16. H. Yoon, A. Guitierrez-Aitken, R. Jambunathan, J. Singh, and P. Bhattacharya, "A 'cold' InP-based tunneling injection laser with greatly reduced Auger recombination and temperature dependence," *IEEE Photonics Technol. Lett.* **7**, 974 (1992).

17. P. Bhattacharya, J. Singh, H. Yoon, X. Zhang, A. Gutierrez-Aitken, and Y. Lam, "Tunneling injection lasers: A new class of lasers with reduced hot carrier effects," *IEEE J. Quantum Electronics* **32**, 1620 (1996).

18. D. Klotzkin and P. Bhattacharya, "Temperature dependence of dynamic and dc characteristics of quantum-well and quantum-dot lasers: A comparative study," *IEEE J. Lightwave Technol.* **17**, 1634 (1999).

19. K. Kamath, D. Klotzkin, and P. Bhattacharya, "Small-signal modulation characteristics of self-organized quantum dot separate confinement heterostructure and tunneling injection lasers," *Proc. 10th IEEE LEOS Annual Meeting*, Vol. 2, San Francisco, CA (1997), pp. 498-499.

20. P. Bhattacharya, X. Zhang, Y. Yuan, K. Kamath, D. Klotzkin, C. Caneau, and R. Bhat, "High-speed tunnel injection quantum well and quantum dot lasers," *Proc. SPIE* **3283**, 702 (1998).

Trends in Semiconductor Laser Design: Balance between Leakage, Gain, and Loss in InGaAsP/InP Multiquantum Well Structures

G. Belenky and L. Shterengas
State University of New York at Stony Brook, Stony Brook, NY 11794, U.S.A.

C. W. Trussell
Night Vision & Electronic Sensors Directorate
10215 Burbeck Road, Ft. Belvoir, VA 22060-5806 U.S.A.

C. L. Reynolds, Jr.
Agere Systems, Breinigsville, PA 18031, U.S.A.

M. S. Hybertsen
Agere Systems, Murray Hill, NJ 07974, U.S.A.

R. Menna
Princeton Lightwave Inc., Cranbury, NJ 08512, U.S.A.

1. Introduction

InGaAsP/InP is one of the basic material systems of modern optoelectronics. Well-developed and mature MOCVD technology allows for high-yield fabrication of lasers in the wavelength range 0.97–2 µm for telecom and other needs. The main peculiarity of devices based on InGaAsP is a relatively small conduction band offset — about 60% of the net band offset is in the valence band. The high mobility of electrons together with a rather small confinement barrier makes carrier leakage through thermionic emission a serious concern.

Modern semiconductor laser designs extensively use doping-profile and waveguide-geometry tuning. In this paper we will consider in detail both of these design tools. Experimental demonstration of their successful application to 1.3-µm and 1.5-µm InGaAsP/InP multiquantum well (MQW) lasers will be shown.

Doping of the interface between the *p*-cladding and the separate-confinement heterostructure (SCH) is a powerful tool for suppressing carrier leakage, leading to high injection efficiency.[1] The introduction of *p*-type doping into the waveguide or even the active region increases device temperature stability and improves high-frequency performance.[2,3] The price paid for these advantages is an increased optical loss due to free-carrier absorption. Alternatively, waveguide broadening minimizes optical loss due to the decreased overlap of the optical mode with the

231

highly doped cladding regions.[4] The trade-off between increased loss and the reduction of carrier leakage from the active region governs laser design. When high modulation bandwidth is an issue, device design is further complicated by carrier transport considerations.[1]

Various semiconductor laser applications place different requirements on the device parameters. Telecom transmitters for digital and analog links need low threshold, high efficiency, low noise, high modulation bandwidth, and highly linear light-current (L–I) characteristics. It is almost impossible to optimize a laser structure for all these parameters simultaneously. However, maximization of the external efficiency and minimization of the threshold current can be achieved for a relatively lightly-doped waveguide but heavily-doped p-cladding/SCH interface.[1] The temperature stability preferred for uncooled device operation is reached with higher doping of the waveguide layer.[2] Active region doping, in turn, increases the differential gain, thus improving the device high-frequency characteristics.[3]

For high-power lasers, small threshold current, large efficiency, and low series resistance are of special importance. Use of the broadened waveguide (BW) approach led to record low values of optical loss, boosting device external efficiency and enabling fabrication of highly efficient high-power devices.[4] The combination of the BW approach with optimization of the doping profile[5] is demonstrated for a 1.5-μm high-power InGaAsP/InP multiple quantum well (MQW) laser. A sharp increase of the Zn concentration near the p-cladding/SCH interface relaxes the L–I rollover, keeping the efficiency at a high level.

This presentation is divided into three sections. Section 2 explains the role of the p-type doping profile in InP-based laser structure performance. Section 3 covers experimental results on 1.3-μm InGaAsP/InP MQW lasers with optimized p-type doping profiles. Section 4 demonstrates design of 1.5-μm InGaAsP/InP MQW lasers in which the BW and optimized Zn doping profile approaches are combined.

2. Role of p-type doping profile in InP-based laser performance

Experimental and modeling results obtained by different groups show that changes of acceptor concentration in different regions of the laser structure significantly affect the device characteristics. We will consider here the effects of the doping of different regions separately.

- *P-type doping of the p-cladding/SCH interface*
  Acceptor centers in the vicinity of the p-cladding/SCH interface in InP-based semiconductor lasers lead to reduced heterobarrier leakage of electrons from the waveguide region into the p-cladding. The energy barrier for thermionic emission controls the amount of heterobarrier carrier leakage. Due to the relatively small conduction band offset at the InGaAsP/InP heterobarrier interface and the small electron effective mass, electron leakage current due to thermal emission into

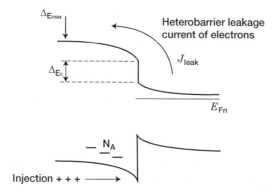

Figure 1. Schematic band diagram of the laser structure near the p-cladding/SCH interface, showing the barrier for thermionic emission of electrons.

the p-cladding can be significant. Hole heterobarrier leakage is not of any concern in this material system due to the large hole effective mass in combination with the higher energy barrier.

The effective barrier for thermionic emission of electrons combines the band discontinuity (ΔE_C) and the local band bending resulting from the modulation doping effect (Fig. 1). The leakage current (J_{leak}) depends exponentially on this energy barrier:

$$J_{\text{leak}} \propto e^{-\left(E_c^{\text{max}} - E_{\text{Fn}}\right)/kT} , \qquad (1)$$

where E_{Fn} is the electron quasi-Fermi level in the waveguide (SCH) near the interface. The local maximum in the conduction band profile (E_C^{max}) includes band bending related to the local doping concentration (N_A) and the influence of injected carriers due to external voltage drop. Higher doping near the interface reduces this additional voltage drop and, consequently, the heterobarrier leakage. Heterobarrier leakage suppression enhances the laser slope efficiency since the injection efficiency increases.

- *P-type doping of the waveguide region*
  High Zn concentration in the waveguide (SCH) region suppresses the effect of the electrostatic band bending.[6] It has been argued that the electrostatic band profile deformation increases device temperature sensitivity. The origin of this effect is separation of the charge in the waveguide. At high injection, electrons can spill over into the SCH region from QWs while holes remain localized within the QW. This separation of charges creates an electrostatic band profile deformation, increasing the effective barrier for the conduction band and decreasing the effective barrier for the valence band. At elevated temperatures, the hole density within the SCH can increase, leading to an increase in loss and recombination in the SCH. As a result, the threshold current rises and the slope

efficiency decreases. Incorporation of acceptors into the waveguide region affects the charge distribution, thus suppressing the accumulation of excess holes in the separate-confinement heterostructure.[2]

- *P-type doping of the active region*
  It was shown theoretically and experimentally that *p*-type doping of the active region in strained MQW InGaAsP/InP lasers can substantially increase the differential gain, leading to greater maximum modulation frequency.[3,7,8] Differential gain or, in other words, gain change per each additional injected electron-hole pair, is controlled by the carrier energy distribution and the relative position of the quasi-Fermi level with respect to states coupled into the laser mode. In most cases, the lasing transition couples states from approximately the bottom of the conduction band to the top of the valence band. The closer are the quasi-Fermi levels to the band edges in the QW, the higher is the differential gain because the population inversion change per each additional electron-hole pair increases. At the transparency current, the quasi-Fermi levels are separated by the bandgap energy. Because the effective mass is about one order of magnitude lower for electrons than for holes, the electron quasi-Fermi level at transparency is in the conduction band while the hole quasi-Fermi level is in the forbidden gap. With an increase of Zn concentration in the active region, the quasi-Fermi levels for electrons and holes shift closer to the energy band edges. As a result, the differential gain increases. Moreover, tuning of the MQW active region doping profile was argued to reduce carrier-transport–related impediments.[1,9]

  Introduction of *p*-type doping into any part of the laser structure that overlaps with the optical field leads to increased optical loss through free-carrier and intervalence-band absorption. These losses cause the threshold to increase and the slope efficiency to decrease, especially serious concerns in high-power lasers.

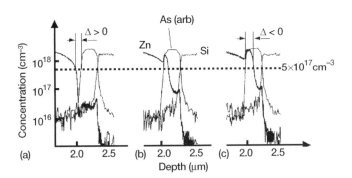

Figure 2. SIMS data on 1.3-μm InGaAsP MQW lasers with low-doped (a), moderately doped (b), and heavily doped (c) Zn profiles.

3.   Optimization of the *p*-type doping in 1.3-μm InGaAsP/InP MQW lasers

In Refs. 1–3, studies of the effect of *p*-type doping on 1.3-μm InGaAsP/InP MQW
laser performance were carried out experimentally. It was shown that there exists
an optimum *p-i* junction placement simultaneously maximizing external efficiency
and minimizing threshold current. To determine the placement of Zn doping
within the laser structure, a setback parameter (Δ) was introduced as the distance
between the Zn doping profile edge at a concentration of 5×10^{17} cm^{-3} and the *p*-
cladding/SCH interface. Figure 2 shows secondary ion mass spectrometry (SIMS)
data for devices with three different doping profiles.[2] The dependence of the
heterobarrier leakage current on Δ is shown in Fig. 3(a). In agreement with the
analysis given in Section 2, heterobarrier leakage current vanishes as *p*-type
doping intrudes deeper into the waveguide region. At the same time, external
efficiency tends to decrease since loss increases with doping due to free-carrier
absorption. Simulation predicts and experiment confirms that Δ = 50 nm is
optimum for minimum threshold current and maximum external efficiency — see
Fig. 3(b).

To minimize the device temperature sensitivity, the value of Δ must be
different. As pointed out in Section 2, the effect of electrostatic band profile
deformation can be suppressed using a doped waveguide. Laser temperature
stability can be characterized by the power-penalty parameter. It is defined as the
change in the output power level caused by ambient temperature change while
current through the device is kept constant:

$$\frac{1}{P}\frac{dP}{dT} = \frac{1}{\eta}\frac{d\eta}{dT} - \frac{1}{T_0(m-1)} \ , \tag{2}$$

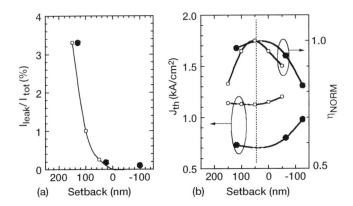

Figure 3. Dependence of the reduced leakage current (a) and of the threshold
current density as well as normalized external efficiency (b) on doping profile.
Open dots: simulation; solid dots: experiment.

where P is the output power, η is the external efficiency, and $m = I/I_{th}$ is the ratio of device current to the threshold value. The second term prevails in the current dependence, as shown in Fig. 4(a). The value of the power penalty decreases with current (Fig. 4). Figure 4(b) shows that the power penalty is minimized at a negative setback value of about –50 nm or, in other words, when the SCH layer is doped.[10] The price for increased temperature stability is decreased efficiency and increased threshold (see Fig. 3(b) for $\Delta = -50$ nm).

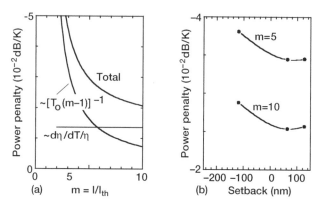

Figure 4. Dependence of the power penalty on pumping (a); power penalty as a function of doping profile for $m = 5$ and $m = 10$ (b).

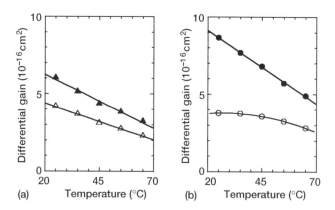

Figure 5. Temperature dependence of the differential gain for moderately doped (open symbols) and doped (solid symbols) Fabry-Perot (a) and distributed feedback (b) lasers.

Finally we will consider the effect of MQW region p-type doping on the device differential gain. Figure 5 shows the temperature dependence of the differential gain for 1.3-μm InGaAsP/InP Fabry-Perot (FP) and distributed feedback (DFB) capped-mesa buried-heterostructure lasers. Results for the devices with two different doping profiles are shown: moderately doped (similar to Fig. 2(b)) and doped (similar to Fig. 2(c)). The shift of the quasi-Fermi levels closer to the band edges with increased p-type doping explains the higher differential gain values in the doped compared to the moderately doped structures. However, the temperature sensitivity of the differential gain is enhanced with doping because the closeness of the quasi-Fermi level to the electronic states coupled to the laser mode also leads to larger changes of the population inversion with temperature.

The temperature dependence of the differential gain for DFB devices is further influenced by detuning (defined as the difference between the wavelength of the gain maximum and the lasing wavelength). For a particular operating carrier density in the active layer, the differential gain is a strong function of wavelength, being smaller on the long wavelength side of the gain peak and larger on the short wavelength side.[11] This dependence explains the larger difference in measured differential gain between the moderately doped (slightly positively detuned) and the heavily doped (negatively detuned) DFB lasers in comparison to the FP lasers. When the lasing wavelength is detuned from the gain peak, then the operating carrier density must be larger, which leads to a lower differential gain overall. For the devices studied, the operating carrier density changes with temperature due both to the intrinsic dependence of gain on temperature and the change in the spectral position of the lasing mode relative to the gain peak. For the devices with negative detuning (the doped lasers) the rise of temperature leads to an additional increase of the operating carrier density due to an increase of the detuning value. In moderately doped devices, the increase of temperature decreases positive detuning, thus suppressing the temperature dependence of the differential gain up to about 40 °C. After this temperature, the effect of negative detuning becomes apparent again.

4. Effect of doping profile on the performance of high-power 1.5-μm InGaAsP MQW BW lasers

The BW design decreases the internal optical loss leading to improved slope efficiency at threshold. The loss decrease is achieved through reduced overlap of the optical field with highly doped cladding layers. Reduction of the mode confinement in the MQW region with waveguide broadening is counterbalanced by the significant optical loss decrease.[4] The devices with BW exhibit improved external efficiency at low injection current. However, at high injection, thermal rollover limits the output power level. It was shown experimentally[12] that the heterobarrier leakage current contributes to the L–I rollover of 1.5-μm InGaAsP/InP narrow-waveguide lasers. In this part of the paper, we present new data showing that the doping of the *p*-cladding/SCH interface of 1.5-μm InGaAsP/InP BW high-power lasers suppresses *L–I* rollover, leading to higher output power.

Broadened-waveguide lasers with three different doping profiles were fabricated. The design of the lasers was similar to the one used in Ref. 4. The structure is MOCVD-grown on an *n*-type InP substrate. The active region contains 3 InGaAsP QWs (4.5 nm) with 1% compressive strain separated by 16-nm InGaAsP barriers. An InGaAsP double-step graded-index SCH provides optical confinement. The total width of the broadened waveguide (BW) region is 710 nm. All layers but the QWs are lattice matched to InP. The waveguide is sandwiched between 1.5-μm InP cladding layers. Lasers with three different Zn doping levels of the *p*-cladding/SCH interface were tested. The corresponding SIMS data for the broad-area 1.5-μm InGaAsP/InP MQW BW lasers are shown in Fig. 6.

The device front and back mirrors were coated for low (3–5%) and high (95%) reflectivity, respectively. In order to minimize thermal effects, low duty cycles (less than 0.1%) and short current pulses (100–200 ns) were used. The two main effects of doping on the *L–I* characteristics are shown in Fig. 7(a). First, the slope efficiency at threshold steadily decreases with doping, suggesting higher optical loss for doped devices. Second, power saturation at high current densities decreases with increased doping levels, which confirms the suppression of heterobarrier leakage.

We measured directly the current dependence of the modal gain spectra of the lasers (Fig. 8). A spatial filtering[13,14] selected on-axis optical modes of the multimode broad-area lasers. Amplified spontaneous emission (ASE) spectra

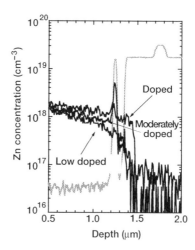

Figure 6. SIMS data for 1.5-μm InGaAsP MQW broadened-waveguide lasers with three different Zn doping profiles.

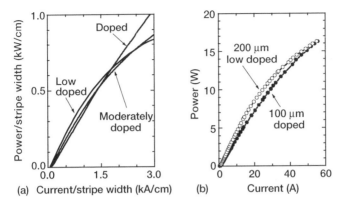

Figure 7. Dependence of the output optical power per stripe width on current per stripe width for devices with three different doping profiles (a). Light-current characteristics for low-doped 200-μm and doped100-μm stripe width lasers (b).

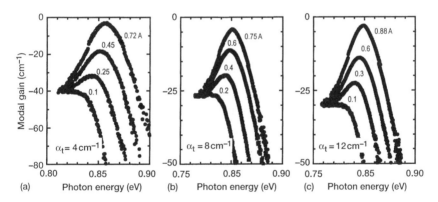

Figure 8. Current dependence of the modal gain spectra for low doped (a), moderately doped (b), and heavily doped (c) lasers.

were monitored using a Fourier transform spectrometer (MAGNA860). Modal gain spectra were obtained from the ASE spectra by the Hakki-Paoli method.[15] In the energy region where modal gain saturates (the range of low photon energies in Fig. 8) the material gain is equal to zero and modal gain is equal to the total optical loss. Subtracting the calculated mirror loss (about 18 cm^{-1} and 36 cm^{-1} for 1-mm and 500-μm cavity length devices, respectively) from the experimentally determined total optical loss, we obtain the internal optical loss value. Lasers with the lowest Zn concentration in the vicinity of the p-cladding/SCH interface have about 4 cm^{-1} internal optical losses, whereas as Zn propagates deeper into the waveguide, the internal loss goes up to 12 cm^{-1} for the doped devices — see Figs. 8(a) and 8(c).

At high injection levels when the barrier for electron thermionic emission from the SCH into the p-type cladding is suppressed by the external voltage, heterobarrier leakage increases and the $L–I$ characteristic saturates. Optimization of the device p-type doping profile allowed us to obtain the same output optical power from 100-μm BW doped lasers as from 200-μm BW low-doped ones, as shown in Fig. 7(b). Far field emission patterns were almost independent of stripe width and doping profile with about 20°×50° divergence for high output power levels. Due to better linearity of the $L–I$ characteristics, the BW doped lasers yielded twice the output optical power density and brightness as BW low-doped devices at 60 A.

Acknowledgments

The authors would like to acknowledge Dmitry Donetsky for valuable discussions and ARO Grant DAAD190010423 for support.

References

1. G. L. Belenky, C. L. Reynolds, Jr., D. V. Donetsky, G. E. Shtengel, M. Hybertsen, M. A. Alam, G. A. Baraff, R. K. Smith, R. F. Kazarinov, J. Winn, and L. E. Smith, "Role of p-doping profile and regrowth on the static characteristics of 1.3 μm MQW InGaAsP/InP lasers: Experiment and modeling," *IEEE J. Quantum Electron.* **35**, 1515 (1999).
2. G. L. Belenky, D. V. Donetsky, C. L. Reynolds, Jr., R. F. Kazarinov, G. E. Shtengel, S. Luryi, and J. Lopata, "Temperature performance of 1.3-μm InGaAsP-InP lasers with different profile of p-doping," *IEEE Photonics Technol. Lett.* **9**, 1558 (1997).
3. G. Belenky, C. L. Reynolds, Jr., L. Shterengas, M. S. Hybertsen, D. V. Donetsky, G. E. Shtengel, and S. Luryi, "Effect of p-doping on temperature dependence of differential gain in FP and DFB 1.3 μm InGaAsP/InP multiple quantum well lasers," *IEEE Photonics Technol. Lett.* **12**, 969 (2000).
4. D. Z. Garbuzov, R. J. Menna, R. U. Martinelli, J. H. Abeles, and J. C. Connolly, "High power continuous and quasi-continuous wave InGaAsP/InP broad-waveguide separate confinement-heterostructure lasers," *Electronics Lett.* **33**, 1635 (1997).
5. I. K Han, S. H. Cho, P. J. S. Heim, D. H. Woo, S. H. Kim, J. H. Song, F. G. Johnson, and M. Dagenais, "Dependence of the light-current characteristics of 1.55-μm broad-area lasers on different p-doping profiles," *IEEE Photonics Technol. Lett.* **12**, 251 (2000).

6. S. Seki and K. Yokoyama, "Electrostatic deformation in band profiles of InP-based strained-layer quantum well lasers," *J. Appl. Phys.* **77**, 5180 (1995).

7. T. Yamanaka, Y. Yoshikuni, K. Yokoyama, W. Lui, and S. Seki, "Theoretical study on enhanced differential gain and extremely reduced linewidth enhancement factor in quantum-well lasers," *IEEE J. Quantum Electron.* **29**, 1609 (1993).

8. I. F. Lealman, M. J. Harlow, and S. D. Perrin, "Effect of Zn doping on modulation bandwidth of 1.55 μm InGaAs/InGaAsP multiple quantum well lasers," *Electronics Lett.* **29**, 1197 (1993).

9. M. A. Alam, M. S. Hybertsen, R. K. Smith, and G. A. Baraff, "Simulation of semiconductor quantum well lasers," *IEEE Trans. Electron Dev.* **47**, 1917 (2000).

10. D. Donetsky, *Temperature Performance of InP and GaSb-based Laser Diodes*, Ph.D. thesis, SUNY at Stony Brook (2000).

11. P. A. Morton, D. A. Ackerman, G. E. Shtengel, R. F. Kazarinov, M. S. Hybertsen, T. Tanbun-Ek, R. A. Logan, and A. M. Sergent, "Gain characteristics of 1.55-μm high-speed multiple-quantum-well lasers," *IEEE Photonics Technol. Lett.* **7**, 833 (1995).

12. L. Shterengas, R. Menna, W. Trussell, D. Donetsky, G. Belenky, J. Connolly, and D. Z. Garbuzov, "Effect of heterobarrier leakage on the performance of high power 1.5-μm InGaAsP MQW lasers," *J. Appl. Phys.* **88**, 2211 (2000).

13. D. J. Bossert and D. Gallant, "Improved method for gain/index measurements of semiconductor lasers," *Electronics Lett.* **32**, 338 (1996).

14. D. V. Donetsky, G. L. Belenky, D. Z. Garbuzov, H. Lee, R. U. Martinelli, G. Taylor, S. Luryi, and J. C. Connolly, "Direct measurements of heterobarrier leakage current and modal gain in 2.3 μm double QW *p*-substrate InGaAsSb/AlGaAsSb broad area lasers," *Electronics Lett.* **35**, 298 (1999).

15. B. W. Hakki and T. L. Paoli, "CW degradation at 300 K of GaAs double-heterostructure junction lasers. II. Electronic gain," *J. Appl. Phys.* **44**, 4113 (1973).

Terahertz Emitters Based on Intersubband Transitions

Q. Hu and B. S. Williams
Dept. of Electrical Engineering and Computer Science
Massachusetts Institute of Technology, Cambridge, MA 02139, U.S.A.

M. R. Melloch
School of Electrical and Computer Engineering
Purdue University, West Lafayette, IN 47907, U.S.A.

J. L. Reno
Sandia National Labs, Dept. 1123, MS 0601
Albuquerque, NM 87185-0601, U.S.A.

1. Introduction

Terahertz (1–10 THz, or 4–40 meV, or 30–300 μm) frequencies are among the most underdeveloped electromagnetic spectra, even though their potential applications are promising for spectroscopy in chemistry and biology, astrophysics, plasma diagnostics, remote atmospheric sensing and imaging, noninvasive inspection of semiconductor wafers, and communications. This underdevelopment is due primarily to the lack of coherent solid-state THz sources that can provide high radiation intensities (greater than a milliwatt). The THz frequency falls between two other frequency ranges in which conventional semiconductor devices have been well developed. One is the microwave and millimeter-wave frequency range, and the other is the near-infrared and optical frequency range. Semiconductor electronic devices that utilize the transport of free charge carriers (such as transistors, Gunn oscillators, Schottky-diode frequency multipliers, and photomixers) are limited by the transit time and parasitic RC time constants. Consequently, the power level of these classical devices decreases as $1/f^4$, or even faster, as the frequency f increases above 1 THz. Semiconductor photonic devices based on quantum-mechanical interband transitions, however, are limited to frequencies higher than those corresponding to the semiconductor energy gap, which is higher than 10 THz even for narrow-gap lead-salt materials. Thus, the frequency range of 1–10 THz is inaccessible for conventional semiconductor devices.

Semiconductor quantum wells are human-made quantum-mechanical systems in which the energy levels can be engineered to be of any value. Consequently, unipolar lasers based on intersubband transitions (electrons that make lasing transitions between subband levels) were proposed for long-wavelength sources as early as the 1970's.[1] This device concept has been realized in the successful development of the quantum-cascade lasers (QCL) at mid-infrared wavelengths.[2] Recent development has extended the operating wavelengths of the QCLs to as long as 24 μm.[3] In spite of these impressive achievements, the THz frequencies below the *Reststrahl* band remain stubbornly unconquered.

Free-carrier absorption increases at long wavelengths as λ^2, which could cause a significant cavity loss at the THz frequencies. By using thick (≥10 μm) active regions or metallic waveguides for mode confinement, in combination with a low doping concentration (<10^{16} cm^{-3}), the cavity losses can be reduced to below 50 cm^{-1}, as verified recently in experiments.[4] Thus, the key to achieve lasing is to obtain a sufficient level of gain to overcome this moderate level of cavity loss.

The intersubband emitters are known to have a large joint density of states, because the two subbands, for example E_3 and E_2, track each other in *k*-space. Thus, ignoring nonparabolicity, electrons emit photons at the same energy regardless of their initial momentum. Therefore, the peak gain is related to the inverted population density $\Delta n = (n_3 - n_2)$ in a simple linear fashion, that is,

$$g = (\Delta n/t)(2e^2\omega/h\varepsilon_r^{1/2}\varepsilon_0 c) z_{ij}^2/\Delta f. \tag{1}$$

In Eq. (1), t is the thickness of the mode confinement region, and thus $\Delta n/t$ is the three-dimensional inverted population density within t, z_{ij} is the radiative dipole moment between the two subbands, and Δf is the FWHM linewidth of spontaneous emission. From Eq. (1), it is obvious that a large peak gain requires a large dipole moment z_{ij}, a narrow emission linewidth Δf, and an appreciable level of population inversion Δn. The optimizations of the first two parameters are usually closely related, as a large dipole moment (accomplished by a strong spatial overlap of the two subband wavefunctions) tends to yield a narrow emission linewidth. Measured emission linewidth as narrow as 0.7 meV has been achieved with the calculated dipole moment greater than 5 nm.[5,6] Even with a moderate level of population inversion of $\Delta n \approx 10^9$ cm^{-2}, the estimated peak gain will be greater than 100 cm^{-1}, which should exceed the cavity loss by a comfortable margin. The fact that THz lasing has not been achieved indicates that it is highly challenging to achieve population inversion between two narrowly separated subband levels.

In this article, we discuss our investigations of two types of electrically pumped intersubband THz emitters. One utilizes electron–LO-phonon scattering, and the other utilizes resonant tunneling to depopulate the lower radiative level in order to achieve population inversion. Related issues, such as the role of interface and confined phonon modes and coupling between subbands near anticrossing (or resonance) will also be discussed. Our investigations of optically pumped intersubband THz sources are discussed elsewhere.[7-9]

2. THz emitters using electron–LO-phonon scattering for depopulation

Following the design principle of the original QCLs, we have designed intersubband THz emitters based on a three-level system. The top two subband levels, E_3 and E_2, form the radiative pair, while the ground state E_1 is below E_2 by more than $\hbar\omega_{LO}$. Since it is energetically allowed, the fast electron–LO-phonon E_2 → E_1 scattering will help to keep the population in E_2 low, and therefore maintain a population inversion between E_3 and E_2. However, because $(E_3–E_2) < (E_2–E_1) \geq \hbar\omega_{LO}$ for THz emitters, it is difficult to implement this three-level system based on an intrawell transition scheme, in which both E_3 and E_2 are primarily located in a single well. The bottom of this well would have to be raised relative to the rest of the structure so that $(E_2–E_1) \geq \hbar\omega_{LO}$. Raising the bottom of the well would require adding aluminum to the GaAs well material, which would cause a significant amount of alloy scattering and result in a broad emission linewidth.

Our design of the three-level systems is based on a scheme in which the radiative transition takes place in a coupled double-well structure. A third well, which is much wider than the coupled wells, contains the ground-state level E_1. In our first design, the wavefunctions of E_3 and E_2 are primarily localized in separate wells, thus the $E_3 \to E_2$ transition is spatially diagonal. This design offers the advantage of a high selectivity in injection into E_3 and removal from E_2, because of the spatial separation of the two wavefunctions. However, the diagonal nature of the radiative transition is quite sensitive to scattering due to interface roughness and alloying in the barrier, and thus the emission showed rather broad spectra ($\Delta f \sim$ 3–5 THz).[10]

It is well known that in a coupled double-well structure, the wavefunctions of the two lowest levels are spatially extended with a strong overlap at the anticrossing. Because of this spatial overlap, both levels are subject to the same interface and alloy scattering, and thus the emission linewidth of the radiative transition between the two levels is reduced. In an improved structure, whose band structure and wavefunctions are shown in Fig. 1, we have taken advantage of this feature to enhance the strength of the radiative $E_3 \to E_2$ transition.[11]

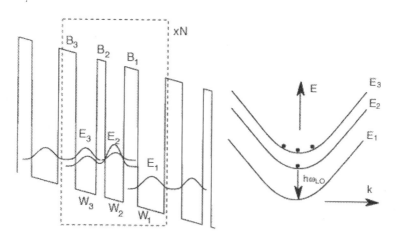

Figure 1. Schematic of a three-level system based on a triple quantum-well structure shown inside the dashed box. On the right is the dispersion relation between the energy and the transverse momentum.

In the structure shown in Fig. 1, the core is a three-well module of GaAs/$Al_{0.3}Ga_{0.7}As$ heterostructures (inside the dashed box), with three barriers B_1 (4.5 nm), B_2 (2.8 nm), B_3 (5.6 nm) and three wells W_1 (8.8 nm), W_2 (5.9 nm), and W_3 (6.8 nm). The collector barrier B_1 is center δ-doped at a level of 6×10^{10} cm^{-2} in order to provide dynamic charges. Under the designed bias of 51 mV per module, the ground state E_1' (not shown) of a previous module is aligned with E_3. Thus, the upper level E_3 can be selectively populated via resonant tunneling. At this bias, the energy separation $E_{32} \approx 11$ meV (corresponding to 2.67 THz), and the dipole moment $z_{32} \approx 3$ nm. The energy separation $E_{21} \approx 40$ meV $\geq \hbar\omega_{LO}$, enabling electron–LO-phonon $E_2 \to E_1$ scattering for depopulation.

Our emission spectra reveal a clear peak due to the $E_3 \to E_2$ intersubband emission. A representative spectrum taken at a device temperature of 5 K is shown in Fig. 2(a), which was taken at the designed bias of 1.6 V (~30×51 mV). The measured peak frequency of 2.57 THz (10.6 meV) is close to the designed value of 11.3 meV. The FWHM linewidth is as narrow as 0.47 THz (1.9 meV). In order to verify the intersubband origin of the measured emission spectra, we have measured the emission spectrum at a high bias of 4.0 V at which the energy levels are severely misaligned. The spectrum is shown in the inset of Fig. 2(a), and it bears little resemblance to the main figure.

Spectra were also taken with the cold stage cooled with liquid nitrogen to 80 K. One taken at a bias of 1.6 V is shown in Fig. 2(b). The main peak is essentially the same as the one measured at 5 K, with a slightly broader linewidth

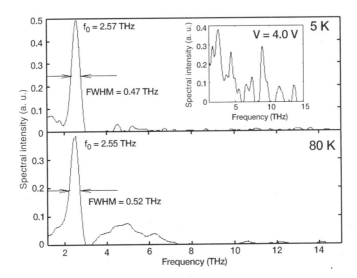

Figure 2. Spectrally resolved THz intersubband spontaneous emission taken at 5 K (a) and 80 K (b) bath temperature under 1.6 V bias. The inset shows the spectrum for 4.0 V bias, clear evidence that emission results from intersubband transitions.

of 0.52 THz (2.14 meV). The linewidth measured at 80 K is expected to be similar to that at 5 K, since nonparabolicity is negligible for THz intersubband emitters. Nevertheless, our experimental verification is encouraging for the development of intersubband THz sources at elevated temperatures.

3.   Role of interface and confined phonon modes

Just as the electronic wavefunctions form discrete subbands in MQW structures, phonon spectra also become discrete and form spatially localized interface and confined modes. Despite this parallel analogy, the bulk LO-phonon mode has been used to calculate the scattering times in QCLs. This practice may be justified for mid-infrared QC lasers, in which $(E_3-E_2) >> \hbar\omega_{LO}$, so that the sum-rule yields the same result as obtained from using the bulk mode. However, as Dutta and Stroscio recently pointed out,[12] such a practice may be questionable for THz intersubband lasers, because $(E_3-E_2) < \hbar\omega_{LO}$ and $(E_2-E_1) \approx \hbar\omega_{LO}$.

We have investigated the role of the complex phonon spectra on the intersubband scattering rates.[13] In the model used in our calculations, the phonon modes are described by the potential $\phi(r)$ resulting from the polarization field created by atomic displacements in a polar semiconductor. Each material layer of index i is described by a dielectric function $\varepsilon_i(\omega)$ as given by the Lyddane-Sachs-Teller relations, which vanishes at the LO-phonon frequencies. For lattice vibrations, since there are no free charges, the phonon potential must satisfy $\varepsilon(\omega)\nabla^2\phi(r) = 0$. Two types of solutions exist: interface modes for which $\nabla^2\phi(r) = 0$, and confined modes for which $\varepsilon(\omega) = 0$.

For the confined modes, $\varepsilon(\omega) = 0$ and therefore $\omega = \omega_{LO}$, where ω_{LO} is the bulk LO-phonon frequency in the layer of interest. Since ω_{LO} changes at the heterointerfaces, the potential must vanish there, and $\phi(r)$ can be described in terms of sine-wave modes in z. For the interface modes, $\varepsilon(\omega) \neq 0$, and the modes have frequencies $\omega \neq \omega_{LO}$. The potential solution is a linear combination of exponential terms peaked at the interfaces, hence the name of "interface mode." For our GaAs/AlGaAs quantum-well structures, there are usually two "GaAs-like" and one "AlAs-like" modes associated with each GaAs/AlGaAs interface. Thus, for the six interfaces in our triple-well structures shown in Fig. 1, there are a total 18 interface modes. The total scattering rate is the sum of the contributions from these 18 interface modes and all the confined modes (30 lowest modes were used in our calculations with the contributions from the higher modes negligible).

We used the transfer matrix approach[14] to account for the electromagnetic boundary conditions and obtain the mode potentials and dispersion relations for the interface modes. As it turns out, the 12 "GaAs-like" modes are clustered around 33–36 meV, close to the bulk GaAs LO-phonon energy. The 6 "AlAs-like" modes are clustered around 45–47 meV, close to the bulk AlAs LO-phonon energy. Special care was taken to ensure a proper normalization of each mode, which was verified by limiting cases.[13]

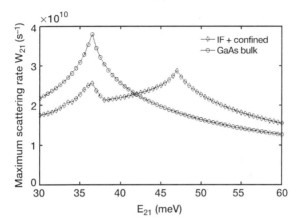

Figure 3. Maximum scattering rate versus subband separation E_{21} for the three-level structure shown in Fig. 1. Maximum scattering rate calculated with GaAs bulk modes is included for comparison.

In order to address the key issue raised in Ref. 12, namely the optimum subband separation E_{21}, we have calculated the maximum total scattering rate as a function of E_{21}, as shown in Fig. 3. The rate shows two peaks, one at ~35 meV due to the "GaAs-like" modes and the other at ~47 meV due to the "AlAs-like" modes. As a comparison, we also include the scattering rate calculated using the bulk GaAs LO-phonon mode. It is clear that the scattering rate is ~30% lower than the bulk mode if $E_{21} \sim 36$ meV, because of the exclusion of the "AlAs-like" modes. However, increasing E_{21} to ~47 meV does not increase the total scattering rate appreciably. The reason is that at $E_{21} \sim 47$ meV, even though all the phonon modes participate in the scattering, the rapid decrease in the strength of the "GaAs-like" modes away from their resonance (at ~36 meV) diminishes most of the benefit gained from including more active phonon modes.

In Fig. 4, we plot the key figure of merit $\tau_3(1-\tau_{21}/\tau_{32})$, which is proportional to the population inversion Δn_{32}, as a function of the scattering time τ_{32}. As E_{21} increases from 36 to 47 meV, the increase in Δn_{32} is marginal at a given τ_{32}. Even this marginal increase should be taken with a grain of salt. As E_{21} increases from 36 to 47 meV, there will be an additional 11 meV energy dissipation per electron. This extra energy dissipation will further raise the electronic temperature. As a result, τ_{32} decreases because of LO-phonon scattering of hot electrons. Figure 4 shows that a reduction in τ_{32} could undo any advantage gained by increasing E_{21} to 47 meV. The conclusion from our analysis is that in electrically pumped THz intersubband emitters, because the barriers are thin and Al concentrations are low ($x \leq 0.3$), the contribution from the higher-energy "AlAs-like" modes only barely makes up for the loss in the strength of "GaAs-like" modes. Thus, increasing E_{21} from 36 to 47 meV will yield at the best only a marginal (if any) improvement in population inversion Δn_{32}.

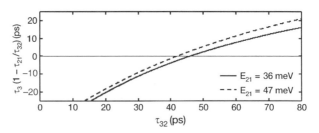

Figure 4. Plot of the quantity $\tau_3(1-\tau_{21}/\tau_{32})$, which is proportional to the population inversion Δn_{32}, versus the lifetime τ_{32} for the structure shown in Fig. 1.

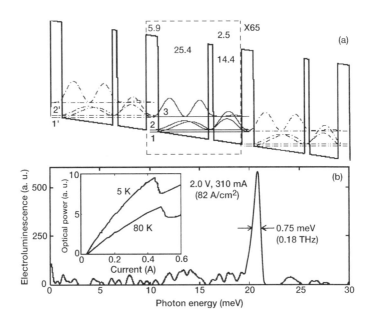

Figure 5. Computed conduction band profile and squared magnitude wavefunctions for the intrawell device (a). Electroluminescence spectrum at 5 K for an applied bias of 2.0 V (b). The inset displays the emitted power versus current.

4.   Intrawell THz emitters using resonant tunneling for depopulation

Figure 5(a) shows the calculated band diagram of another type of THz intersubband emission structure that we investigated recently.[6] In this structure, the radiative transition takes place between two subband levels primarily located in one quantum well. We consider this scheme "intrawell" as opposed to the interwell scheme discussed earlier. It is known that the intrawell scheme yields a larger dipole moment and a narrower emission linewidth, because the subband separation is less sensitive to impurity and interface roughness scattering.[5]

The core of the structure shown in Fig. 5(a) is a coupled double-well module shown inside the dashed box. Sixty-five nominally identical modules are cascade connected. Under the designed bias of ~20 mV/module, the lower level E_2 in the wide well is aligned with the lowest level E_1 in the narrow well, which in turn is aligned with the upper level in the wide well of the following module. The radiative transition takes place between E_3 and E_2, which have a strong spatial overlap as can be seen in Fig. 5(a). The depopulation of E_2 is facilitated through "resonant tunneling" from E_2 to E_1, which deserves special attention and will be discussed in the following. The calculated dipole moment $z_{32} \approx 6$ nm at the designed bias, as a result of the strong radiative coupling in this intrawell scheme.

Figure 5(b) shows a measured emission spectrum taken at a bias of 2.0 V, which is somewhat greater than the designed bias (65×20 mV ≈ 1.3 V). The spectrum shows a clear peak at 21 meV (~5.04 THz), with a FWHM linewidth as narrow as 0.75 meV (0.18 THz). The measured emission frequency is close to the calculated subband separation $E_{32} \approx 18.5$ meV, and the narrow linewidth indicates the high quality of the MBE growth.

The term "resonant tunneling" usually refers to electron transport between two subband levels close in energy. There has been a long debate on whether this process should be analyzed as intersubband scattering between two spatially extended wavefunctions (the scenario of coherent resonant tunneling), or as tunneling between two spatially localized states (the scenario of incoherent sequential tunneling). This question is not merely academic, but is crucial for the successful development of intersubband THz lasers, as the two transport mechanisms yield very different depopulation rates. For coherent resonant tunneling, the electron transport is facilitated by fast (<1 ps) electron-electron intersubband scattering. For incoherent sequential tunneling, the depopulation rate is determined by the barrier transparency and can be much slower than in the first scenario.

In the coherent picture, the energy difference between any pair of two subbands has a finite minimum value, known as the anticrossing gap Δ, which characterizes how strongly the two subbands are coupled. In contrast, in the incoherent scenario, the two subband wavefunctions are spatially localized in different wells. Their energy levels can be degenerate and thus may be arbitrarily close at resonance. Based on the coherent model, we have calculated Δ_{21} (the minimum value of $|E_2-E_1|$) to be ~2.5 meV. Even though current-field (I–B) magneto-tunneling spectroscopy has been used successfully to resolve subband separations of ~20 meV,[15,16] the resolution of the anticrossing gaps of only a few meV requires the measurement of conductance $G = dI/dV$ to enhance the energy resolution and sensitivity.

Fig. 6 shows both I–B and G–B curves measured in a voltage range near the E_2–E_1 anticrossing. The I–B curves show a single-period oscillation, revealing an energy separation of ~19 meV, close to the calculated intersubband spacing. The G–B data show a closely separated double resonance, as highlighted by the double arrows. The dependence of these conductance peaks on bias is plotted in Fig. 6(c), and it displays a typical anticrossing behavior between a Stark-shifted peak and another fixed at ~6.0 T (~21 meV). This peak at 21 meV peak corresponds to the 3 → 2 transition, in agreement with the emission data. The Stark-shifted peak corresponds to the 3 → 1 transition, as the two wavefunctions are spatially more separated than those of E_3 and E_2. The minimum separation of the two conductance peaks is ~1.7 meV, which is close to the calculated anticrossing gap of $\Delta_{21} \approx 2.5$ meV. The overall dependence of the conductance peaks on the bias agrees reasonably well with the calculated energy differences E_{31} and E_{32}, as shown in Fig. 6(c). Our work shows that in the particular structure shown in Fig. 5(a), the $E_2 \to E_1$ depopulation should be modeled as intersubband scattering between two spatially extended states.

In conclusion, we have investigated intersubband THz emission structures with the depopulation of the lower radiative level assisted by either electron–LO-phonon or electron-electron scattering. Our theoretical analysis shows that population inversion can be achieved in both structures and both should be pursued experimentally. Inclusion of the complex phonon spectra yields only a marginal difference from calculations using the bulk phonon mode.

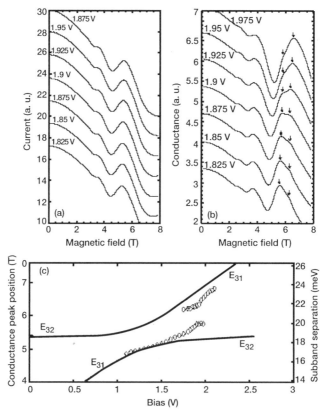

Figure 6. Current (a) and conductance (b) versus magnetic field for several biases. Positions of the double peaks in *G–B* plots versus the bias (c). The solid lines are the calculated subband energy separations E_{32} and E_{31}.

Acknowledgments

We would like to thank B. Xu and H. Callebaut for their participation in this work, and K. Kempa for his insightful comments. This work is supported by AFOSR, NASA, NSF, and ARO. Sandia is a multiprogram laboratory operated by Sandia Corporation, a Lockheed Martin Company, for the United States Department of Energy under Contract DE-AC04-94AL85000.

References

1. R. F. Kazarinov and R. A. Suris, "Possibility of amplification of electromagnetic waves in a semiconductor with a superlattice," *Sov. Phys. Semicond.* **5**, 707 (1971).
2. J. Faist, F. Capasso, D. L. Sivco, C. Sirtori, A. L. Hutchinson, and A. Y. Cho, "Quantum cascade laser," *Science* **264**, 477, (1994).
3. R. Colombelli *et al*, "Far-infrared surface-plasmon quantum-cascade lasers at 21.5 and 24 μm wavelengths," *Appl. Phys. Lett.* **78**, 2620 (2001).
4. M. Rochat, M. Beck, J. Faist, and U. Oesterle, "Measurement of far-infrared waveguide loss using a multisection single-pass technique," *Appl. Phys. Lett.* **78**, 1967 (2001).
5. M. Rochat, J. Faist, M. Beck, U. Oesterle, and M. Ilegems, "Far-infrared (λ = 88 μm) electroluminescence in a quantum cascade structure," *Appl. Phys. Lett.* **73**, 3724 (1998).
6. B. S. Williams, H. Callebaut, Q. Hu, and J. Reno, "Magnetotunneling spectroscopy of resonant anticrossing in terahertz intersubband emitters," *Appl. Phys. Lett.* **79**, 4444 (2001).
7. I. Lyubomirsky and Q. Hu, "Energy level schemes for far-infrared quantum well lasers," *Appl. Phys. Lett.* **73**, 300 (1998).
8. I. Lyubomirsky, Q. Hu, and M. R. Melloch, "Measurement of far-infrared intersubband spontaneous emission from optically pumped quantum wells," *Appl. Phys. Lett.* **73**, 3043 (1998).
9. H. Callebaut and Q. Hu, "Design analysis of interband-pumped intersubband THz lasers," submitted to *Appl. Phys. Lett.* (2001).
10. B. Xu, Q. Hu, and M. R. Melloch, "Electrically pumped tunable THz emitters based on intersubband transition," *Appl. Phys. Lett.* **71**, 440 (1997).
11. B. S. Williams, B. Xu, Q. Hu, and M. R. Melloch, "Narrow-linewidth terahertz intersubband emission from three-level systems," *Appl. Phys. Lett.* **75**, 2927 (1999).
12. M. Dutta and M. A. Stroscio, "Comment on 'Energy level schemes for far-infrared quantum well lasers," *Appl. Phys. Lett.* **74**, 2555 (1999); see also comment by Q. Hu and I. Lyubomirsky, *Appl. Phys. Lett.* **74**, 3065 (1999).
13. B. S. Williams and Q. Hu, "Optimized energy separation for phonon scattering in three-level terahertz intersubband lasers," *J. Appl. Phys.* **90**, 5504 (2001).

14. S. G. Yu, K. W. Kim, M. A. Stroscio, G. J. Iafrate, J.-P. Sun, and G. I. Haddad, "Transfer matrix method for interface optical-phonon modes in multiple-interface heterostructure systems," *J. Appl. Phys.* **82**, 3363 (1997).

15. J. H. Smet, C. G. Fonstad, and Q. Hu, "Magnetotunneling spectroscopy in wide InGaAs/InAlAs double quantum wells," *Appl. Phys. Lett.* **63**, 2225 (1993).

16. J. Ulrich, R. Zobl, W. Schrenk, G. Strasser, K. Unterrainer, and E. Gornik, "Terahertz quantum cascade structures: Intra- versus interwell transition," *Appl. Phys. Lett.* **77**, 1928 (2000).

The Future of Photovoltaics

Martin A. Green
Special Research Centre for Third Generation Photovoltaics
University of New South Wales, Sydney, 2052, Australia

1. Introduction

Many have thought about how photovoltaic technology may evolve in the future. Since the early days of terrestrial photovoltaics, a common perception has been that "first generation" silicon wafer-based solar cells would be replaced by a "second generation" of lower cost thin-film technology, probably also involving a different semiconductor. Historically, CdS, a-Si, $CuInSe_2$, CdTe and, more recently, thin-film Si have been regarded as key thin-film candidates. Since any mature solar cell technology must evolve to the stage where costs are dominated by those of the constituent materials, be it silicon wafers or glass sheet, it is argued that photovoltaics must evolve, in its most mature form, to a "third generation" of high-efficiency thin-film technology. By high-efficiency, what is meant is energy conversion values double or triple the 15–20% range presently targeted. Tandem cells provide the best known example of such high-efficiency approaches, where efficiency can be increased merely by adding more cells of different bandgap to a stack. However, a range of other more "parallelled" approaches are possible that offer similar efficiency to an infinite stack of tandem cells.

Most solar cells presently sold are based on silicon wafers, so-called "first generation" technology. As this technology has matured, costs have become increasingly dominated by material costs, namely those of the silicon wafer, the strengthened low-iron glass cover sheet, and those of other encapsulants. This trend is expected to continue as the photovoltaic industry continues to mature. A recent study[1] of costs of manufacturing in greatly increased 500 MW/year production volume suggests material costs would account for over 70% of total manufacturing costs. This situation favors high-efficiency processing sequences that produce solar cells with high energy conversion efficiency, provided these can be implemented without unduly complicating cell processing.[1] Nonetheless, module energy conversion efficiency above 16% is not contemplated in this future scenario.[1]

For the past 15 years, the industry has seemed on the verge of switching to a "second generation" of thin-film cell technology. Regardless of semiconductor employed, thin films offer prospects for a quantum reduction in material costs by eliminating the silicon wafer. Thin films also offer other advantages, such as the increase in the unit of manufacturing from a silicon wafer ($\sim$100 cm^2) to a glass sheet ($\sim$1 m^2), about 100 times larger. With time, this "second generation" technology may largely bridge the present gap with regard to conversion efficiency between it and "first generation" product.

As thin-film "second generation" technology matures, costs will become progressively dominated by those of the constituent materials, in this case, the top cover sheet and other encapsulants. There will be a lower limit on such costs (say, US \$30/m^2) which, when combined with likely cell efficiency (15% or 150 W$_p$/m^2), determines the lower limit on photovoltaic module costs (US \$0.20/W$_p$) and electricity generation costs (US \$0.02–0.04/kW hr), assuming module costs are half system costs and a range of insolation and cost-of-money scenarios.

To progress further, conversion efficiency must be increased substantially. The Carnot limit on the conversion of sunlight to electricity is 95% as opposed to the theoretical upper limit of 33% for a single-junction cell, such as a silicon wafer and most present thin-film devices. This comparison suggests the performance of solar cells could be improved 2–3 times if different concepts were used to produce a "third generation" of high performance, low-cost photovoltaic product.

What are the prospects for thin-film cells based on new concepts capable of "third generation" performance? Fortunately, with the likely evolution of new materials technology over the coming decades, quite good! In the following, possible third-generation approaches are reviewed and implementation strategies suggested. New concepts such as thermophotonic conversion are also introduced, suggesting that there is scope for new approaches on which to base "third generation" devices.

2.   Efficiency losses in a standard solar cell

Losses in a standard solar cell are shown in Fig. 1. A key loss process is thermalization (process 1), whereby the photoexcited pair quickly loses energy in excess of the bandgap. A low energy red photon is just as effective as a much higher energy blue photon. This loss alone limits conversion efficiency of a cell to about 44%.

Another important loss process is recombination of the photoexcited electron-hole pairs (process 4). This loss can be kept to a minimum by using material with high lifetimes for the photogenerated carriers, which can be ensured by eliminating all unnecessary defects. The lifetime is then determined by radiative recombination in the cell, the inverse to photoexcitation. As shown in 1960,[1] this symmetry between light absorption and light emission can be used to derive quite fundamental limits on achievable solar cell performance. This approach revisits "black body" radiation, the topic that stimulated the birth of quantum mechanics.

Radiation from the sun approximates that from a "black body" held at a warm 6000 K, the temperature of the sun's photosphere. The energy distribution of this radiation is described by a formula developed by Planck. In Shockley and Queisser's approach,[1] the cell is also modeled as a black body, but at the typical terrestrial temperature of 300 K. They realised that Planck's formula

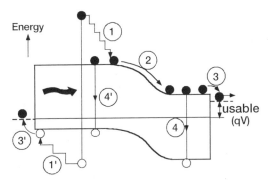

Figure 1. Loss processes in a standard solar cell: thermalization loss (1); junction-voltage (2) and contact-voltage (3) losses; recombination loss (4).

would need to be modified for a device where light was generated by recombination between electrons and holes at different potentials. When open-circuited, the voltage of the ideal cell builds up so that the number of above bandgap photons emitted as part of this voltage-enhanced radiation balances those in the incoming sunlight. At voltages below open-circuit, the number emitted is less, the difference between incoming and outgoing being due to electrons flowing through the cell terminals.

In this way, Shockley and Queisser showed that the performance of a standard cell was limited to 31.0% efficiency for an optimal cell with a bandgap of 1.3 eV. This figure is lower than the 44% previously mentioned since the output voltage of the cell is less than the bandgap potential, with the difference made up by voltage drops at the contact and junction.

These drops can be reduced if sunlight is focused to increase the photon density striking the cell. Under the maximum possible sunlight concentration (46,200 times!), the limiting efficiency increases to 40.8%. However, only direct sunlight can be focused in this way. This is not an issue above the earth's atmosphere. However, sunlight is scattered by this atmosphere so that only about 75% of the light reaching the earth's surface is direct.

As the efficiency under maximal concentration gives the highest numerical value and also applies to the conversion of direct light even when unconcentrated, this efficiency is a useful figure in comparing the ultimate potential of any given approach. This efficiency is also more directly comparable with results from classical thermodynamics.

For example, the conversion efficiency of energy from a source at 6000 K with a sink temperature of 300 K is limited to 95% by the Carnot efficiency, given by $(1 - T_{sink}/T_{source})$. However, this value does not count the photons emitted by the cell as a waste, since it assumes they get back to the sun, helping it to maintain its temperature! A limit that regards these photons as a loss while assuming the process is reversible, as in the Carnot limit, is 93.3%.

3.   Tandem cells

The key recombination loss (process 4 of Fig. 1) largely can be eliminated if the energy of the absorbed photon is just a little higher than the cell bandgap. This leads to the tandem cell, where multiple cells are used with different bandgaps, each converting a narrow range of photon energies close to its bandgap.

Fortunately, just stacking the cells with the highest bandgap cell uppermost as in Fig. 2 automatically achieves the desired filtering. Performance increases as the number of cells in the stack increases, with a direct sunlight conversion efficiency of 86.8% calculated for an infinite stack of independently operated cells.[2]

Having to operate each cell independently is a complication best avoided. Usually cells are designed with their current outputs matched so that they can be connected in series. This constraint reduces achievable performance. More importantly it makes the design very sensitive to the spectral content of the sunlight. Once the output current of one cell in a series connection drops more than about 5% below that of the next worst, the best approach for overall performance is to short-circuit the low output cell, otherwise it will consume power, rather than generate power.

Tandem cells are now in commercial production. Double- and triple-junction cells based on GaInP/GaAs/Ge have been developed for use on spacecraft with terrestrial efficiencies approaching 30%. Quadruple-junction devices with efficiencies approaching 40% are presently under development. Tandem cells are also used to improve the performance and reliability of terrestrial amorphous silicon cells with stabilized efficiencies up to 12% confirmed for triple-junction cells based on the Si:Ge:H alloy system. Modules with efficiencies up to 6-7% are available incorporating double- and triple-junction devices.

4.   Multiple electron-hole pairs

If, instead of giving up their excess energy as heat, the high-energy electron-hole pair used it to create additional pairs, higher efficiency would be possible. Evidence for the creation of more than one pair by high-energy photons is

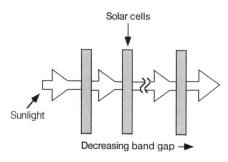

Figure 2. Tandem cell approach.

documented and attributed to impact ionization by the photoexcited carriers.

The limiting efficiency for an idealized cell capable of taking full advantage of this impact ionization effect is calculated as 85.4% for a cell of bandgap approaching zero. This design allows, on energy grounds at least, many electron-hole pairs to be generated by each incident photon. In reality, the measured effect to date is so weak as to produce negligible improvement in device performance. Competing processes for the relaxation of the high-energy photoexcited carriers are too efficient.

A more recent idea is based on Raman luminescence. Raman scattering is a generic term applied to the inelastic scattering of photons (scattering that results in a change in photon energy and also, usually, in direction). Formally, the scattering involves creation of a "virtual" electron-hole pair by the photon in a process that conserves momentum but not necessarily energy. The virtual pair remains viable for a short time determined by the energy imbalance. In Raman luminescence during this period, the virtual pair relaxes by emitting a photon of an energy that differs from that of the original photon by the energy of the generated electron-hole pair.

An analysis of the efficiency of cells based on Raman luminescence gives identical bounds to those based on impact ionization. In principle, 85.4% efficiency is possible from such cells. The difference may prove to be in the practicality of implementation.

5.   Hot-carrier cells

When photoexcited carriers collide elastically with one another, no energy is lost. It is inelastic collisions with the atoms of the cell material that result in an energy loss (through phonon emission). In principle, if such atomic collisions can be avoided during the time it takes a photogenerated carrier to traverse the cell, the energy loss associated with process 1 of Fig. 1 can be avoided.

The various time constants can be appreciated by imagining a direct-bandgap cell illuminated by a short pulse of monochromatic light such as from a laser.

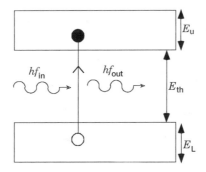

Figure 3. Device based on Raman luminescence.

Such a pulse would create electrons in the conduction band and holes in the valence band of distinct energy and momentum as in Fig. 4. Collisions of these carriers occur in less than a picosecond, tending to smear this distribution.

The peaked distributions will become broader and tend towards a Boltzmann type of distribution. If carriers collide elastically only with carriers of the same type, no energy is lost from this group. The temperature of the "hot carrier" distribution will be determined by the total number of carriers created by the laser pulse and the total energy given to each carrier type. Different temperatures are possible for electrons and holes unless they are efficient at sharing their energy.

In the next phase, collisions with the lattice atoms become important. These result in energy loss (phonon emission). During this phase, the number of electrons and the number of holes remain constant, but the average energy and carrier temperature decrease due to this loss. The temperature of electrons and holes equalize and both are reduced towards that of the host material. Finally, recombination in the semiconductor becomes important. The distributions of electrons and holes retain the same general shape, determined by the ambient temperature, but the number of carriers at each energy is reduced until finally reaching the levels they had prior to the laser pulse.

A standard cell is designed to collect the carriers before they get too far into the latter recombination stage of this decay sequence. A hot-carrier cell has to catch them before the carrier cooling stage. Carriers either have to traverse the cell very quickly or cooling rates have to be slowed in some way. Special contacts to prevent the contacts from cooling the carriers may also be required.[3,4] The limiting efficiency of this approach lies between the values of 85.4% and 86.8% of the previous sections. However, to reach this limiting efficiency carrier cooling rates have to be reduced sufficiently or radiative recombination rates sufficiently accelerated that the latter is faster than the former.

6. Multiband cells

Standard cells rely on excitations between the valence and conduction bands. A recent analysis[5] has shown advantages if a third band, nominally an impurity band,

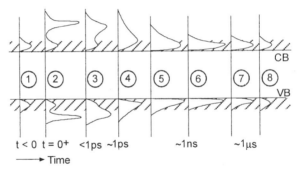

Figure 4. Carrier energy relaxation after a short high-intensity laser pulse at $t = 0$.

is included in the analysis (Fig. 5).

This theory has been extended to an *n*-band cell and additional implementation approaches discussed. These approaches include using excitations between minibands in superlattices, if phonon relaxation processes can be controlled; using semiconductors with multiple narrow bands, such as I-VII and I_3-VI compounds; and using high concentrations of impurities such as rare-earths to form multiple impurity bands in wide bandgap semiconductors.

The limiting efficiency for an *n*-band cell is identical to the 86.8% figure for a large stack of tandem cells. However, the effective cell connections in the *n*-band approach show much more redundancy than a series-connected tandem cell (Fig. 6), suggesting that this approach may be more tolerant to spectral variations in sunlight.

Recent work has also already resolved a controversy as to whether an idealized cell incorporating multiple quantum wells can exceed the efficiency of an idealized standard cell. The structure of Fig. 7 shows a multiple-quantum-well cell that meets all the requirements, in principle, to attain limiting 3-band cell performance. The previous question is now answerable in the affirmative!

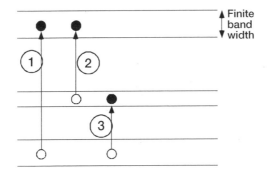

Figure 5. Three-band solar cell.

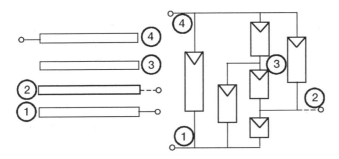

Figure 6. Four-band cell and equivalent circuit.

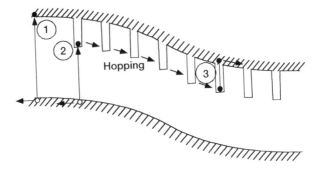

Figure 7. Multiple-quantum-well solar cell meeting the constraints of three-band theory.

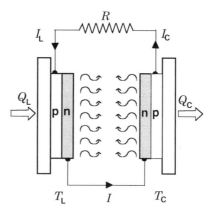

Figure 8. Thermophotonic conversion.

The author has recently extended the multi-band approach to devices with discrete mid-gap impurity levels and shown that most of the benefits also transfer to such "impurity photovoltaic" devices, if properly designed.

7. Thermophotovoltaic and thermophotonic devices

Thermophotovoltaics is a well-established branch of photovoltaics where the light from a heated body other than the sun is used as the illumination. A development of this approach has been dubbed "thermophotonics." In this case, the exponentially enhanced light output of a device where the light is generated by band-to-band recombination is used to advantage. Figure 8 shows the basic arrangement, which is nearly symmetrical. Two diodes acting as solar cells/light emitters face each other, connected by a load. Heat is supplied to one to heat it hotter than the other and heat is extracted from the other to cool it. The devices are optically coupled but thermally isolated. The combination is able to convert heat supplied to the hotter device to electricity in the load with an efficiency approaching the Carnot efficiency, in principle.

The heated device acts as an emitter of narrow bandwidth light that has an energy within kT of the bandgap energy. This near-monochromatic light can be converted very efficiently by the cell. Moreover, light emitted by the cell is recycled back to help drive the light-emitting diode. Since the same current flows in the cell and source diode, the voltage across the diode will be smaller than that across the cell when the diode is at higher temperature. This difference results in power dissipation in the load.

With on-going evolution in device design, both experimental solar cells and light-emitting diodes are approaching the stage where internal recombination can be limited by radiative processes, a prerequisite for this scheme. If used to convert solar radiation in conjunction with a thermal absorber, energy conversion efficiency up to 85.4% is obtainable in principle. Alternatively, the approach could be used for maximally efficient conversion of fossil fuels or waste heat, particularly when the heat is available at low temperature.

8. Conclusion

The author's prognosis is that material costs will, with the fullness of time, push photovoltaic evolution in the direction of the highest possible efficiency. Work to date suggests there is scope for improving solar cell performance by exploring approaches capable of giving efficiencies closer to thermodynamic limits. Low dimensional structures seem to show some promise due to the small dimensions and new features offered.

The new Centre for Third Generation Photovoltaics has commenced operation at the University of New South Wales in early 2000 supported by the Australian Research Council (ARC), with a 9-year program to bring the most promising of these approaches to the "proof-of-concept" level.

References

1. W. Shockley and H. J. Queisser, "Detailed balance limit of efficiency of *pn* junction solar cells," *J. Appl. Phys.* **32**, 510 (1961).
2. A. Marti and G. L. Araujo, "Limiting efficiency for photovoltaic energy conversion in multigap systems," *Sol. Energy Mater. Sol. Cells* **43**, 203 (1996).
3. R. T. Ross and A. J. Nozik, "Efficiency of hot-carrier solar energy

converters," *J. Appl. Phys.* **53**, 3813 (1982).

4. P. Wurfel, "Solar energy conversion with hot electrons from impact ionizaton", *Sol. Energy Mater. Sol. Cells* **46**, 43 (1997).

5. A. Luque and A. Marti, "Increasing the efficiency of ideal solar cells by photon induced transitions at intermediate levels," *Phys. Rev. Lett.* **78**, 5014 (1997).

Infrared Detectors Based on InAs/GaSb Superlattices

M. Razeghi and Y. Wei
Center for Quantum Devices, Dept. of Electrical and Computer Engineering
Northwestern University, Evanston, IL 60208, U.S.A.

G. Brown
Air Force Research Laboratory
Materials Directorate, WPAFB, OH 45433-7707, U.S.A.

1. Introduction

Since the discovery of infrared radiation by Herschel in 1800,[1] many different types of infrared detectors have been developed. Based on different detecting mechanisms and varying in manufacturing cost, they are widely used in industry, medical, and military areas.[2] In general, infrared detectors can be divided into two categories, photon detectors and thermal detectors. Photon detectors can be further divided into intrinsic, extrinsic, free carrier, and quantum well/superlattice detectors. Thermal detectors include ferroelectric/pyroelectric detectors and bolometers.

Type II InAs/GaSb detectors fall into the category of quantum well/superlattice detectors. The development of this type of detector is based on its flexibility to tune the spectral range to cover about 2-µm and longer wavelengths; the unique properties that other material systems do not have, including reduced Auger recombination rate,[3-5] stable material forms, highly uniform growth,[6] and very high-speed response.[7,8] Extrinsic detectors have to

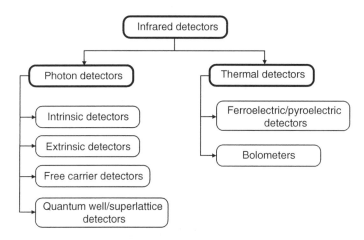

Figure 1. Categorization of infrared detectors.

operate at very low temperatures that require multistage cryocoolers, and are not for high-speed applications. Mercury cadmium telluride (MCT) detectors are the most widely used intrinsic detectors today. However, as the wavelength goes to the VLWIR range, growth nonuniformity and problems caused by high mercury incorporation are drawbacks. Thermal detectors operate at or above room temperature, but their speeds are low.

Quantum well infrared photodetectors (QWIPs) operate at very low temperatures. The idea of type-II band alignment was first put forth by Sai-Halasz and Esaki in the 1970's.[9–11] High quality material growth of type-II InAs/GaSb superlattices has become available only in recent years via MBE and MOCVD. The quality of the superlattices has been improved dramatically during the past few years and high-performance single-element detectors have achieved similar or better performance than MCT detectors.[12–19] In the meantime, other groups are working on type-II optoelectronic devices in the areas of modulators,[20] detectors,[21,22] and lasers.[23,24]

The InAs/GaSb superlattice we choose has a type-II misaligned band line-up. The line-up can be described conveniently by the model solid theory.[25] The valance band of GaSb lies above the conduction band of InAs. The effective bandgap can be smaller than that of either of the two kinds of bulk materials. Calculations have shown that the band gap can range from zero to about 600 meV, which is ideal for devices within the infrared range. The correlations of electron states in different layers form minibands. The minimum of the conduction bands and the maximum of the valance bands lie at the center of the Brillouin zone of the superlattice. This direct bandgap feature is of crucial importance for optoelectronic devices. Due to the misaligned bandgap, the InAs layers become quantum wells for electrons, while the GaSb layers become quantum wells for holes. The effective mass of electrons is much smaller than that of holes. This difference leads to the sensitivity of the effective superlattice bandgap to the InAs

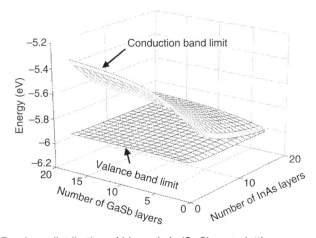

Figure 2. Bandgap distribution of binary InAs/GaSb superlattices.

layer thickness, and the insensitivity to GaSb layer thickness. These observations are confirmed by our empirical tight binding model (ETBM)[26,27] calculations of the superlattice bandgap over a wide range of different layer thicknesses for the binary system InAs/GaSb. A series of InAs/GaSb superlattices can have the same bandgap, but entirely different energy dispersion in the reciprocal space, as illustrated in Fig. 2. Therefore we can engineer the energy bands.

The type-II binary system has a clear advantage in that it is more immune to the interface roughness than the ternary system InAs/GaInSb. Calculations show that the higher the indium mole fraction is, the more sensitive the bandgap is to the layer thickness.[28] Comparison of the results of three different groups shows clear correlations of the indium mole fraction in the GaInSb layers with band broadening. We have achieved the lowest bandgap broadening using the binary system. This result is important for multicolor detectors.

2.   Theory of InAs/GaSb superlattice detectors

The general purpose of modeling a detector is to maximize its detectivity and achieve background limited infrared photodetector (BLIP) performance at higher temperature. This goal remains one of the most important aspects of developing detectors. However, as focal plane array (FPA) implementation becomes feasible, the improvement of the resistivity (R_0A) becomes more and more important.

There are already comprehensive treatments for both photoconductors and photodiodes in the literature.[29] We will briefly discuss the modeling of each type for completeness, and then continue with the modeling of type-II superlattice detectors.

- *Noise mechanisms in photoconductors*
  The major intrinsic noise in photoconductors includes Johnson-Nyquist noise, generation-recombination noise, and $1/f$ noise. The Johnson-Nyquist noise has the following form:

$$V_{J-N} = \sqrt{4kTR\Delta f} , \qquad (1)$$

where k is the Boltzmann constant, T is the absolute temperature, R is the resistance of the detector, and Δf is the bandwidth. For near-intrinsic photoconductors, the generation-recombination noise can be expressed as

$$V_{gr} = \frac{2V_B}{\sqrt{lwt}} \left(\frac{l+b}{bn+p} \right) \sqrt{\frac{np}{n+p}} \sqrt{\frac{\tau \Delta f}{1+\omega^2\tau^2}} , \qquad (2)$$

in which V_B stands for the bias voltage; l, w and t are the length, the width, and the thickness of the photoconductor; τ is the excess carrier lifetime; n and p are the carrier concentrations for electrons and holes; $b = \mu_e/\mu_h$ is the ratio of electron mobility μ_e and hole mobility μ_h; and ω is the operating frequency of the device. The $1/f$ noise is in the form

$$I_{1/f} = \sqrt{\frac{KI_B^{\alpha}\Delta f}{f^{\beta}}} \ , \tag{3}$$

where α and β are exponents, f is the operating frequency, I_B is the bias current, and K is a constant.

- *Noise mechanisms in photodiodes*

The shot noise of the diode current is the main source of internal noise for photodiodes. The $1/f$ noise will be absent when the diode is operated under zero bias. Under the zero-bias condition and thermal equilibrium, the Johnson-Nyquist noise can be written as

$$I_n^2 = 2q(I_D + 2I_S)\Delta f = 4qI_S\Delta f = \frac{4kT}{R_0}\Delta f \ . \tag{4}$$

Here $R_0 = kT/qI_S$ is the differential resistance of the photodiode under zero bias, T is the diode temperature, q is the electron charge, I_D is the diode current, and I_S is the saturation current under reverse bias.

- *Background noise for photoconductors*

The background photon flux Φ_b will affect the carrier concentrations in photoconductors and thus affect the generation-recombination noise. The change in the carrier concentrations can be written as follows:

$$n \rightarrow n_0 + \frac{\eta\Phi_b\tau}{t} \ , \tag{5}$$

$$p \rightarrow p_0 + \frac{\eta\Phi_b\tau}{t} \ , \tag{6}$$

in which n_0 and p_0 are the carrier concentrations under thermal equilibrium, η is the quantum efficiency, and τ is the excess carrier lifetime. At low enough temperature and sufficient bias voltage, the generation-recombination noise will dominate the Johnson-Nyquist noise and BLIP performance will be achieved. For n-type (p-type) material, the BLIP condition is $\eta\Phi_b\tau/t >> p_0$ ($\eta\Phi_b\tau/t >> n_0$).

- *Background noise for photodiodes*

The noise caused by the background photon flux will contribute an independent photocurrent for photodiodes, thus an additional shot noise will be added to the total noise current:

$$I_n^2 = 2q(I_D + 2I_S + I_{ph})\Delta f \ , \tag{7}$$

in which $I_{ph} = q\eta A\Phi_b$ (where A is the detector area). When $2I_S >> I_{ph}$, the detector is thermal noise limited, whereas when $2I_S << I_{ph}$, the detector is background limited (BLIP).

- *Detectivity*

The detectivity is in fact the signal-to-noise ratio expressed in a specific form so as to normalize the size of the detector and the frequency bandwidth. There are different forms of definition for detectivity, but the most widely used forms are

$$D^* = \frac{R_V}{V_n}\sqrt{A\Delta f} \text{ and } D^* = \frac{R_I}{I_n}\sqrt{A\Delta f}, \qquad (8)$$

in which R_V is the voltage responsivity, R_I is the current responsivity, V_n is the noise voltage, and I_n is the noise current. Different forms should be equivalent. Series resistance R_S will decrease the external quantum efficiency of the detector by a factor of $R_0/(R_0+R_S)$, and the measured detectivity will be decreased by a factor of $[R_0/(R_0+R_S)]^{1/2}$. Parallel resistance (surface leakage resistance) R_{surf} will not affect the measured quantum efficiency, but will increase the current noise, and lead to a decrease of detectivity by a factor of $[R_{surf}/(R_0+R_{surf})]^{1/2}$.

- *Modeling of InAs/GaSb superlattices*

The modeling of the superlattice is a fundamental step to design type-II superlattice detectors. We can predict the bandgap, the electron and hole wavefunctions, and the band diagram. More information can be extracted. We have developed two different kinds of techniques for this purpose. One is the well-developed $k \cdot p$ model,[30] and the other one is the empirical tight binding model (ETBM).[26] There exist other models such as the bond-orbital model (BOM) and the linear muffin-tin orbital model (LMTO) for superlattices.

Within the $k \cdot p$ model, we use the envelope function approximation to separate the crystal periodicity from the heterostructure energy envelope. Aperiodic heterostructures can be modeled. The band line-up is modeled by model solid theory and it has shown indirectly very good agreement with extracted data from experiments. For the empirical tight binding model, the superlattice electron wavefunction is built atom after atom. Very small period superlattice and specific interfaces can be modeled accurately. Strain is taken into account by the well-known d^2 scaling rule.[31] The band structure along arbitrary directions within the entire Brillouin zone can be modeled. Typical band diagrams of a type-II superlattice are shown in Fig. 3.

Figure 3(a) shows the calculation from the $k \cdot p$ model, while Fig. 3(b) shows that from the ETBM calculations. An aperiodic $k \cdot p$ modeling of the wave functions is shown in Fig. 4. The allowed bound energy states and corresponding wavefunctions can be calculated numerically. The electrons and holes are less populated near the substrate (or the buffer layer).

The superlattice is assumed to have an averaged effect as an equivalent bulk material for the device modeling. This approach has been shown to agree with experimental results very well. We optimize the device parameters to give the maximum detectivity.[28]

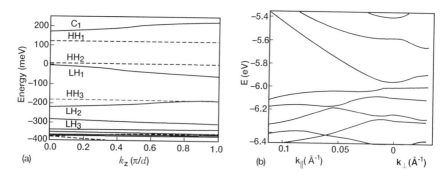

Figure 3. Band-structure of a 21-InAs/13-GaSb superlattice based on the **k·p** model (a).[28] Empirical tight binding model (ETBM) calculation of the band-structure for the same 21-InAs/13-GaSb superlattice (b).

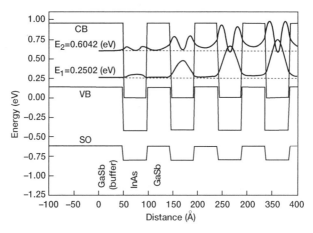

Figure 4. Calculation of the electron wavefunction distribution for an aperiodic structure based on the **k·p** model.[28]

3.   Experimental results

The superlattices are grown by MBE (Intevac Modular Gen II). The binary system we use brings very high bandgap uniformity. We have grown superlattices with cutoff wavelength from 10 μm to over 20 μm. As shown in Fig. 5, the 90% to 10% cutoff energies are all below $2kT$, where T represents the detector temperature, mostly liquid nitrogen temperature. The optical response of a photodiode at 9.5 K is shown in Fig. 6. The cutoff wavelength is around 25 μm. The marked transitions are in good agreement with the calculated values from a four-band **k·p** model.

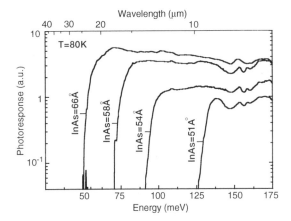

Figure 5. Spectral response of photodiodes with different InAs/GaSb superlattices as the active layer. The thickness of the GaSb layer is 4 nm for all of the superlattices, while the thickness of the InAs layer is shown for each device. The 90%–10% cutoff energies of all of the devices are below $2kT$.

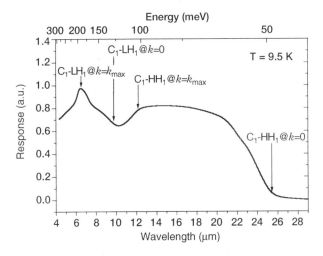

Figure 6. The optical spectral response of a photodiode at $T = 9.5$ K. The arrows indicate transition energies at $k_\perp = 0$ and at the edge of the Brillouin zone in the growth direction k_{max}, as calculated with a four-band model.

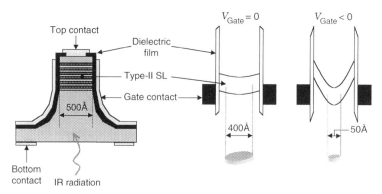

Figure 7. Schematic diagram of the gated pillars with an adjustable lateral confinement.

7. Conclusion

The current progress towards high-performance type-II detectors of practical interest has shown very promising results. Lattice-matched type-II InAs/GaSb superlattices with excellent crystalline quality and surface morphology can now be readily grown. Well-established detector processing techniques make reliable devices and repeatable performance. The realization of focal-plane arrays on type-II InAs/GaSb superlattices becomes feasible, although surface passivation needs to be studied further for small detectors. The background doping level needs to be decreased below 10^{15} cm^{-3} to achieve a higher BLIP performance temperature in the VLWIR range. Anti-reflection coatings also needed to be developed and this move will increase the detectivity by about 30%. On the other hand, integration of a micro-lens would increase the optical area of the detector significantly and improve the detectivity easily by an order of magnitude. Nanometer-sized infrared detectors may also show wavelength tunability and much higher-temperature performance. A schematic diagram of the structure of nanometer-sized detectors is shown in Fig. 7. Simple calculations show that with a moderate voltage bias, the cut-off wavelength is tunable within the 8–12 µm wavelength region. All of this research may lead to the realization of high-quality focal-plane arrays for type-II detectors based on InAs/GaSb superlattices.

Acknowledgments

The authors would like to thank Dr. H. Mohseni for his contribution to this work. We would like to thank Dr. Y. S. Park of ONR, Dr. D. Johnstone and Dr. W. Mitchel of Air Force, for their support, interest, and encouragement. This work is partially supported under Contracts N00014-99-1-0630 and F49620-01-1-0087.

References

1. W. Herschel, "Investigation of the powers of the prismatic colours to heat and illuminate objects; with remarks, that prove the different refrangibility of radiant heat. To which is added, an inquiry into the method of viewing the sun advantageously, with telescopes of large apertures and high magnifying powers," *Phil. Trans. Roy. Soc. London* **90**, 255 (1800).
2. R. Hudson, *Infrared System Engineering*, New York: Wiley, 1969.
3. E. R. Youngdale, J. R. Meyer, C. A. Hoffman, and F. J. Bartoli, "Auger lifetime enhancement in InAs-Ga$_{1-x}$In$_x$Sb superlattices," *Appl. Phys. Lett.* **64**, 3160 (1994).
4. G. Zegrya and A. Andreev, "Mechanism of suppression of Auger recombination processing in type-II heterostructures," *Appl. Phys. Lett.* **67**, 2681, (1995).
5. H. Mohseni, V. I. Litvinov, and M. Razeghi, "Interface-induced suppression of the Auger recombination in type-II InAs/GaSb superlattices," *Phys. Rev. B* **58**, 15378 (1998).
6. H. Mohseni, A. Tahraoui, J. Wojkowski, M. Razeghi, G. J. Brown, W. C. Mitchel, and Y. S. Park, "Very long wavelength infrared type-II detectors operating at 80 K," *Appl. Phys. Lett.* **77**, 1572 (2000).
7. H. Mohseni and M. Razeghi, "First demonstration of high-speed uncooled type-II superlattices for long wavelength infrared detection," *Proc. ISDRS* (1999), pp. 563-564.
8. H. Mohseni, Y. Wei, and M. Razeghi, "High performance type-II InAs/GaSb superlattice photodiodes," *Proc. SPIE* **4288**, 191 (2001).
9. G. A. Sai-Halasz, R. Tsu, and L. Esaki, "A new semiconductor superlattice," *Appl. Phys. Lett.* **30**, 651 (1977).
10. G. A. Sai-Halasz, L. L. Chang, J. M. Welter, and L. Esaki, "Optical absorption of In$_{1-x}$Ga$_x$As-GaSb$_{1-y}$As$_y$ superlattices," *Solid State Commun.* **27**, 935 (1978).
11. G. A. Sai-Halasz, and L. Esaki, "InAs-GaSb superlattice energy structure and its semiconductor-semimetal transition," *Phys. Rev. B* **18**, 2812 (1978).
12. H. Mohseni, E. Michel, M. Razeghi, W. Mitchel, and G. Brown, "Growth and characterization of InAs/GaSb type-II superlattices for long-wavelength infrared detectors," *Proc. SPIE* **3287**, 30 (1998).
13. H. Mohseni and M. Razeghi, "Growth and characterization of InAs/GaSb type-II superlattice for 8–12 µm room temperature detectors," *Proc. Electrochem. Soc.* **98-21**, 170 (1998).
14. H. Mohseni, J. Wojkowski, and M. Razeghi, "Uncooled InAs/GaSb type-II infrared detectors grown on GaAs substrate for the 8–12 µm atmospheric window," *IEEE J. Quantum Electronics* **35**, 1041 (1999).
15. H. Mohseni, A. Tahraoui, J. Wojkowski, M. Razeghi, W. Mitchel, and A. Saxler, "Growth and characterization of very long wavelength type-II infrared detectors," *Proc. SPIE* **3948**, 145 (2000).

16. H. Mohseni, J. Wojkowski, A. Tahraoui, M. Razeghi, G. Brown, and W. Mitchel, "Growth and characterization of type-II non-equilibrium photovoltaic detectors for long wavelength infrared range," *Proc. SPIE* **3948**, 153 (2000).

17. H. Mohseni and M. Razeghi, "Long wavelength type-II photodiodes operating at room temperature," *IEEE Photonics Technol. Lett.* **13**, 517 (2001).

18. H. Mohseni, M. Razeghi, G. Brown, and Y.S. Park, "High-performance InAs/GaSb superlattice photodiodes for the very long wavelength infrared range," *Appl. Phys. Lett.* **78**, 2107 (2001).

19. M. Razeghi and H. Mohseni, "Miniaturization: Enabling technology for the new millennium," *Optoelectonics Rev.* **9**, 101 (2001).

20. H. Xie, and W. I. Wang, " Infrared electroabsorption modulation at normal incidence in asymmetrically stepped AlSb/InAs/GaSb/AlSb quantum wells," *J. Appl. Phys.* **76**, 92, (1994).

21. J. L. Johnson, L. A. Samoska, A. C. Gossard, J. Merz, M. D. Jack, G. R. Chapman, B. A. Baumgratz, K. Kosai, and S. M. Johnson, "Electrical and optical properties of infrared photodiodes using the $InAs/Ga_{1-x}In_xSb$ superlattice in heterojunctions with GaSb," *J. Appl. Phys.* **80**, 1116 (1996).

22. F. Fuchs, U. Weimer, W. Pletschen, J. Schmitz, E. Ahlswede, M. Walther, J. Wagner, and P. Koidl, " High performance $InAs/Ga_{1-x}In_xSb$ superlattice infrared photodiodes," *Appl. Phys. Lett.* **71**, 3251 (1997).

23. B. H. Yang, D. Zhang, Rui Q. Yang, C.-H. Lin, S. J. Murry, and S. S. Pei, "Mid-infrared interband cascade lasers with quantum efficiencies > 200%," *Appl. Phys. Lett.* **72**, 2220 (1998).

24. C. L. Felix, J. R. Meyer, I. Vurgaftman, C. H. Lin, S. J. Murry, D. Zhang, and S. S. Pei, "High-temperature 4.5-μm type-II quantum-well laser with Auger suppression," *IEEE Photonics Technol. Lett.* **9**, 734, (1997).

25. C. G. Van de Walle, "Band lineups and deformation potentials in the model-solid theory," *Phys. Rev. B* **39**, 1871 (1989).

26. J. C. Slater and G. F. Koster, "Simplified LCAO method for the periodic potential problem," *Phys. Rev.* **94**, 1498 (1954).

27. D. N. Talwar, J. P. Loehr, and B. Jogai, "Comparative study of band-structure calculations for type-II $InAs/In_xGa_{1-x}Sb$ strained-layer superlattices," *Phys. Rev. B* **49**, 10345 (1994).

28. H. Mohseni, *Type-II InAs/GaSb Superlattices for Infrared Detectors*, Ph.D. Thesis, Northwestern University, 2001.

29. A. Rogalski, *Infrared Photon Detectors*, Bellingham: SPIE Press, 1995.

30. E. O. Kane, *Physics of III-V Compounds*, New York: Academic Press, 1966.

31. W. A. Harrison, *Electronic Structure and the Properties of Solids*, San Francisco: W. H. Freeman and Company, 1980.

Solid-State Lighting

A. Zukauskas
Institute of Materials Science and Applied Research
Vilnius University, Sauletekio al. 9-III, LT-2040 Vilnius, Lithuania and
Rensselaer Polytechnic Institute, Troy, NY 12180, U.S.A.

M. S. Shur
Dept. of Electrical, Computer and Systems Engineering
Rensselaer Polytechnic Institute, Troy, NY 12180, U.S.A.

R. Gaska
Sensor Electronic Technology, Inc., 21 Cavalier Way, Latham, NY 12110, U.S.A.

1. Introduction

Breakthroughs in artificial light sources — a piece of burning wood invented more than 500,000 years ago, gas lighting (1772), the Agrand lamp with a tubular wick (1874, see Fig. 1), electric lighting (1876, see Fig. 2), the Edison bulb (1879), and

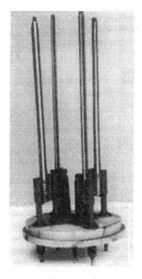

Figure 1. Agrand lamp with tubular wick (English patent No. 1425 of 1784).

Figure 2. The first electric lighting device: Yablochkov candle (1876).

Figure 4. Blue LED on silicon developed by the University of South Carolina and SET, Inc.

Figure 3. Replica of the Edison incandescent lamp.

fluorescent lamps (1938) — have led to the development of modern lighting sources, which are tungsten incandescent and compact fluorescent lamps for residential use, fluorescent lamps for work environments, and "ugly" sodium lamps for street lighting. Figure 3 shows a replica of the Edison bulb, which is not so different from present-day incandescent bulbs.

Today, lighting consumes approximately 2000 TWh annually, which means that about 21% of all energy use is in lighting.[1] Perhaps half of this energy can be saved by switching to efficient and cold solid-state lighting sources. Projected cumulative financial savings from solid-state lighting might reach $115 billion by the year 2020. Solid-state lighting will use visible and ultraviolet (UV) light-emitting diodes (LEDs) that are projected to reach lifetimes exceeding 100,000 hours. The efficiency of white LEDs (now up to approximately 20 lm/W, already twice that of incandescent lamps) is expected to reach 60 lm/W by year 2010. From traffic lights to road signs, from automobile tail-lights to outdoor displays, from landscape to accent lights, solid-state light sources have already arrived as harbingers of the next lighting revolution. Solid-state lighting relies on AlGaInP- and AlInGaN-based semiconductors that can be grown using the mature technology of metalorganic chemical vapor deposition (MOCVD). AlInGaP materials provide for emitters in the red to yellow region. A revolution in lighting technology is emerging with the GaN-based green, blue, and UV emitters initiated by the pioneering work of Pankove,[2] Akasaki,[3] and Nakamura.[4] As an example, Fig. 4 shows a blue GaN-based LED fabricated on a silicon substrate.

In this review we shall discuss the present status of solid-state lighting and make projections of future developments. A more detailed discussion of many issues related to solid-state lighting may be found in our upcoming book.[5]

2.   Current status and future prospects

Figure 5 shows the luminous efficiency for different light sources. Fluorescent lights have some advantage over solid-state lamp (LED) technology. However, fluorescent lights have some other disadvantages. They flicker, are difficult to dim, produce noise, lose power in the electrical ballast, have hazards from the glass parts, and require special disposal means since they contain mercury.

The full advantages of solid-state lighting become apparent on comparing the expected lifetime of different light sources, illustrated in Fig. 6. Whereas white solid-state lighting is still to be developed and perfected, color solid-state sources are already very much better than conventional sources of colored light (see Fig. 7). There are, however, many challenges for the deployment of this technology. The issues include efficient light generation, efficient light extraction, color rendering (*i.e.* the quality of light), and, finally, the especially important issue of cost.

For many years, most manufactured LEDs have employed a primitive design comprising a planar structure on an absorbing substrate encapsulated into an epoxy dome. Such a design results in an extremely poor light extraction efficiency $\eta_{opt} \sim$ 4%, much smaller than typical internal quantum efficiencies. Since the overall

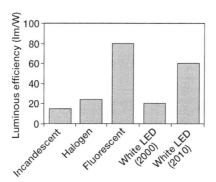

Figure 5. Luminous efficiency of light sources.

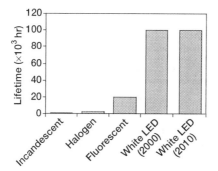

Figure 6. Expected lifetime of light sources.

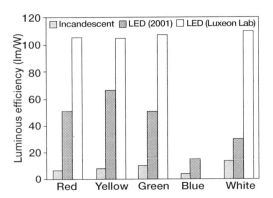

Figure 7. Luminous efficiency of color-filtered incandescent and LED sources.

luminous efficiency of such LEDs is just a few percent, they are able to compete with other light sources only for applications in color indicator lamps and miniature numeric displays, where small dimensions and extended lifetimes are the main advantages. Lighting applications require LEDs with the greatly improved efficiency that is typical for modern high-brightness LEDs.

The main physical reason that makes light extraction difficult is a large ratio of the refractive index of the semiconductor to that of the surrounding medium. Consequently, a major part of the light generated inside the chip is reflected back into the semiconductor. High-brightness LED designs implement additional means for easy photon escape.

An example of such a design is the geometrically deformed LED that was reported by Krames et al.[6] in 1999 and Holcomb et al.[7] in 2000. The devices are made from an epitaxial AlGaInP structure wafer-bonded to a thick GaP substrate. By using a beveled dicing blade, truncated-inverted-pyramid (TIP) chips with sidewall angles of 35° with respect to the vertical were fabricated. The TIP geometry improves light extraction by redirecting totally internally reflected photons from the sidewalls to the top surface or from the top surface to the sidewalls at small incidence angles. At present, the AlGaInP TIP LED is the record semiconductor visible-light source. In the orange region (611 nm), it exhibits the highest reported luminous efficiency, exceeding 102 lm/W with a peak luminous flux of 60 lm. In the red region (652 nm), an external quantum efficiency as high as 55% is achieved under dc operation. Measured luminous efficiencies for AlGaInP/GaP rectangular LEDs, TIP LEDs, and conventional lamps are compared in Fig. 8.

Creating sources of white light is the ultimate goal of solid-state lighting technology. The most challenging application for LEDs is the replacement of conventional incandescent and, probably, even fluorescent lamps. Attempts to upconvert the long-wavelength emission of infrared (IR) LEDs to broader visible

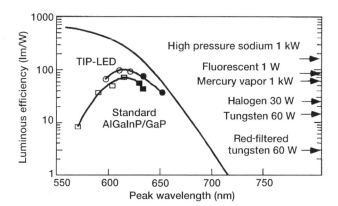

Figure 8. Luminous efficiencies for standard AlGaInP/GaP LEDs, AlGaInP/GaP TIP-LEDs, and various lamps (after Ref. 8). The envelope curve shows luminous efficiency for 100% external quantum efficiency.

spectra were undertaken years ago. However, practical white LEDs became feasible only after the development of high-brightness blue AlInGaN emitters.[9] Based on short-wavelength LEDs, white LEDs that exploit the mixture of two or three colors (dichromatic and trichromatic LEDs, respectively) are being developed. In view of the potential applications, the designs of these solid-state light emitters aim at the combination of high efficiency and high color rendering.

There are basically two ways to produce white LEDs. The first approach is to mix light of different colors emitted by different chips. A white multichip LED (MC-LED) should comprise at least two electroluminescent emitters. Another way is to downconvert the emission from a blue or UV LED to a longer wavelength using phosphors (in the case of a blue LED, a part of the initial emission is used as a component of the white light as well). The number of phosphors involved in a phosphor-conversion LED (PC-LED) may vary depending on the required device characteristics.

The more efficient way to obtain white light is to mix different colors emitted by a few primary LED chips.[10,11,12] The multichip solid-state lamp should be superior in efficiency and longevity to a phosphor-conversion white LED. Besides, the multichip lamp features more flexibility. For instance, pulse-width modulation, a technique widely used in video displays, can adjust the chromaticity of the lamp by varying the average contribution of emissions from the primary LEDs. State-of-the-art red-to-yellow AlInGaP LEDs and blue-green AlInGaN LEDs exhibit quantum efficiencies in excess of 50%[6] and 20%,[13] respectively. Even polychromatic solid-state lamps based on this existing technology are promising sources of white light that are capable of surpassing conventional bulbs and tubes in performance. Improvements in LED technology are expected to make such light sources economically viable, opening a huge potential market for this new "disruptive" technology.

The spectral power distribution (SPD) $S(\lambda)$ of a polychromatic solid-state lamp consists of relatively narrow lines generated by the primary LEDs. An infinite number of combinations of the primary emissions can yield white light of a predetermined chromaticity. In order to maximize the desired quantitative and qualitative characteristics of the light emission, optimization of the SPD is required. Such optimization relies on two figures of merit: the luminous efficacy K and the general color rendering index (CRI). The luminous efficacy represents the amount of visual stimulus per unit power of the radiant flux, whereas the general CRI (R_a) describes the ability of a source to properly reproduce the colors of the illuminated objects. The former is determined by the spectral sensitivity function for photopic vision specified by the International Commission on Illumination (CIE). The latter is obtained by estimating the colorimetric shift with respect to a reference source for the specified reflectivity spectra of the test sample, with the chromatic adaptation of the eye taken into account. Generally, the luminous efficacy and the CRI are somewhat incompatible and the optimization involves a trade-off between these two figures of merit.

Recently, we reported on an optimization technique for the SPD of a white multichip lamp for any number of primary LEDs with arbitrary individual spectra, and the use of this technique for the optimization of quadrichromatic and quintichromatic LED systems.[14,15] The purpose of the optimization is to find the peak wavelengths and the relative emission intensities from primary LEDs that yield the highest values of the objective function $\sigma K + (1-\sigma)/R_a$, where σ is the weight that controls the trade-off between efficacy and color rendering ($0 \leq \sigma \leq 1$).

Figure 9 depicts the calculated optimal boundaries of the phase distribution (K, R_a) obtained for 2, 3, 4, and 5 primary LEDs at the color temperature of 4870

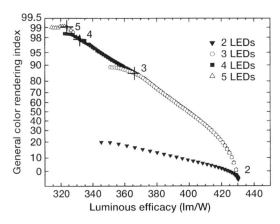

Figure 9. Optimal boundaries of white polychromatic solid-state lamps for 2, 3, 4, and 5 LEDs. Crosses mark the points that are suggested as practical upper limits of CRI for each number of primary LEDs (see spectra in Fig. 10).

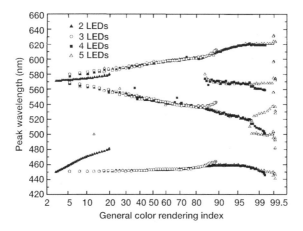

Figure 10. Peak-wavelength variation with color rendering index for optimal solid-state lamps composed of different numbers of primary LEDs.

K (direct sunlight). The width of the individual LED lines was set to 30 nm, which is an average value for AlInGaP and AlInGaN high-brightness LEDs. The contours are seen to partially overlap and a joint optimal boundary is formed. Typically, in the overlap regions, the lamps with a larger number of LEDs exhibit spectra with merged emission lines, so that the actual number of peaks is reduced to that of the lamp with the smallest possible number of LEDs (Fig. 10).

Figure 10 shows the dependences of the peak wavelengths on general CRI for lamps composed of different numbers of primary LEDs. With small exceptions, the dependences are seen to contain continuous regions and abrupt branching to higher numbers of the primary LEDS at certain values of R_a. The trichromatic system evolves from the dichromatic one at R_a values slightly above zero. The fourth branch of wavelengths evolves at approximately 84 points of the general CRI. Finally, at $R_a > 97$, quintichromatic solutions emerge.

The analysis of the joint optimal boundaries and relevant peak wavelengths provides guidelines for assembling polychromatic solid-state lamps. The joint boundary is formed of segments, each of which requires a certain minimal number of primary LEDs. One can see that the introduction of an additional primary source is reasonable at particular points of the joint boundary and that the increase of the color rendering index at the expense the luminous efficacy is continuous in spite of the discrete variation of the number of LEDs. The crosses in Fig. 8 mark suggested practical limits to the highest values of the general CRI for each number of the primary LEDs. It is obvious that the dichromatic lamp makes sense only for the rightmost points of the joint boundary, *i.e.* when the highest efficacy is required with the general CRI close to zero. Although this lamp can yield a CRI up to 20 points, the trichromatic solution offers higher efficacies for $R_a \geq 4$. Similarly, the trichromatic lamp becomes less efficient in comparison with the quadrichromatic lamp when a value of the general CRI in excess of 85 points is required. The quadrichromatic lamp features the SPD with the highest efficacy for CRI values between 85 and 98 points. Introduction of a fifth LED marginally improves the general CRI by one additional point.

Figure 11 presents SPDs of white solid-state lamps composed of 2, 3, 4, and 5 LEDs. The SPDs refer to the suggested practical limits of the color rendering index for each number of primary LEDs (points marked by crosses in Fig. 9). Generally, with increasing n, the empty regions of the spectra are gradually occupied. Finally, the quintichromatic system exhibits a quasi-continuous spectrum. Despite the negligible benefit in color rendering, the quintichromatic lamp (and lamps with a larger number of primary LEDs) is capable of providing supreme-quality white light that might be required for special applications (e.g., for lighting sources for individuals with color vision defects).

It is worth discussing the feasibility of current technology furnishing multichip solid-state lamps with primary LEDs of the required peak wavelengths. As one can see from Fig. 10, a trichromatic lamp with moderate color-rendering ability ($50 < R_a < 85$) requires LEDs with peak wavelengths around 450, 540, and 600 nm. This kind of lamp can exhibit high performance, since it might be composed of efficient blue (450 nm) and green (540 nm) InGaN LEDs and orange (600 nm) AlInGaP LEDs that are available through mature MOCVD technology. For instance, by using InGaN and AlInGaP LEDs with 20% and 50% wall-plug

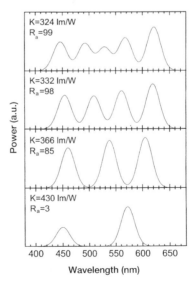

Figure 11. Spectral power distributions of white polychromatic solid-state lamps composed of 2, 3, 4, and 5 LEDs for the points marked by crosses in Fig. 9. Luminous efficacy and general CRI are indicated at each spectrum.

efficiencies, an optimal-SPD trichromatic white lamp with wall-plug performance of 94 lm/W and color rendering of 85 points can be designed. By reducing the value of general CRI to 50 points, the performance may exceed 100 lm/W. Even better performance can be obtained by optimizing the luminous efficiency instead of the luminous efficacy, implying that trichromatic solid-state lamps with luminous efficiency higher than that of fluorescent tubes (typically, around 80 lm/W) are quite plausible.

Implementation of the quadrichromatic lamp, which is capable of providing for "deluxe" lighting with the general CRI in the range 85–98 points might encounter problems associated with the low efficiency of yellow-green LEDs emitting around 570 nm. The quantum efficiencies of 570 nm LEDs offered by both InGaN and AlInGaP technologies do not yet exceed 2%.[16] This low value is due to the direct-to-indirect band crossover in the AlGaInP materials system and to the deterioration of InGaN emissive properties at high In molar fractions.[17] The 570-nm "abyss" is the main obstacle that might hinder the implementation of high-quality multichip solid-state lamps. The same problem is characteristic of the quintichromatic lamp, where the 570 nm line also is important (see Fig. 11). Therefore, our results point out the technological problem that has to be solved for greatly improving the prospects of "deluxe" solid-state lighting.

3.   Applications of solid-state lighting

Owing to improved efficiency, flexibility, device longevity, and environ-mental considerations, solid-state lighting technology gradually penetrates into numerous application fields. In addition to the inherent robustness, small weight, and low maintenance cost of LED lamps, an important property is their low heat production that allows designers to use low-cost low-temperature plastic materials for fixtures and secondary optics. Solid-state local illumination became economically justified with the development of red high-brightness LEDs.   One of the first demonstrations of cost-efficient illumination was the decorative Christmas tree lighting developed by the Lighting Research Center at Rensselaer Polytechnic Institute in 1991.[18] The installation featured four-LED clusters instead of 7-W miniature incandescents.

Since that time, colored illumination has undergone considerable development.   The price charged by the AlGaInP LED suppliers to OEM manufacturers dropped to $0.06/lm in 2000. For a 610-nm LED with 100-lm/W luminous efficiency and a lifetime of 100,000 hours, this rate translates to a price of light of $1.6/Mlm·h that equals the price of light for fluorescent triphosphor and low-pressure sodium lamps and is much lower than that for incandescent lamps. Since plasma-based tubes are not available for wattage below 5 W, AlGaInP LEDs are rapidly occupying the niche of low-wattage red-to-orange illumination. For example, such illumination is required for outlining the contours of buildings and other structures for decorative and advertising purposes.   Contour lighting systems, which are made of LED chains and LED-backlighted plastic tubes, are gradually replacing neon tubes. Despite a higher cost, green, blue, and white LEDs are also penetrating this niche for decorative needs. An example is a 70-m bridge in Duisburg, Germany, illuminated with 140 vertical strips of white LEDs.

Another important application of colored LED illumination is low-wattage security and landscape lighting. All kinds of stairs, cellars, passageways, aisles, park/garden walks, and lanes can be cost-efficiently illuminated by LED fixtures to provide orientation and psychological safety.

Since LED outputs can be confined to narrow angles, inexpensive low-wattage spotlighting can be implemented for all colors. For instance, durable LED-based pocket flashlights have much longer battery life and feature no shock-breakable filament in comparison with their incandescent counterparts. LED spotlights may be useful in museum and gallery applications, causing none of the deterioration of artwork due to heat and UV light that are drawbacks of conventional incandescent and halogen sources. The same considerations are applicable for spotlighting merchandise in retail shops.

LED-based spotlighting and illumination of compact spaces find many applications in automotive interior lighting, such as map-reading lamps incorporated into rearview mirrors. Manufacturers are gradually converting to all-LED interior lighting. For instance, since 2001, all Porsche models are equipped with LED interior orientation lights. LEDs provide illumination of the central console, cockpit, ignition lock, light switch, and door latches.

High-power LED-based local illumination such as floodlighting is also under consideration. A breakthrough of LED-based technology in general lighting is expected to occur when the luminous efficiency of solid-state sources approaches 200 lm/W,[19] more than twice the luminous efficiency of fluorescent lamps. For polychromatic multichip LED systems, such performance can be attained in the near future. However, the breakthrough can occur even at lower efficiencies if environmental and energy-saving considerations become central.

What's next? Are there any other physical principles that can surpass the recombination of electrons and holes in light emission efficiency? Can white light produced by LEDs maintain all quality and quantity standards? How deeply will LEDs penetrate into lighting technology? What amount of effort is required to completely implement solid-state lighting? One of the answers to these questions was given by Professor Nick Holonyak, Jr. who stated that "... it is vital to know that the LED is an ultimate form of lamp, in principle and in practice, and that its development indeed can and will continue until all power levels and colors are realized."[20]

References

1. A. Lidow, "The power conversion process as a prosperity machine. II. Power semiconductor road maps," *Proc. 14th Ann. Applied Power Electronics Conf.,* Vol. 1, Piscataway, NJ: IEEE Press, 1999, p. 10.

2. J. I. Pankove and T. D. Moustakas, eds., *Gallium Nitride (GaN) I*, Vol. 50 in R. K. Willardson and E. R. Weber, eds., *Semiconductors and Semimetals*, New York: Academic Press, 1998.

3. H. Amano, N. Sawaki, I. Akasaki, and Y. Toyoda, "Metalorganic vapor phase epitaxial growth of a high quality GaN film using an AlN buffer layer," *Appl. Phys. Lett.* **48**, 353 (1986).

4. S. Nakamura, T. Mukai, and M. Senoh, "Candela-class high-brightness InGaN/AlGaN double-heterostructure blue-light-emitting diodes," *Appl. Phys. Lett.* **64**, 1687 (1994).

5. A. Zukauskas, M. S. Shur, and R. Gaska, *Introduction to Solid-State Lighting*, New York: Wiley, 2002.

6. M. R. Krames, M. Ochiai-Holcomb, G. E. Höfler, C. Carter-Coman, E. I. Chen, *et al.*, "High-power truncated-inverted-pyramid $(Al_xGa_{1-x})_{0.5}In_{0.5}P/GaP$ light-emitting diodes exhibiting >50% external quantum efficiency," *Appl. Phys. Lett.* **75**, 2365 (1999).

7. M. O. Holcomb, M. R. Krames, G. E. Hofler, C. Carter-Coman, E. Chen, *et al.*, "High-power truncated-inverted-pyramid $(Al_xGa_{1-x})_{0.5}In_{0.5}P$ light-emitting diodes," *Proc. SPIE* **3938**, 77 (2000).

8. M. Holcomb, P. Grillot, G. Höfler, M. Krames, and S. Stockman, "AlGaInP LEDs break performance barriers," *Compound Semicond.* 7 (April), 59 (2001).

9. S. Nakamura and G. Fasol, *The Blue Laser Diode: GaN Based Light Emitters and Lasers*, Berlin: Springer, 1997.

10. G. Bogner, A. Debray, G. Heidel, K. Hoehn, U. Muller, and P. Schlotter, "White LED," *Proc. SPIE* **3621**, 143, 1999.

11. R. Mueller-Mach and G. O. Mueller, "White light emitting diodes for illumination," *Proc. SPIE* **3938**, 30 (2000).

12. A. Zukauskas, M. S. Shur, and R. Gaska, "Light-emitting diodes: Progress in solid-state lighting," *MRS Bull.* **26**, 764 (2001).

13. J. J. Wieret, D. A. Steigerwald, M. R. Krames, J. J. O'Shea, M. J. Ludowise, *et al.*, "High-power AlGaInN flip-chip light-emitting diodes," *Appl. Phys. Lett.* **78**, 3379 (2001).

14. A. Zukauskas, F. Ivanauskas, R. Vaicekauskas, M. S. Shur, and R. Gaska, "Optimization of multichip white solid-state lighting source with four or more LEDs," *Proc. SPIE* **4425**, 148 (2001).

15. A. Zukauskas, R. Vaicekauskas, F. Ivanauskas, R. Gaska, and M. S. Shur, "Optimization of white polychromatic semiconductor lamps," *Appl. Phys. Lett.* **80**, 234 (2002).

16. M. R. Krames, G. Christenson, D. Collins, L. W. Cook, M. G. Craford, *et al.*, "High-brightness AlGaInN light-emitting diodes," *Proc. SPIE* **3938**, 2 (2000).

17. T. Mukai, M. Yamada, and S. Nakamura, "Characteristics of InGaN-based UV/blue/green/amber/red light-emitting diodes," *Jpn. J. Appl. Phys.* **38**, 3976 (1999).

18. S. Peralta and H. Ruda, "Applications for advanced solid-state lamps," *IEEE Ind. Appl. Mag.* **4**, 31 (1998).

19. R. Haitz, F. Kish, J. Tsao, and J. Nelson, "Another semiconductor revolution: This time it's lighting!" *Compound Semiconductor* **6** (February), pp. 34, 36-37 (2000).

20. N. Holonyak, Jr., "Is the light emitting diode (LED) an ultimate lamp?" *Am. J. Phys.* **68**, 864 (2000).

Reduction of Reflection Losses in Nonlinear Optical Crystals by Motheye Patterning

A. Zaslavsky and C. Aydin
Div. of Engineering and *Physics Dept.*
Brown University, Providence, RI 02912, U.S.A.

G. J. Sonek
Optical Switch Corporation, 65 Wiggins Avenue, Bedford, MA 01730, U.S.A.

J. Goldstein
Air Force Research Laboratory, Wright-Patterson AFB, OH 45433, U.S.A.

1. Introduction

Modern microfabrication capabilities have opened a number of new and unexpected avenues for creating novel electronic and optoelectronic devices or modifying the properties of existing ones. Some of these avenues are slated for a major impact on mainstream technology: two classic examples are self-assembled quantum dot lasers in optoelectronics[1] and silicon-on-insulator transistors[2,3] in VLSI circuitry. Other less-heralded approaches are intended for niche applications: the performance of a well-established device can sometimes be enhanced by including a microfabrication step. This contribution will discuss an unusual technique — the so-called "motheye patterning" — that applies the capabilities of interference lithography (IL) and submicron processing to reduce surface reflection losses in nonlinear crystals operating in the mid-infrared.

Nonlinear optical crystals are essential for mid-IR generation using techniques such as optical parametric amplification, optical parametric oscillation, and second harmonic and sum-frequency generation. Mid-IR sources have applications to infrared countermeasures, remote sensing of chemical and biological hazards, and windshear detection. All of these techniques involve the pumping of the nonlinear crystal with high-intensity and high-fluence lasers. Surface reflection losses are extremely undesirable, as they reduce the intensity of the pump-frequency light coupled into the crystal and the output mid-IR frequency light coupled out of the crystal. The standard approach for reducing reflection losses depends on dielectric antireflection (AR) coatings deposited on both the input and output surfaces of the crystal. However, under high-intensity illumination, dielectric AR coatings are plagued by poor adhesion, thermal mismatch, surface contamination, delamination effects under thermal cycling, and absorption at the coating-crystal interface that can lead to thermal runaway and catastrophic damage. All of these problems arise from the mismatch between the AR dielectric and the nonlinear crystal — if the AR coating had the same thermal, optical, and mechanical properties as the crystal, then the threshold for laser damage would be governed by the crystal itself. Hence, integrating the AR coatings directly into the crystal substrate should reduce reflection losses and increase the damage threshold, leading to improved performance. Motheye patterning provides a microfabrication-based technique for precisely this integration. Below we discuss the concept and implementation of motheye patterning on an exemplary nonlinear optical crystal — zinc germanium phosphide (ZGP) — as well as the obtained reduction in surface reflection losses.

2.   Motheye patterning: concept and implementation

The motheye antireflection patterning works by creating a region of gradually varying effective refractive index between air and the ternary nonlinear crystal. The motheye is a periodic sub-wavelength surface relief structure that can produce nearly zero reflectance over a large range of wavelengths and fields of view.[4] The basic motheye geometry is illustrated in Fig. 1(a): a repeated pattern of tapered features is etched into the surface, with the pattern period Λ designed to be smaller than the smallest wavelength of interest. The result is a "transition" region that

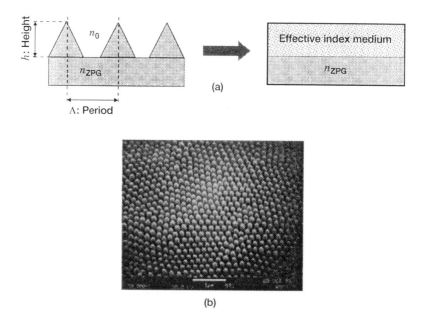

(a)

(b)

Figure 1. A schematic diagram of an etched motheye surface relief structure (a) that acts as a region of graded index of refraction between air and the nonlinear optical crystal (in our case, zinc germanium phosphide). A high-magnification photograph of a real moth's eye (b).[4]

continuously grades the refractive index n from that of the incident medium (in our case air) to that of the nonlinear crystal. Provided the motheye pattern is etched deeply enough with an optimal sidewall profile,[4] light incident on a motheye-patterned substrate should suffer no Fresnel reflection losses. Previously, motheye patterning has been employed to reduce reflection losses in diamond[5] and diamond-coated Ge,[6] (gradient-index AR coatings have also been reported for ZnSe[7]). Since the motheye AR structures are patterned in the nonlinear crystal itself, there is no thermal or physical degradation as in the case of coated dielectric films. These structures are ideal for optical applications where high-power lasers are used and reflection losses are an issue. The specific period, depth, and cross-sectional geometry of the pattern determine the antireflective properties of the motheye, as well as the lower and upper cutoff wavelengths of operation. Curiously, the "motheye" designation arose by analogy with the physiology of the night-flying moth: the periodic rod-like structure shown in Fig. 1(b) reportedly reduces reflection from the eye itself, making the moth less visible to predators.[4]

The nonlinear optical crystal we chose for our motheye patterning was ZGP ($ZnGeP_2$). This material is characterized by a large nonlinear coefficient, natural birefringence, broad optical transparency (1–12 μm), phase match ability over a broad spectral range, high thermal conductivity (180 mW/cm·K), good mechanical properties, and high laser-damage threshold (> 10^3 W/cm^2 cw, 10^6–10^{10} W/cm^2 pulsed),[8] so it is a popular material for the generation of mid-IR radiation. In a typical nonlinear process, a ZGP crystal is pumped by a high-energy laser source (e.g. λ_{IN} = 2.09 μm) on the input crystal surface and emits longer-wavelength broadband tunable radiation (λ_{OUT} = 3–6 μm) from the other (output) surface.[9]

We fabricated our motheye patterns by combining interference lithography (IL), capable of producing periodic patterns with submicron period over a large area, with chlorine chemistry reactive ion etching (RIE) using hardened photoresist as an etch mask. The desired motheye parameters — period Λ and etch height h, see Fig. 1(a) — are obtained from the following expression:[4]

$$\Lambda \leq \frac{\lambda}{\left(n_{ZGP} + n_0 \sin\theta_{MAX}\right)}, \tag{1}$$

where n_{ZGP} = 3.17 and n_0 = 1 label the ZGP and air indices of refraction respectively (since ZGP is birefringent, the larger $n_{ZGP} \sim 3.17$ extraordinary index value should be used), while θ_{MAX} is the acceptance angle. This expression shows that the periodicity of the surface relief structure must be smaller than the shortest wavelength of operation in order to avoid significant diffraction and surface-scatter effects. The $\sin\theta_{MAX}$ term takes into account the possibility that the incident laser beam may have a finite convergence angle. To take into account the finite bandwidth of the pump and signal beams, we included an additional bandwidth margin in the output wavelength specification, e.g. 4.18 ± 0.3 μm. The required height (etch depth) of the motheye structures was taken to be 40% of the longest operating wavelength, i.e. $h \geq 0.4\lambda_{OUT}$.[4] As a result, for the output λ_{OUT} = 4.18 μm we obtain Λ = 1.17 μm and h = 1.8 μm.

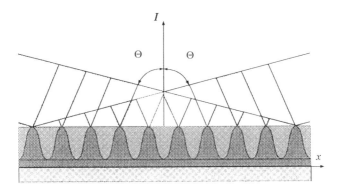

Figure 2. Schematic illustration of interference lithography. Two exposures, with the sample rotated 90° before the second, are needed to produce a regular 2D checkerboard array of features (a hexagonal array is possible with three exposures).

Figure 3. Ideal motheye pattern sidewall profile, described by a fifth-order polynomial,[9] calculated using the actual $n_{ZGP} \sim 3.17$ value.

In order to achieve the necessary repeated features with submicron periodicity, we employed crossed two-beam IL,[10,11] illustrated in Fig. 2. For a given motheye periodicity Λ, the IL parameters are obtained easily from the geometry of Fig. 2. The motheye pattern was subsequently transferred into ZGP using a chlorine-chemistry RIE process, with $SiCl_4$ providing the etching radicals. We developed a process that etched ZGP at a rate of 45 nm/min with an acceptable etch selectivity *vs.* photoresist of 1.3:1 (at 10 mT pressure, 30 W power and dc bias in the neighborhood of −500 V for our home-built parallel-plate electrode RIE system).

The final relevant issue for motheye patterning is the sidewall profile, which serves to grade the effective index of refraction. The ideal index profile has a fifth-order-polynomial (quintic) dependence on depth with a concave sidewall,[12] illustrated in Fig. 3 for the actual refractive index mismatch between air and ZGP. The actual etched sidewall profile depends on the etch anisotropy and selectivity, which respectively determine the mask undercutting and erosion. Two etched sidewall profiles corresponding to the Λ = 1.17 μm motheye, intended for the output surface of the ZGP crystal, are shown in the SEM photographs of Fig. 4.

The differences in the sidewall profile arise from the etch anisotropy, which is a function of the RIE process parameters — we have obtained sidewalls ranging from smoothly sloped and rounded, as in Fig. 4 on the left, to pyramidal as in Fig. 4 on the right. Although neither of the obtained sidewall profiles matches the ideal shown in Fig. 3, the pyramidal shape on the right in Fig. 4 comes closer. Our samples did not achieve the targeted etch depth, etching to only ~1 μm (vs. the intended 1.8 μm). This result indicates that the initial resist layer had a suitable profile, but insufficient thickness, and etched faster than expected. Nonetheless, the reduction in Fresnel reflection losses obtained in these ZGP samples was pronounced, indicated the robustness of the motheye patterning approach.

3.   Reduced reflection losses due to motheye patterning

These motheye-patterned output surfaces (in crystals with uncoated input surfaces) showed a significant increase in transmission over uncoated crystals, shown in Fig. 5. In the best case of the motheye with a straight pyramidal sidewall of Fig. 4(b), the overall transmission reached a maximum of 67% for wavelengths above 3.8 μm. We note that the largest theoretically achievable overall transmission for a ZGP crystal with only one surface patterned is 73%, due to the Fresnel reflection losses at the other interface,[13] not too far from our result in Fig. 5.

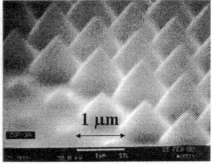

Figure 4. SEM photographs of motheye-patterned ZGP surfaces with smoothly sloped and rounded (left) and straight pyramidal (right) sidewall profiles. The sidewall profile depends on the etch anisotropy and varies with etching parameters.

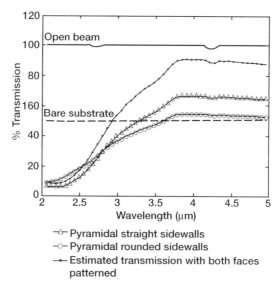

Figure 5. Comparison of the FTIR transmission spectra of ZGP crystals with output faces uncoated *vs.* motheye-coated by pyramidal (straight-sloped) and rounded profile patterns. The line with solid dots shows the estimated transmission spectrum for ZGP with motheye patterning on both input and output faces.

Thus, despite the evident imperfections of our motheye patterning process, the AR properties of the patterned surface are quite good. The estimated transmission curve for a ZGP crystal patterned on both surfaces using our current process is also shown in Fig. 5. The patterning of the input face ($\lambda_{IN} = 2.09\ \mu m$) in that case would, from Eq. (1), require $\Lambda = 607$ nm (well within the reach of modern IL) and $h = 856$ nm (from $h \geq 0.4\lambda_{IN}$). Thus, motheye patterning both the input and output faces with an optimized process should result in a reduction of Fresnel losses comparable to dielectric antireflection coatings. Also from Fig. 5, we find that the broadband transmission is nearly constant above the 3.8-μm cut-off wavelength, but drops off dramatically for $\lambda < 3.8\ \mu m$ because of diffraction effects. Other rounds of etching tests provided evidence that etching parameters and results are quite reproducible. Furthermore, in testing for surface contamination by Auger spectroscopy, we found that Cl, Si, and O are confined to within ~1–2 monolayers at the surface. Finally, as long as the etching runs were interrupted to prevent excessive sample heating, our ZGP material did not exhibit sub-bandgap absorption due to excess vacancies[14] or other damage mechanisms (*e.g.* ion bombardment effects).

4. Conclusions

We have succeeded in demonstrating that motheye surface relief structures of the proper period can be etched into ZGP crystals. It was shown that these structures effectively achieve antireflection functionality by increasing single-surface transmittance. Our motheye etching study revealed that:

1) the required surface periodicity can be achieved with the resist geometry formed by a crossed two-beam exposure, but improved etch selectivity of ZGP *vs.* photoresist is required to achieve the desired motheye etch depth;

2) even though our motheye patterns were shallower than optimal, they did significantly reduce reflection losses; and

3) there is negligible surface contamination from the etching process.

Future work should include motheye patterning of both crystal surfaces and greater motheye etch depth either by achieving better lithographic contrast in the photoresist (*e.g.* by using triple-exposure IL) or by using other masking materials (*e.g.* oxide or metal). Then the laser damage threshold in motheye patterned ZGP crystals should be measured and compared to those with dielectric antireflection coatings. Obviously, our technique could be extended to other ternary nonlinear optical crystals where Fresnel losses are a problem.

Acknowledgments

This work was funded in part by the U.S. Air Force (STTR contract #F33615-99-C-5428). The processing facilities at Brown University are supported by the Brown MRSEC Center (DMR-0079964).

References

1. See, for example, N. N. Ledentsov, "Long wavelength quantum dot lasers: From promising to unbeatable," in this volume.

2. A. J. Auberton-Hervé and M. Bruel, "Why can Smartcut® change the future of microelectronics," in: Y.-S. Park *et al.*, eds., *Frontiers in Electronics: From Materials to Systems*, Singapore: World Scientific, 2000, pp. 131-146.

3. See, for example, F. Allibert, J. Prétet, S. Cristoloveanu, and A. Zaslavsky, "Future silicon-on-insulator MOSFETs: Chopped or genetically modified?", in this volume.

4. S. J. Wilson and M. C. Hutley, "The optical properties of motheye antireflection surfaces," *Opt. Acta* **29**, 993 (1982).

5. A. B. Harker and J. F. DeNatale, "Diamond gradient index motheye anti-reflection surfaces for LWIR windows," *Proc. SPIE* **1760**, 261 (1992).

6. J. F. DeNatale, P. J. Hood, J. F. Flintoff, and A. B. Harker, "Fabrication and characterization of diamond motheye antireflective surfaces on Ge," *J. Appl. Phys.* **71**, 1388 (1992).

7. H. Sankur and W. H. Southwell, "Broadband gradient-index antireflection coating for ZnSe," *Appl. Optics* **23**, 2770 (1984).

Future Trends in Microelectronics

8. Y. V. Rud, "Optoelectronic phenomena in zinc germanium diphosphide," *Semicond.* **28**, 633 (1994).
9. K. L. Vodopyanov, "Parametric generation of tunable infrared radiation in ZnGeP$_2$ and GaSe pumped at 3 μm", *J. Opt. Soc. Am.* **10**, 1723 (1993).
10. H. I. Smith, "Holographic lithography," *Proc. IEEE* **62**, 1361 (1974).
11. S. H. Zaidi and S. R. J. Brueck, "Multiple exposure interferometric lithography," *J. Vac. Sci. Technol. B* **11**, 658 (1993).
12. William Southwell, "Pyramid-array surface-relief structures producing antireflection index matching on optical surfaces," *J. Opt. Soc. Am. A* **8**, 549 (1991).
13. D. E. Zelmon, E. A. Hanning, and P. Schunemann, "Refractive index measurements and new Sellmeier coefficients of zinc germanium phosphide from 2-9 microns with implications for phase matching in optical parametric oscillators," in: M. O. Manasreh, B. J. H. Stadler, I. Ferguson, and Y.-H. Zhang, eds., *Infrared Applications of Semiconductors III, Materials Research Society Symposium Proceedings*, Vol. 607, Boston, 1999, pp. 451-457.
14. S. D. Setzler, P. G. Schunemann, T. M. Pollak, M. C. Ohmer, J. T. Goldstein, *et al.*, "Characterization of defect-related optical absorption in ZnGeP$_2$," *J. Appl. Phys.* **86**, 6677 (1999).

Growth of III-Nitrides on Si(111) and GaN Templates: Challenges and Prospects

M. A. Sánchez-García, E. Calleja, E. Muñoz, F. Calle,
F. B. Naranjo, J. L. Pau, A. Jiménez, S. Fernández, and J. Ristic
Dpto. Ingeniería Electrónica and
Institute for Optoelectronic Systems and Microtechnology (ISOM)
ETSI Telecomunicación, Universidad Politécnica de Madrid
Ciudad Universitaria s/n, 28040 Madrid, Spain

1. Introduction

The growth and fabrication of optoelectronic devices based on III-nitrides (GaN, AlN, and InN) has been a very active field during the last decade.[1] The announcement in 1997 by Nakamura *et al.* of continuous-wave blue laser emission at room temperature with a lifetime above 10,000 hours[2] clearly demonstrated all the capabilities of these materials. Most of this device-quality material has been obtained by metalorganic growth techniques (MOCVD) using sapphire or 6H-SiC substrates. However, the growth of very high quality material on sapphire relies on sophisticated processes, like the epitaxial lateral overgrowth (ELO),[3] whereas the use of SiC substrates is still hampered by their high price. Recently, there is an increasing interest in the growth of III-nitrides on other alternative and inexpensive substrates, like Si(111),[4–6] which also offers the possibility of integration, as well as on thin SiC layers synthesized on top of Si(111) wafers.[7,8] This last choice will add the benefits of bulk SiC at a much lower price.

A second issue might be raised regarding the growth technique employed. Molecular beam epitaxy (MBE) generally leads to low efficiency optoelectronic III-nitride devices when growing heteroepitaxially on sapphire or SiC substrates. However, the use of GaN templates (*i.e.* homoepitaxy) will allow the advantages of the MBE technique to be exploited for clearly improved device performance. One of these advantages is the ultra high vacuum (UHV) environment plus the use of high-purity materials, which minimize the presence of contaminants in the grown layers, in particular carbon, oxygen, and hydrogen. Another important aspect is the low growth temperature, as compared with MOCVD techniques, especially for the growth of InGaN-based material with high In content. The lower growth temperature prevents the thermal desorption of In from the surface of the layer, allowing a higher In mole fraction incorporation and minimizing the In segregation process. Finally, MBE offers very high reproducibility and abruptness control of heterointerfaces, which are key issues to achieve high quality superlattices such as distributed Bragg reflectors (DBRs), and high electron mobility transistors (HEMTs).

The fabrication of GaN-based optoelectronic devices, like ultraviolet (UV) light-emitting diodes (LEDs) and UV photodetectors, grown by MBE on Si(111) substrates, and the demonstration of the device performance improvement when growing homoepitaxially by MBE on GaN templates are the main issues covered in this work.

2. Experimental details

All the structures presented in this paper were grown in a commercial MBE system (MECA2000) equipped with a radio-frequency plasma source from Oxford Applied Research to activate the N_2. Standard Knudsen effusion cells were employed for the rest of the elements (Ga, Al, In, Si, Be, and Mg). The substrate temperature was monitored with an optical pyrometer. The growth chamber also includes a reflection high-energy electron diffraction (RHEED) system for observing *in-situ* the surface reconstruction during growth. More details about the system configuration are given in Ref. 9.

It has been reported previously that the effective group-III/group-V ratio at the surface of the substrate during growth is the most critical parameter controlling the properties (morphological, optical, and electrical) of the layer.[10] Ratios below 1 (*i.e.* N-rich conditions) lead to the formation of nanocolumns (60 to 150 nm diameter) of very high crystal quality, as determined by the intense and narrow (full width at half maximum, FWHM, of 2 meV) excitonic low-temperature photoluminescence (PL) emissions.[6] As the III/V ratio increases towards the stoichiometric value (*i.e.* III/V ratio of 1) the morphology evolves to a compact structure followed by a decrease in the PL efficiency. These changes in growth morphology have been observed independently of the substrate employed and the type of buffer layer used.[6] For device processing it is very critical to maintain the growth conditions slightly above stoichiometry (*i.e.* slightly Ga-rich conditions) in order to avoid the columnar morphology.

3. Growth and device fabrication on Si(111) substrates

The Si(111) substrates were cleaned prior to loading into the MBE system using a modified RCA procedure and mounted in In-free molybdenum holders. After outgassing at 850 °C for 20 minutes, a clear 1×1 RHEED pattern with prominent Kikuchi lines is observed, verifying desorption of the native oxide. Below 750 °C, the typical 7×7 reconstruction from Si(111) can be observed.

Due to the high lattice (17%) and thermal expansion coefficient (37%) mismatches between Si(111) and GaN, the use of an AlN buffer layer is necessary to improve the quality of the GaN films. However, prior to the growth of this buffer layer, a coverage of Al is performed to avoid the formation of an amorphous Si_xN_y layer.[11] Figure 1(a) shows a typical GaN/Si(111) interface without the pre-deposition of Al, where a 2–3 nm thick amorphous layer is clearly

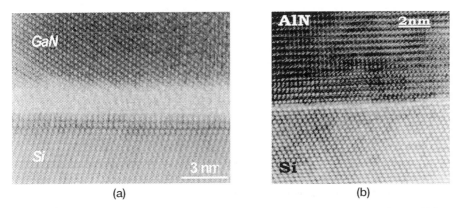

Figure 1. High resolution TEM photographs of the GaN/Si(111) interface without initial Al deposition (a) and the AlN/Si(111) interface with Al predeposition (b).

observed. The interface becomes abrupt and clean when depositing Al prior the activation of the nitrogen plasma (Fig. 1(b)). The AlN buffer layers were grown at 800 °C using a III/V ratio close to stoichiometry, as established in a previous work.[12]

The optimization of the growth parameters (III/V ratio and substrate temperature) and the use of an AlN buffer layer yield high quality GaN layers with intense and excitonic low temperature photoluminescence — see Fig. 2(a). Layers are under biaxial tensile strain, as is expected from the difference in thermal expansion coefficients between Si and GaN. The RMS surface roughness is below 5 nm and XRD values of 8.5 arcmin are obtained, using a $\theta/2\theta$ configuration with open detector — see Fig. 2(b).

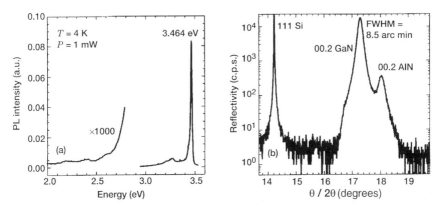

Figure 2. Low-temperature PL spectra (a) and $\theta/2\theta$ scan (open-detector) (b) of a GaN/AlN/Si(111) layer.

The residual carrier concentration of these layers is not easy to determine by standard Hall measurements due to an Al/Si interdiffusion process at the AlN/Si interface.[13] A value of 2×10^{17} cm^{-3} has been obtained using a Schottky barrier. The density of dislocations, as determined by plan-view TEM, is in the range $6 \times 10^9 - 1 \times 10^{10}$ cm^{-2}. N-type doping is achieved with Si, reaching electron concentrations of 2×10^{19} cm^{-3}. P-type doping has been attempted with Be, Mg, and C, yielding a much lower optical activation energy for Be (90 meV) than for the other acceptors,[14] although there is no evidence of real p-type conduction. The low Be substitutional solubility in GaN may be a limiting factor.[15]

The quality of the grown material on Si(111) substrates is confirmed via the growth and fabrication of two common optoelectronic devices, namely, a UV-LED and a UV-photodetector. The LED consists of a single GaN:Mg/AlGaN heterojunction with the following growth sequence: Si(111), 50 nm of AlN (at 820 °C), 0.7 μm Al$_{15}$Ga$_{85}$N n.i.d. (at 760 °C), 0.8 μm GaN:Mg (at 760 °C). The residual n-type concentration of the unintentionally doped AlGaN, measured by means of Ni/Au Schottky barriers, was 5×10^{17} cm^{-3}. The hole concentration on the p-type side was determined using similarly Mg-doped GaN layers grown on sapphire substrates leading to a value of 1.2×10^{17} cm^{-3}. Mesa structures were defined by reactive ion etching (RIE) with SF$_6$ plasma. A Ni/Au contact was deposited on the GaN:Mg layer, while Ti/Al was used for the n-type contact on the AlGaN.[16] Room temperature electroluminescence (EL) spectra under forward bias are shown in Fig. 3 for different injection currents together with a schematic of the LED structure. The dominant component of the spectrum is a near band-edge ultraviolet (UV) emission centred around 365 nm for low current injection (15 mA), with a FWHM value of 8 nm.

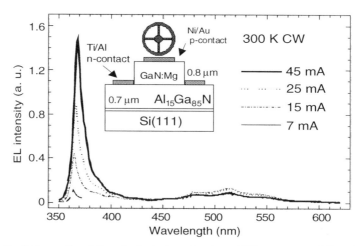

Figure 3. Schematic of the single heterostructure LED grown on Si(111) and its room temperature EL spectra at different injection currents.

The optical power output of the LED was estimated using a pyroelectric radiometer, with a high-pass filter to eliminate the heat radiation from the diode, but without any focusing or converging optics. A value of 1.5 μW was obtained for 35-mA/15-V LED operation. Although the optical quality of the device is still far from the state of the art (*i.e.* blue-violet devices fabricated on sapphire by MOVPE) it represents an improvement for UV-LEDs grown by MBE on silicon substrates.[17]

Schottky-based UV-photodetectors were fabricated using Si-doped AlGaN layers with an electron concentration around 1×10^{18} cm^{-3}. An extended Ti/Al (300-Å/700-Å) bilayer was used as an ohmic contact, while 100-Å thick Au disks were employed as semitransparent Schottky barriers. Figure 4 shows the schematic of this photodetector grown on Si(111) and its responsivity curve indicating a visible/UV contrast ratio higher than 4 orders of magnitude. The normalized responsivity of similar devices grown on sapphire is also shown. The same UV/VIS contrast is obtained for both devices, although that grown on sapphire shows a more abrupt edge. The shoulder observed in the responsivity of the device grown on Si(111) is also present in PL measurements, suggesting that it must be related to the presence of defects in the AlGaN layer. Once again, this result confirms the feasibility of fabricating GaN-based devices on Si(111) substrates with similar properties to those of devices grown on sapphire.[18]

4.   Prospects for the growth on Si(111) substrates

The growth of III-nitrides on thin SiC layers synthesised on top of a Si(111) substrate is an attractive alternative to the growth directly on Si(111), since one

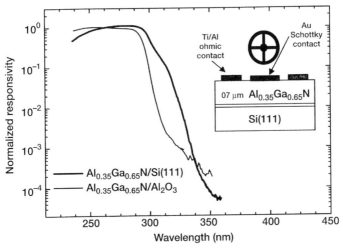

Figure 4. Schematic of the Schottky-based AlGaN photodetector grown on Si(111) and the normalized responsivity compared with a similar device grown on a sapphire substrate.

may benefit from the good lattice match present between SiC and GaN, at a much lower price compared to bulk SiC. In our case, the synthesis of SiC layers is achieved by means of a multiple C ion implantation at 500 °C plus a thermal annealing process at 1150 °C for 6 hours in N_2 atmosphere. Figure 5(a) shows x-ray diffraction rocking curves of the SiC layer before and after the annealing process, indicating a clear improvement. The SiC layer formed is embedded inside the Si(111) wafer, but with the top layers being not stoichiometric, *i.e.* with an excess of C. Reactive ion etching (RIE) with $SF_6:O_2$ is employed to etch this non-stoichiometric layer until stoichiometric SiC is reached (3 minutes of etching is needed to completely remove the non-stoichiometric zone).

The growth of GaN layers was performed at 700–730 °C directly on bare SiC pseudosubstrates with different etching times, *i.e.* 1 and 3 minutes, corresponding to a non-stoichiometric surface and a stoichiometric one, respectively. XRD rocking curves of these GaN layers lead to a FWHM value of 53 arcmin for the layer grown on top of the non-stoichiometric surface and 21 arcmin for the stoichiometric one. Clearly, the stoichiometry of the initial SiC surface plays an important role in terms of the mosaicity of the GaN layer. A second analysis consisted of determining the optical properties of GaN layers grown on stoichiometric SiC layers either directly on the bare substrate or with the use of an AlN buffer layer. Figure 5(b) shows the low temperature PL spectra of these samples, with a similar intensity of the dominant band-edge emission and with no evident effect of the AlN buffer. Both of them show comparable (even larger) intensity than a reference GaN layer grown on Si(111) with AlN buffer layer (also shown in Fig. 5(b)). This result might be surprising, taking into consideration the high level of mosaicity present in the layers grown on SiC as compared with the one grown on Si(111). We suggest that this behaviour might be explained in terms

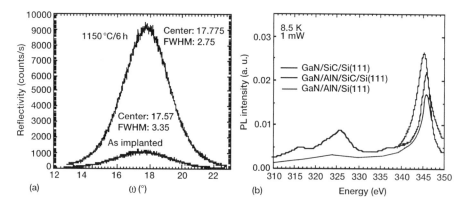

Figure 5. XRD rocking curve of the C-implanted Si substrate before and after annealing (a); low-temperature PL spectra of GaN layers grown on stoichiometric SiC with and without AlN buffer, and a reference GaN/AlN/Si(111) sample (b).

of the size of the microcrystals forming in each GaN layer. TEM measurements must be performed to confirm this hypothesis.

5. Growth and device fabrication on GaN templates

The templates used consisted of 1.5- to 2.0-μm thick GaN layers grown by MOCVD on sapphire. They were Ti-coated (back side) to monitor the growth temperature and cleaned with organic solvents before loading in the MBE system. After being outgassed at 650 °C for 30 min, a bright 1×1 wurtzite RHEED pattern was clearly observed. The MBE growth on the template starts with a thin (200 nm) GaN layer that allows accurate control of the substrate temperature (700 °C) and the stoichiometric III/V ratio by means of the RHEED pattern. Once these conditions are achieved, a clean 2×2 surface reconstruction appears, indicative of a smooth surface and two-dimensional (2D) growth.

It has been proven that MBE grown layers on top of MOCVD GaN templates can replicate or even improve the characteristics of the initial template.[19] Figure 6 shows the low-temperature PL spectra of a GaN template (with no MBE layer grown on top) and that of a GaN epilayer grown on top of that template. Although the intensities of the dominant PL emission at 3.481 eV are very similar in both cases, there is an improvement in terms of FWHM, from 17 meV for the template down to 13 meV for the GaN epilayer. There is also a substantial decrease (close to one order of magnitude) in the intensity of the emission centred around 3.25 eV for the case of the MBE-grown GaN epilayer, which is assigned to carbon contamination. Another characteristic of the MOCVD-grown material is the appearance of the yellow band, which can also be observed in the spectrum of the template but is absent in the MBE-grown GaN layer.

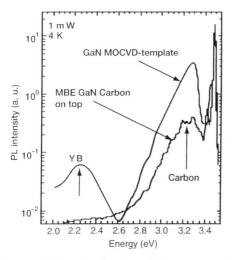
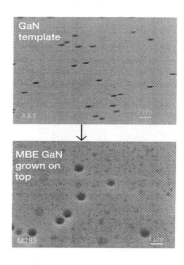

Figure 6. Low-temperature PL spectra and SEM photographs of a GaN template and an MBE-GaN layer grown on top of that same template.

The morphological properties of the GaN template are also replicated. Figure 6 shows scanning electron microscopy (SEM) photographs of the surface of a GaN template and the one corresponding to an MBE-GaN layer grown on top of it. The same hexagonal defects present initially on the template are maintained throughout the growth of the MBE layer. This result indicates that in order to exploit fully the MBE technique, smooth (defect-free) surfaces are needed as the starting point, and that means high-quality GaN templates.

Once the growth parameters are optimized (*i.e.* substrate temperature and III/V ratio), the final assessment of the material quality is performed directly on devices having similar or improved characteristics to those grown by MOCVD on sapphire or SiC. In this respect two complementary structures are presented: an AlGaN/GaN-based distributed Bragg reflector (DBR) and a multi-quantum well (MQW) InGaN-based LED. Both of these structures can be combined to fabricate a resonant cavity LED (RCLED). Finally, the fabrication of a HEMT structure is also described.

The DBR structure presented in this work consists of a 10-period AlGaN/GaN superlattice, where both the Al content (30%) and thickness of each layer (55.5 nm for AlGaN, 52.8 nm for GaN) are designed to obtain a maximum reflectivity centred at 510 nm. During the growth of this structure, the substrate temperature is changed abruptly from 680 °C (for the GaN layers) to 730 °C (for the AlGaN layers) and *vice versa* without any growth interruption. Figure 7(a) shows a cross-sectional transmission electron microscopy (TEM) image of this DBR revealing a periodic structure with clean and abrupt interfaces and very good uniformity. The threading dislocation crossing the whole structure originates at the GaN template and it does not affect the reflectivity nor the periodicity of the reflector. All the dislocations observed come from the template, which indicates a good MBE regrowth process. Figure 7(b) shows the measured reflectivity spectrum of this mirror, obtaining a maximum reflectivity of 48% at 511 nm, in very good

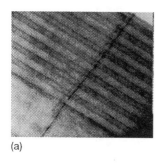

(a)

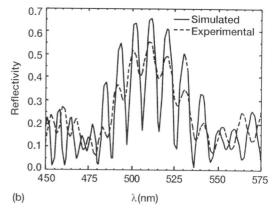

(b)

Figure 7. Cross-section TEM image of a 10-period AlGaN/GaN DBR (a) and its reflectivity spectra (simulated and measured) (b).

agreement with the simulated spectrum (also shown in Fig. 6(b)).

The second device grown on GaN templates is a green-emitting LED using an InGaN-based MQW structure as the active layer of the device. The first step consisted in successfully incorporating enough In (>15%) in the InGaN QW layers. For this task, MBE represents a clear advantage over MOCVD, since the growth temperatures for the latter are much higher (700 °C) than the ones used in MBE (540–600 °C). Initial calibrations on InGaN bulk layers confirmed the In incorporation covering a range between 20 and 30%, from PL and XRD (both symmetric and asymmetric reflections). A second important issue lies in the interface quality of these QWs. Again, the UHV environment and precise growth rate control offered by MBE play an important role.

Figure 8(a) shows a high-resolution TEM (HRTEM) photograph of an InGaN/GaN MQW structure showing clean and abrupt interfaces. The MQW low-temperature PL emission can be tuned between 2.4 eV (516 nm) and 2.7 eV (460 nm) simply by varying the thickness of the InGaN well from 1 to 3 nm and maintaining the GaN barrier thickness fixed at 6 nm. The LED structure is formed by a thin (200 nm) undoped GaN buffer (common to all the structures grown on GaN templates); a 0.8 μm thick GaN:Si layer, a 5-period $In_{0.28}Ga_{0.72}N$/GaN MQW with nominal thicknesses of 20 and 60 Å respectively, and a 0.3 μm thick GaN:Mg layer. A substrate temperature of 660 °C was employed for the growth of the initial thin GaN, the GaN:Si and the GaN:Mg layers. The MQW structure was grown at 550 °C, interrupting the growth and decreasing the substrate temperature after the growth of the first GaN barrier. During the growth of the last GaN barrier the substrate temperature was increased again to 660 °C. Figure 8(b) shows the room-temperature continuous-wave electroluminescence (EL) spectrum of one of these devices, measured at 10 mA / 6.1 V, with an emission centred at 502 nm.

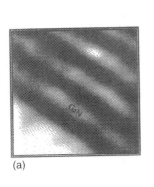

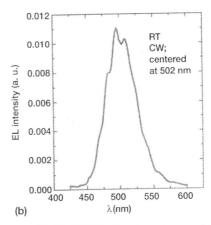

Figure 8.   HRTEM image of an InGaN/GaN MQW structure (a) and room-temperature electroluminescence spectrum of an InGaN-based MQW LED (b).

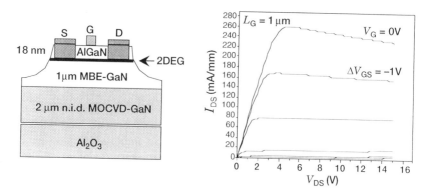

Figure 9. GaN/AlGaN HEMT schematic and current-voltage characteristic.

Finally, the growth and fabrication of an AlGaN/GaN-based HEMT structure on GaN templates is presented. For the growth of this structure a high-resistivity GaN template is needed to avoid parallel conduction, which degrades the optimal performance of the device. The schematic of the device is shown in Fig. 9(a). The growth of an unintentionally doped GaN layer by MBE takes place at 660 °C while the AlGaN layer is grown typically at 720 °C. The source and drain contacts are made with Ti/Al/Au while Pt/Ti/Au is used for a T-shaped gate 1 μm long. A typical transistor output characteristic is shown in Fig. 9(b), with a trans-conductance value of 93 mS/mm and a saturation current of 260 mA/mm.

6.   Conclusions

In this paper we have presented high-quality Ga(Al)N layers grown on Si(111) substrates using AlN buffer layers, leading to the fabrication of optoelectronic devices (UV SHLED and UV Schottky-based photodetector). Synthesis of SiC layers on Si(111) substrates by a multiple C ion implantation allows the growth of GaN films on this type of pseudosubstrate, with similar or even higher luminescence efficiency than equivalent GaN layers grown on Si(111), in spite of the polycrystalline nature of the SiC layer. MBE growth of III-nitrides on MOCVD-GaN templates replicates, or even improves, the quality of the initial template, exploiting the advantages of this growth technique. High quality AlGaN/GaN Bragg reflectors are grown on GaN templates, with abrupt interfaces and free of cracks at the surface, exhibiting high reflectivities. In incorporation in InGaN bulk layers is achieved covering a range of 20–30% and InGaN/GaN MQW-based green LEDs structures are obtained.

Acknowledgments

The authors wish to thank Dr. P. Gibart's group at CHREA (Valbonne, France) and Prof. A. Morante´s group at the Universidad of Barcelona for the provision of MOCVD-GaN templates and SiC/Si(111) substrates, respectively. Partial financial support was provided by IST ESPRIT 1999-10292 AGETHA Project.

References

1. S. Nakamura, "InGaN-based violet laser diodes," *Semicond. Sci. Technol.* **14**, R27 (1999).
2. S. Nakamura, M. Senoh, S. Nagahama, N. Iwasa, T. Yamada, T. Matsushita, H. Kiyoku, Y. Sugimoto, T. Kozaki, H. Umemoto, M. Sano, and K. Chocho, "InGaN/GaN/AlGaN-based laser diodes with modulation-doped strained-layer superlattices," *Jpn. J. Appl. Phys.* **36**, L1568 (1997).
3. T. S. Zheleva, O. H. Nam, M. D. Bremser, and R. F. Davis, "Dislocation density reduction via lateral epitaxy in selectively grown GaN structures," *Appl. Phys. Lett.* **71**, 2472 (1997).
4. F. Semond, B. Damilano, S. Vézian, N. Gradjean, M. Leroux, and J. Massies, "GaN grown on Si(111) substrate: From two-dimensional growth to quantum well assessment," *Appl. Phys. Lett.* **75**, 82 (1999).
5. S. A. Nikishin, V. G. Antipov, S. Francoeur, N. N. Faleev, , G. A. Seryogin, V. A. Elyukhin, H. Temkin, T. I. Prokofyeva, M. Holtz, A. Konkar, and S. Zollner, "High-quality AlN grown on Si(111) by gas-source molecular-beam epitaxy with ammonia," *Appl. Phys. Lett.* **75**, 484 (1999).
6. E. Calleja, M. A. Sanchez-Garcia, F. J. Sanchez, F. Calle, F. B. Naranjo, E. Muñoz, S. I. Molina, A. M. Sanchez, F. J. Pacheco, and R. García, "Growth of III-nitrides on Si(111) by molecular beam epitaxy. Doping, optical, and electrical properties," *J. Cryst. Growth* **201-202**, 296 (1999).
7. A. J. Steckl, J. Devrajan, C. Tran, and R. A. Stall, "SiC rapid thermal carbonization of the (111)Si semiconductor-on-insulator structure and subsequent metalorganic chemical vapor deposition of GaN," *Appl. Phys. Lett.* **69**, 2264 (1996).
8. D. Wang, Y. Hiroyama, M. Tamura, and M. Ichikawa, "Growth of hexagonal GaN on Si(111) coated with a thin flat SiC buffer layer," *Appl. Phys. Lett.* **77**, 1846 (2000).
9. M. A. Sánchez-García, E. Calleja, E. Monroy, F.J. Sánchez, F. Calle, E. Muñoz, and R. Beresford, "The effect of the III/V ratio and substrate temperature on the morphology and properties of GaN and AlN layers grown by molecular beam epitaxy on Si(111)," *J. Crystal Growth* **183**, 23 (1998).
10. M. A. Sánchez-García, F. J. Sánchez, F. B. Naranjo, F. Calle, E. Calleja, E. Muñoz, U. Jahn, and K. H. Ploog, "Crystal morphology and optical emissions of GaN layers grown on Si (111) substrates by molecular beam epitaxy," *MRS Internet J. Nitride Semicond. Res.* **3**, 32 (1998).

of GaN layers grown on Si (111) substrates by molecular beam epitaxy,"*MRS Internet J. Nitride Semicond. Res.* **3**, 32 (1998).

11. M. A. Sánchez-García, E. Calleja, F. J. Sánchez, F. Calle, E. Monroy, D. Basak, E. Muñoz, C. Villar, A. Sánz-Hervás, M. Aguilar, J. J. Serrano, and J. M. Blanco, "Growth optimization and doping with Si and Be of high quality GaN on Si(111) by molecular beam epitaxy," *J. Electron. Mater.* **27**, 276 (1998).

12. M. A. Sanchez-Garcia, E. Calleja, E. Monroy, F. J. Sanchez, F. Calle, E. Muñoz, A. Sanz-Hervas, C. Villar, and M. Aguilar, "Study of high quality AlN layers grown on Si(111) substrates by plasma-assisted molecular beam epitaxy," *MRS Internet J. Nitride Semicond. Res.* **2**, 33 (1997).

13. E. Calleja, M. A. Sánchez-García, D. Basak, F. J. Sánchez, F. Calle, P. Youinou, E. Muñoz, J. J. Serrano, J. M. Blanco, C. Villar, T. Laine, J. Oila, K. Saarinen, P. Hautojarvi, C. H. Molloy, D. J. Somerford, and I. Harrison, "Effect of Ga/Si interdiffusion on optical and transport properties of GaN layers grown on Si(111) by molecular beam epitaxy," *Phys. Rev. B* **58**, 1550 (1998).

14. F. J. Sánchez, F. Calle, M. A. Sánchez-García, E. Calleja, E. Muñoz, C. H. Molloy, D. J. Somerford, J. J. Serrano, and J. M. Blanco, "Experimental evidence for Be shallow acceptor in GaN grown on Si (111) by molecular beam epitaxy," *Semicond. Sci. Technol.* **13**, 1130 (1998).

15. F. Bernardini, V. Fiorentini, and A. Bosin, "Theoretical evidence for efficient *p*-type doping of GaN using beryllium," *Appl. Phys. Lett.* **70**, 2990 (1997).

16. M. A. Sánchez-García, F. B. Naranjo, J. L. Pau, A. Jiménez, E. Calleja, and E. Muñoz, "Ultraviolet electroluminescence in GaN/AlGaN single heterojunction light emitting diodes grown on Si(111)," *J. Appl. Phys.* **87**, 1569 (2000).

17. S. Guha and N. A. Bojarczuk, "Ultraviolet and violet GaN light emitting diodes on silicon," *Appl. Phys. Lett.* **72**, 415 (1998).

18. J. L. Pau, E. Monroy, E. Muñoz, F. B. Naranjo, F. Calle, M. A. Sanchez-Garcia, and E. Calleja, "AlGaN photodetectors grown on Si(111) by molecular beam epitaxy," *J. Cryst. Growth* **270**, 548 (2001).

19. M. A. Sanchez-García, F. B. Naranjo, J. L. Pau, A. Jiménez, E. Calleja, E. Muñoz, S. I. Molina, A. M. Sánchez, F. J. Pacheco, and R. García, "Properties of homoepitaxial and heteroepitaxial GaN layers grown by plasma-assisted MBE," *Phys. stat. sol. (a)* **176**, 447 (1999).

4 THE FUTURE WAY BEYOND SILICON: OTHER PARADIGMS

Contributors

Quantum Computing: A View from the Enemy Camp

M. I. Dyakonov
Laboratoire de Physique Mathématique
Université de Montpellier II, 34095 Montpellier, France

1. Introduction

We are witnessing an overwhelming rush towards "quantum computing." Quantum information research centers are opening all over the globe, funds are generously distributed, and breathtaking perspectives are presented to the layman by enthusiastic scientists and journalists. Many people feel obliged to justify whatever research they are doing by claiming that it has some relevance to quantum computing. The impression is created that quantum computing is going to be the next technological revolution of the 21st century.

The comments below reflect my personal frustration with this state of affairs. I will attempt to dispel the general enthusiasm and put the subject into a proper perspective. No references to original work will be given. The reader may consult the excellent review by Steane[1] on quantum computing with an historical survey of the development of the basic ideas and references (see also Ref. 2). The skeptical remarks due to Landauer[3] and to Haroche and Raimond[4] are also recommended. The so-called quantum cryptography is beyond the scope of this presentation.

Quantum information theory is a respectable branch of mathematics, within which outstanding results, such as the famous Shor's algorithm and error-correction codes, were obtained during the last decade. Still, quoting Steane,[1] who has contributed much to the field, "The quantum computer is first and foremost a machine, which is a theoretical construct, like a thought experiment, whose purpose is to allow quantum information processing to be formally analyzed."

The natural questions are, first, do we need a quantum computer, and, second, can we build a practically working quantum computer in the foreseeable future? Although in my opinion the answer to both questions is negative, I will concentrate on the second one. I will argue that quantum computing relies on storing and processing information within a physical system with a large number of continuous degrees of freedom, and that the requirements for the precision with which this physical system should be manipulated are absolutely unrealistic.

In fact, a quantum computer is a complex system of $n = 10^3$–10^5 particles with two states each. While theoretically it may be quite enlightening to think about a complete description of the system by a grand wave function with its 2^n independent amplitudes, in practice it would be rather difficult to manipulate this wave function by controlled unitary transformations.

A classical example may help to better understand my point. Take an ensemble of many oscillators, say a cubic centimeter of NaCl with its 10^{22} vibrational modes, or an array of LC circuits, and consider the following idea. Let us store information by imposing prescribed amplitudes and phases for all of our oscillators. We can then process this information by switching on external fields and interactions between oscillators. This system would be a classical analog computer, and everyone can see immediately that such a proposal does not make sense for many obvious reasons. Nonetheless, this classical system can very well mimic a quantum computer, although one would need 2^n classical oscillators to simulate a quantum computer with n spins. In spite of this fundamental difference, the basic idea of employing many continuous degrees of freedom for information processing is the same. From the practical point of view, *this* property, not the quantum laws, makes all the difference with a digital computer. It is by no means easier to manipulate the state of 10^5 classical oscillators with the required speed and precision, virtually eliminating relaxation processes within this system, than to fulfill the same requirements for 10^5 quantum spins. In fact, the requirements for a quantum system are more stringent.

One of the main reasons for the interest in a quantum computer is that it should be able to solve problems which are beyond the limits of classical computation. The number of such known problems, like factorizing numbers greater than 10^{130}, is currently between 2 and 10, depending on whether one counts similar problems as distinct, or not. Certainly, the mere fact that such problems exist is of considerable theoretical interest. However, from the practical point of view, their existence may be of interest to a relatively small minority involved in cracking cryptography codes. Can this goal really justify years of efforts by an army of researchers?

In the meantime, many proposals for practical implementations of quantum computing have been advanced, mostly by theorists. Some think that spins of electrons confined in quantum dots and interacting via the two-dimensional electron gas in the regime of the quantum Hall effect will do the job. Others find it more natural to use anyons, hypothetical objects with a statistics intermediate between bosons and fermions that could exist in a two-dimensional space. Less exotic ideas, like using conventional NMR techniques, were also put forward (see Section 3 for a discussion of this idea). Most of the ongoing experimental work is concerned with manipulating and measuring the states of 1–3 individual spins, or atoms. Thus there is some hope that within the next 20 years we will be able to factor the number 6, or even 15, by using Shor's algorithm! It is for the reader to judge whether more exciting possibilities may exist.

2.   What is a quantum computer and how does it work?

The main idea is to replace the classical bit by the quantum bit, or qubit, a quantum two-level system, which can be thought of as the spin-up and spin-down states of a spin-1/2 particle. These states are denoted as $|1\rangle$ and $|0\rangle$, respectively.

The general state of the spin is described by the wave function Ψ, which is a superposition of the two states

$$\Psi = a|0\rangle + b|1\rangle . \tag{1}$$

The complex amplitudes a and b satisfy the normalization condition $|a|^2 + |b|^2 = 1$, so that Ψ is defined by two real parameters. Classically speaking, these could be chosen as two angles describing the orientation of the spin vector. More precisely, it is the orientation of the quantum mechanical average of the spin vector S (which would obey the classical Bloch equation in an external magnetic field). The components of S are expressed through a and b in Eq. (1) as:

$$S_z = (|b|^2 - |a|^2)/2, \quad S_x = (ab^* + a^*b)/2, \quad S_y = (ab^* - a^*b)/2i . \tag{2}$$

It is often stressed that a qubit can contain infinitely more information than a classical bit. The reason is that the classical yes/no switch is replaced by an object with continuous degrees of freedom, the two polar angles determining the orientation of the spin vector. This object certainly remains quantum, since one obtains the results given by Eq. (2) only in measurements averaged over an ensemble of spins, all of which are described by the wave function of Eq. (1). An individual measurement of the spin component along any direction will always give values $+1/2$ or $-1/2$ with certain probabilities. Thus, the probability of having a projection $\pm 1/2$ on the x-axis is $1/2 \pm S_x$.

Quantum computing is based on controlled manipulation of the quantum state of n spins, or n qubits, where, if one wants to have a real computer, not just a demonstration toy, n should be on the order of 10^3-10^5 (the origin of this number will be explained below). Let us begin with a two-spin or two-qubit system. The general wave function is a superposition of four available states

$$\Psi = a|00\rangle + b|01\rangle + c|10\rangle + d|11\rangle \tag{3}$$

with four independent amplitudes, restricted only by the normalization condition. To get a feeling of what this state describes, we must again calculate the average values of the components of S_1 and S_2, but also *correlations* between the two spins, so that we will need average values of products $S_{1x}S_{2y}$, etc. The number of such independent physical quantities is exactly equal to the number of independent real parameters in Eq. (3). Thus, Eq. (3) generally describes a state in which the two spins are correlated, the nature of correlation being defined by the values of amplitudes.

Finally, a general state of n spins is a superposition of 2^n basis states, like $|01001110...\rangle$. Each such state can be labeled by a single number x, which has the corresponding representation in the binary code. Symbolically the wave function can be written as

$$\Psi = \Sigma A_x|x\rangle , \tag{4}$$

where the sum is over all values of x, from 0 to $(2^n - 1)$, each x represents a certain state of n spins (example: $|5\rangle$ stands for $|101\rangle$, if $n = 3$), and A_x are the corresponding amplitudes. In such a state, higher-order correlations between all of the n spins may exist.

Now we can have a definition of a quantum computer, which was basically put forward by Deutsch in 1985 (quoted from Steane[1]). A quantum computer is a set of n qubits in which the following operations are experimentally feasible:

1. Each qubit can be prepared in some known state $|0\rangle$.

2. Each qubit can be measured in the basis $\{|0\rangle, |1\rangle\}$.

3. A universal quantum gate (or set of gates) can be applied at will to any fixed-size subset of qubits.

4. The qubits do not evolve other than via the above transformations.

To get just a glimpse of the beautiful ideas in quantum computing, consider the ingenious trick that is at the heart of quantum algorithms. Suppose we have an integer variable x, between 0 and 2^n-1, and a function $f(x)$, whose values are also integers in the same interval (for simplicity). Assume that there exists an efficient (classical) algorithm for calculating $f(x)$ for a given x, but that if n is large, say 1000, calculation for *all* x would take quite a lot of time. The quantum computer, in a certain sense, performs all these calculations much faster, during a time which is not exponential, but polynomial in n. (I note in passing, that the *difficulty* of building a quantum computer increases exponentially with n, which is a kind of Nature's revenge.)

Here is how it works. Take two sets of n qubits each. One will serve to store the values of x, while the other is for the values of $f(x)$. Start with the initial state $|0\rangle|0\rangle$, which means that all of the spins in both sets are in their down positions. Apply a unitary transformation to change the initial state to

$$\Psi = 2^{-n/2} \Sigma \, |x\rangle \, |0\rangle, \tag{5}$$

where again the sum is over all x, *i.e.* over all 2^n possible states of the first set. These states enter Eq. 5 with identical amplitudes, while the second set remains unchanged in all terms.

Now apply another unitary transformation, U_f, which acts on any pure state $|x\rangle|0\rangle$ as

$$U_f|x\rangle \, | \, 0\rangle = |x\rangle \, | \, f(x)\rangle, \tag{6}$$

where, as before, $| \, f(x)\rangle$ stands for the state of n qubits of the second set, corresponding to the binary representation of the value of $f(x)$. Equation 6 means simply that U_f calculates $f(x)$ for a given x. However, when applied to the superposition in Eq. 5, this transformation gives

$$U_f\Psi = 2^{-n/2} \Sigma \, |x\rangle \, | \, f(x)\rangle. \tag{7}$$

Thus we have obtained a state of the two sets of spins that contains information on the values of $f(x)$ for *all* x. This property is called quantum parallelism. The important point is that in order to construct the transformation U_f, one does not need to pre-calculate the values of $f(x)$ for all x (in which case the procedure would be senseless). Instead, U_f should contain the description of the algorithm to calculate $f(x)$, which is something much simpler. It is the hardware that performs automatically all the 2^n applications of this (classical) algorithm.

If one measures the state of the system described by Eq. 7, one will obtain one of the pairs $(x, f(x))$ with equal probabilities. So, this state is not useful by itself. The idea of quantum algorithms (see Ref. 1 for details) is to perform further unitary transformations with the final goal to enhance the probability of the state containing the answer to our problem.

3. Discussion of feasibility

Certainly, the ideas of quantum computing, which were only very briefly sketched in the previous section, are beautiful and appealing. However it should be fully realized that the practical implementation of these ideas requires complete control over all the intimate details of a many-particle quantum system. I will now discuss the principles of a quantum computer, outlined above, from the practical point of view.

Clearly, these mathematical postulates cannot be fulfilled exactly in any real experimental set-up. So, we must have an idea of the required precision and also of the number n that are needed to outperform the classical computer. If specially designed error-correction codes are used, an estimate for the tolerable noise level gives the order of magnitude of 10^{-5} per qubit per gate.[1] This estimate could be somewhat exaggerated, since only a simplified model of relaxation was used (uncorrelated errors in each qubit). The values of n of practical interest are defined by the maximal numbers that we should be able to store in order to beat the classical computer and also by the large quantity of additional qubits required for error correction (without error correction, a fantastic precision of 10^{-13} is required). From these considerations an estimate $n = 10^3 - 10^5$ is obtained.[1]

While the basic points of the following discussion are independent of the physical nature of qubits, to be more specific, we will use the term "spins" instead of "qubits." Thus, we must have some $10^3 - 10^5$ spins and the operation of our quantum computer depends on the possibility of the following:

1. We must be able to put them in the spin-down state with a precision 10^{-5}. No problem about that: if these are electronic spins we can have this polarization in a magnetic field of 1 T at temperatures below 100 mK. However, this magnetic field will create some problems with postulate 4, to be discussed below. (Here it is assumed that the precision required for the initial polarization, as well as for measurement of the system state, is of the same order as the estimated tolerable noise level. It should be noted that while physically the spins are equivalent, from the computational point of view it is not so: a mistake in the first spin changes the corresponding number x by 2^{n-1}, while a mistake in the nth spin changes it only by 1. This difference does not matter if we do

everything *exactly*, but it seems that it should be taken into account when one is concerned with the effects of errors.)

2.   We must be able to measure simultaneously the state of each spin, *i.e.* its projection on the direction of the magnetic field, presumably with the same precision. This formidable task may be compared with measuring simultaneously *all* the velocities of a thousand atoms in a gas cell. It is very difficult, though not impossible, to measure the state of a single spin. This measurement cannot be done directly, but rather through the influence of our spin on some process, such as photon emission (in this case we must measure the polarization of a single photon with a precision of 10^{-5}, which is unheard of), or transport properties, which involve complex physics and additional interactions. Any measurement scheme will be based on spin-orbital or exchange interaction with the environment, which will make it more difficult to satisfy postulate 4. However, the problem of measuring one spin is infinitesimal compared to the problem of measuring simultaneously 10^5 or even 10^2 spins (remember, with a precision 10^{-5}!).

3.   We must have a universal quantum gate, that is, a device that can transform an arbitrary state of our system into another state of our choice via a unitary transformation of the system wave function.

This universal quantum gate is reminiscent of an old Russian joke. The story goes that during World War II an inventor appeared with an idea of extreme military value. Since at the time quantum cryptography did not yet exist, the inventor insisted that they take him to the very top, that is, to Stalin.

> – *So, tell me what is it about?*
> – *It's simple, comrade Stalin. You will have three buttons on your desk, a green one, a blue one, and a white one. If you press the green button, all the enemy ground forces will be destroyed. If you push the blue button, the enemy navy will be destroyed. If you push the white button, the enemy air force will be destroyed.*
> – *OK, it sounds nice, but how will it work?*
> – *Well, it's up to your engineers to figure it out! I'm just giving you the idea.*

Quantum-mechanically, anything that may happen in this world is a unitary transformation. Thus the wave function of the destroyed ground force is related to that of the intact ground force by a unitary transformation, and the same is true for the navy and the air force. The required apparatus is illustrated in Fig. 1. Before dismissing this analogy, the reader is invited to consider the difficulty of constructing a classical equivalent of the universal quantum gate, which must be

Figure 1. A schematic view of the Supreme Commander's work desk after the installation of the universal military transformation device.

easier. Is there any hope of ever learning to transform an arbitrary state of 10^5 classical oscillators into another prescribed state with a precision 10^{-5}?

It was proved that an arbitrary transformation could be decomposed into a sequence of elementary transformations that influence either individual spins, or all possible pairs of spins. The total number of such operations needed to beat the classical computer in factorizing numbers is estimated as 10^{10}. For each transformation, such as U_f, a special Hamiltonian of our spin system must be designed and implemented. This requirement means that magnetic fields should be applied individually and independently to each spin, but also $n(n-1)/2$ interactions between any two spins should be switched on and off at will.

There are some quite trivial difficulties. An individual spin can be manipulated by applying either static magnetic fields or pulses of high frequency fields under the conditions of spin resonance. Each spin must have an individual set of coils producing such fields, and each spin must be shielded from fields acting on other spins with enormous precision, which means that the inter-spin distance should be large enough (there should also be some space for the coils). Then how can we make them interact? If the field of 1 T that is needed to provide the initial polarization is present all the time, so that we can use spin resonance techniques, there are other problems. It is easy to reverse the spin, by applying a $180°$-pulse, however it is virtually impossible to put the spin in a fixed direction in the xy-plane, because of the continuous fast precession around the direction of the magnetic field. (This point will be explained in other terms below).

This problem makes it impossible to achieve even the first step of quantum algorithms, the transformation, leading to the wave function in Eq. (5). It can be easily seen that this wave function is the joint eigenfunction of all operators S_{1x}, S_{2x}, ... S_{nx}. In other words, Eq. 5 describes a state in which all the spins point in the x-direction, while in the initial state they pointed in the z-direction.

4. Any evolution of the system, apart from the transformations imposed by our quantum gates, should be avoided. Relaxation, or "decoherence" as a source of errors, was studied in many works in the framework of a simplified model of uncorrelated stochastic noise (the effects of noise on different qubits, or on the same qubit at different times, are assumed to be uncorrelated). New and profound ideas of quantum error-correction were proposed,[1,2] with the resulting estimate of 10^{-5} for the required precision, as was used above. There are trivial problems that were not treated. Suppose that the magnetic fields applied to individual spins, or interactions between pairs of spins do not have *exactly* the values they should have, which certainly will be the case for any practical device. This discrepancy would mean that the errors *are* correlated: each time a certain spin, or a certain pair of spins, is manipulated, the same type of error will occur.

The list of similar problems could be enlarged easily. However, there is a more fundamental reason why in practice the postulate 4 will be violated. So long as the energies of states $|0\rangle$ and $|1\rangle$ are different, which is the case for two-level atoms or spins in an external magnetic field, the general superposition in Eq. (4) will not remain unchanged, even if there is no relaxation and no quantum gates are applied. The reason is simple: each amplitude A_x will acquire a time-dependent phase $k\Omega t$, where $\Omega = E/\hbar$, E is the energy difference, t is the time, and k is the number of qubits in the $|1\rangle$ state within a state x. (For example, $k = 3$ for the state $|1011\rangle$, corresponding to $n = 4$, $x = 11$). This time-dependent phase difference is the quantum-mechanical way of saying that each spin performs a fast precession around the direction of the magnetic field. Thus, our 2^n states will be divided into groups with the same phase (same values of k), which are physically similar but computationally quite different (for example, states $|1000\rangle$ ($x = 8$) and $|0001\rangle$ ($x = 1$) fall into one group).

For atomic hyperfine levels, the frequency $\Omega = E/\hbar$ falls in the 10^{10}–10^{12} s^{-1} range, and it is impossible to fix the relative phase of states with different k on such a time scale. As a result, the relative phases will have random values at the moments when our gates are applied, which will completely disorganize the performance of our quantum computer. For electron spins in a magnetic field of 1 T, the corresponding frequency Ω is on the order of 10^{11} s^{-1}. Even if we move the fundamental frequency to an accessible range, say 100 MHz, by lowering the field to 0.001 T and making the temperature 0.1 mK, it still will not help. There always will be higher harmonics with the difference in k on the order of $n = 10^{3}$–10^{5}, whose coherence it would be impossible to preserve. The existence of the higher harmonics with frequencies $p\Omega$ ($p = 1, ..., n$) has a simple physical explanation:

they are the frequencies of evolution of the p-spin correlations of the type $<S_{1x} S_{2x}$ ... $S_{px}>$. Since each spin makes a precession in the xy-plane with a frequency Ω, such a product will have all the harmonics up to $p\Omega$.

It appears that the only way to avoid this fast and unwanted evolution of our wave function is to have qubits with degenerate or quasi-degenerate states. This approach means that we will have to abandon the spin-resonance method and switch off the initial field of 1 T. This switching will take a long time, during which some relaxation could take place. Also the switching-off process will, most probably, perturb our spin system. Besides, we will have to do this switching on and off each time we need the state $|00000...\rangle$ for the next step of calculations. After we have made the spin states degenerate, we will have to use quasi-static magnetic fields to independently rotate individual spins in the desired direction with the desired precision (please, design the experimental set-up).

It seems that at this stage it would be pointless to discuss numerous additional problems related to the possibilities of controlling interactions between any two spins, which is indispensable for quantum computing.

4.   The classical limit of a quantum computer

There is no impenetrable wall between quantum and classical physics. Quantum effects may gradually disappear when some parameter is changed. Under certain conditions, even such small objects as electrons are very well described by Newton's laws. For a solid-state physicist, a familiar example is provided by magnetotransport phenomena. At room temperature, transport is well described by the classical Drude theory. As the temperature is lowered, small Shubnikov-de Haas oscillations appear, which are due to the quantization of the electron energy in a magnetic field. In the strong magnetic field limit one enters the extreme quantum Hall effect regime. Note that the condition for a classical description to be valid is *not* the requirement that the Landau energy spacing be smaller than $\Delta E = \hbar/\tau$, where τ is the relaxation time. Their ratio is a classical parameter $\omega_c\tau$, where ω_c is the cyclotron frequency. This parameter, which enters the classical Drude formulas, may be quite large and still the classical description will hold. The true criterion is the ratio between the Landau spacing and the thermal energy, kT. The classical description fails when this ratio becomes large enough.

In view of these remarks, the following questions seem to be justified. What is the classical limit of a quantum computer? What are the conditions for achieving this limit? These are difficult questions, which were never addressed properly by the quantum computing community. Apparently, most people in the field believe that the classical limit of a quantum computer is a digital computer, employing switching between spin-up/spin-down states as a classical bit, and that this limit is achieved when the coherent superposition of states in Eq. (4) is destroyed by relaxation processes. This limit is supposed to occur when the energy spacing between the spin-up and spin-down states, due to an applied magnetic field, becomes less than $\Delta E = \hbar/\tau$. (Another way of expressing this idea is to say that the classical limit is achieved when the off-diagonal elements of the density matrix disappear. This statement is not correct. First, a matrix that is diagonal in one basis may become non-diagonal in another basis. Secondly, it is the non-diagonal density matrix that transforms into the classical distribution function of our system in the classical limit, see, for example the Wigner density matrix.) Just as in the example above, the quantity $\Omega\tau$, where Ω is the spin precession frequency in the external magnetic field, is a *classical* parameter (it does not contain the Planck constant) describing the quality of the spin resonance, so that quantum *vs.* classical behavior cannot be chosen depending on the value of this parameter. One need not go to the over-damped situation ($\Omega\tau < 1$) to insure classical behavior.

In my opinion, the classical limit of the quantum computer is a classical analog machine that stores information in the mutual orientations of a large number of classical magnetic moments (spins). Evidently, the practical possibility of building such a machine is questionable, to say the least. However it is much, much easier than to build a true quantum computer.

To shed more light on these questions, let us for a moment replace the two-state qubit ($S = 1/2$) by a "many-state qubit" (high spin $S \gg 1$). For example for $S = 9/2$, which is the spin value for the ^{113}In nucleus, one has $(2S+1) = 10$ states. This change will not affect the basic ideas for quantum computing, except that now we will have to abandon the binary code for writing numbers in favor (for example) of the familiar decimal code. It is well known that for high quantum numbers the behavior of any quantum system approaches that of a corresponding classical system. Thus, under application of external fields, various interactions, and so on, the hardware of our high-spin quantum computer should behave very similarly to a classical system of spins. However, under ideal conditions it should still remain a quantum computer.

The paradox is resolved by taking into account the effects of finite temperature, finite precision of our unitary transformations, or finite accuracy of our measurements. As soon as the precision becomes such that we cannot distinguish between neighboring states m and $m+1$, where m is the projection of S on a chosen axis (though we can still distinguish between states for which $|m_1-m_2| \gg 1$), we will see purely classical behavior. This path is the usual one from quantum towards classical mechanics. A tennis ball in a rectangular one-dimensional potential well obeys textbook quantum mechanics. For high energy levels the wave function is a standing wave, but since the wavelength is smaller than the precision of any imaginable instrument, we will see the classical distribution function, which is uniform throughout the well. (Note again that the classical limit does not require that the energy spacing be smaller than $\hbar/\tau$, or in other words, that the tennis ball density matrix be diagonal.) But at this point the quantum computer will stop working as such, since we will be making random mistakes of ± 1 in each digit of all the numbers written in our $(2S+1)$-code. Instead, we will have an analog machine consisting of classical spins.

In this context, some remarks concerning the proposals to use NMR techniques for quantum computing are in order. At first glance, the advantage of nuclear spins is in their extremely long relaxation times, due to the very small value of nuclear magnetic moments and consequently to their poor coupling to fluctuating magnetic fields. However, for the same reason it is extremely difficult to obtain full nuclear polarization and it is practically impossible to measure the state of an individual nuclear spin. In order to overcome these difficulties, it is suggested to use not an individual spin, but rather a cluster of, say, 10^{15} nuclear spins, which may be weakly polarized. A thousand such clusters are supposed to play the role of the qubits.

In fact, the idea is completely misconceived. A weakly polarized cluster of many nuclear spins behaves like a classical spin and cannot be represented by a wave function given by Eq. (1). Accordingly, a measurement of the spin per nucleus will always give the small average value with a 100% probability (more precisely, with an uncertainty $10^{-15/2}$). The situation would not change even if the nuclear polarization were to reach 99.99%, which is the reason one actually does not need quantum mechanics to understand most of the phenomena pertaining to NMR or, more generally, to the average spin of large ensembles. Many clusters would work as a classical analog machine, described above, but by no means like a quantum computer.

5. Conclusions

We should make a clear distinction between the quantum computer of the abstract information theory and the working quantum computer, between a thought experiment and a real experiment. For a mathematician, rotating a vector in a huge Hilbert space is a routine algebraic operation, while doing it in practice may be possible or not, depending on the required precision, the dimension of the space, and other conditions imposed by the real physical world.

Modern ideas linking quantum mechanics and information theory are quite fascinating, providing better insight into both theories, but this advance is not related in any way to the possibility of building a working quantum computer.

The practical performance of a quantum computer is based on manipulating on a microscopic level and with an enormous precision a many-particle physical system with continuous degrees of freedom. Obviously, for large enough systems, either quantum or classical, this task becomes impossible, which is why such systems belong to the domain of statistical, not microscopic physics. The question is whether a system of $n = 10^3$–10^5 quantum spins, the number needed to beat the classical computer in solving a limited special class of problems, is large enough in this sense. Can we ever learn to control the 2^n amplitudes defining the quantum state of such a system? Based on the previous analysis, my answer is *no, never*.

There is an intermediate link between a quantum and a digital computer, which is a classical analog machine, such as a system of 10^5 oscillators or classical spins. Why not try to build such a machine first, and then, when we see that it works, try to accomplish the orders-of-magnitude more difficult task of constructing a quantum computer with 10^5 quantum spins? I do not believe that even such a classical analog machine of sufficient complexity will ever work. The fashion of quantum computer building will gradually die away, the sooner the better. Research in atomic and spin physics is interesting and useful on its own and need not be justified by irresponsible projects and promises.

Acknowledgments

I am indebted to Andrew Steane for illuminating correspondence. I have also benefited from discussions with Sergei Meshkov, Gérard Mennessier, and Serge Luryi.

References

1. A. Steane, "Quantum computing," *Reports Prog. Phys.* **61**, 117 (1998); LANL e-print quant-ph/9708022.
2. D. Bouwmeester, A. Ekert, and A. Zeilinger, eds., *The Physics of Quantum Information*, Berlin: Springer, 2000.
3. R. Landauer, "The physical nature of information," *Phys. Lett. A* **217**, 188 (1996).
4. S. Haroche and J-M. Raimond, "Quantum computing: Dream or nightmare?" *Physics Today* **49**, August (1996), p. 51.

Entanglement and Quantum Gate Operations with Spin-Qubits in Quantum Dots

John Schliemann
Dept. of Physics, The University of Texas, Austin, TX 78712, U.S.A.

Daniel Loss
Dept. of Physics and Astronomy, University of Basel
Klingelbergstrasse 84, CH-4056 Basel, Switzerland

1. Introduction

Quantum entanglement is one of the most intriguing features of quantum mechanics.[1-4] In the beginning of modern quantum theory, the notion of entanglement was first noted by Einstein, Podolsky, and Rosen,[5] and by Schrödinger.[6] While in those days quantum entanglement and its predicted physical consequences were (at least partially) considered as an unphysical property of the formalism (a "paradox"), the modern perspective on this issue is very different. Nowadays quantum entanglement is seen as an experimentally verified property of nature providing a resource for a vast variety of novel phenomena and concepts such as quantum computation, quantum cryptography, or quantum teleportation.

While the basic notion of entanglement in pure quantum states of bipartite systems (Alice and Bob) is theoretically well understood, fundamental questions are open concerning entanglement in mixed states (described by a proper density matrix)[7-10] or the entanglement of more than two parties.[8,11-17] The most elementary example for entanglement in a pure quantum state is given by a spin singlet composed of two spin-1/2 objects (qubits) owned by Alice and Bob (A and B, respectively):

$$(1/\sqrt{2}) \; [|\!\uparrow\rangle_A \otimes |\!\downarrow\rangle_B - |\!\downarrow\rangle_A \otimes |\!\uparrow\rangle_B] \; . \tag{1}$$

The state of such a combined system cannot be described by specifying the states of Alice's and Bob's qubits separately. It is a standard result of quantum information theory[1] that this property does not depend on the basis chosen in Alice's or Bob's Hilbert space. As we shall see below, the entanglement of such a quantum state (quantified by an appropriate measure) is invariant under (independent) changes of basis in both spaces.

Physically measurable consequences of quantum entanglement of the above kind arise typically (but not exclusively) in terms of two-body correlations between the subsystems. In this case the effects of entanglement typically can be cast in terms of so-called Bell inequalities,[18] whose violation manifests the presence of entanglement in a given quantum state. Using this formal approach the physical existence of quantum entanglement (as opposed to classical correlations) has been unambiguously verified for the polarization states of photons by Aspect and co-workers.[19] Moreover, quantum entanglement is an essential ingredient of algorithms for quantum computation,[2,3] in particular for Shor's algorithm for decomposing large numbers into their prime factors.[20] This problem is intimately related to public-key cryptography systems such as RSA encoding, which is widely used in today's electronic communication.

Among the many proposals for experimental realizations of quantum information processing, solid-state systems have the advantage of offering the prospect for integration of a large number of quantum gates into a quantum computer once the single gates and qubits are established. Recently, a proposal has been put forward involving qubits formed by the spins of electrons living on semiconductor quantum dots.[21-25] In this scenario, the indistinguishable character of the electrons leads to entanglement-like quantum correlations that require a description different from the usual entanglement between distinguishable parties (Alice, Bob, ...) in bipartite (or multipartite) systems. In such a case the proper statistics of the indistinguishable particles has to be taken into account.

In this article we give an elementary introduction to the notion of quantum entanglement between distinguishable parties and review the aforementioned proposal for quantum computation with spin-qubits in quantum dots. The indistinguishable character of the electrons whose spins realize the qubits gives rise to further entanglement-like quantum correlations. We summarize recent results on the characterization and quantification of these quantum correlations, which are analogues of quantum entanglement between distinguishable parties.[24,26-30]

2. Quantum entanglement between distinguishable parties

We now give an introduction to the basic concepts of characterizing and quantifying entanglement between distinguishable parties. We concentrate on pure states (*i.e.* elements of the joint Hilbert space) of bipartite systems. We then comment only briefly on the case of mixed states (described by a proper density operator), and entanglement in multipartite systems.

One of the most prominent examples of an *entangled state* was already given in the previous section, namely a spin singlet built up from two qubits. More generally, if Alice and Bob own Hilbert spaces H_A and H_B with dimensions m and n, respectively, a state $|\psi\rangle$ is called *nonentangled* if it can be written as a product state,

$$|\psi\rangle = |\alpha\rangle_A \otimes |\beta\rangle_B , \qquad (2)$$

with $|\alpha\rangle_A \in H_A$, $|\beta\rangle_B \in H_B$. Otherwise $|\psi\rangle$ is entangled. The question arises whether a given state $|\psi\rangle$, expressed in some arbitrary basis of the joint Hilbert space $H = H_A \otimes H_B$, is entangled or not, *i.e.* whether there are states $|\alpha\rangle_A$ and $|\beta\rangle_B$

fulfilling Eq. (2). Moreover, one would like to quantify the entanglement contained in a state vector.

An important tool to investigate such questions for bipartite systems is the biorthogonal Schmidt decomposition.[1] It states that for any state vector $|\psi\rangle \in H$ there exist bases of H_A and H_B such that

$$|\psi\rangle = \sum_{i=1}^{r} z_i \left(|a_i\rangle \otimes |b_i\rangle \right) , \qquad (3)$$

with coefficients $z_i \neq 0$ and the basis states fulfilling $\langle a_i|a_j\rangle = \langle b_i|b_j\rangle = \delta_{ij}$. Thus, each vector in both bases for H_A and H_B enters at most only one product vector in the above expansion. As a usual convention, the phases of the basis vectors involved in Eq. (3) can be chosen such that all z_i are positive. The expression (3) is an expansion of the state $|\psi\rangle$ into a basis of product vectors $|a\rangle \otimes |b\rangle$ with a minimum number r of nonzero terms. This number ranges from one to $\min\{m, n\}$ and is called the *Schmidt rank* of $|\psi\rangle$.

With respect to arbitrary bases in H_A and H_B, a given state vector can be represented by

$$|\psi\rangle = \sum_{a,b} M_{ab} |a\rangle \otimes |b\rangle , \qquad (4)$$

with an $m \times n$ coefficient matrix M. Under unitary transformations U_A and U_B in H_A and H_B, respectively, M transforms as

$$M \rightarrow M' = U_A M U_B^T , \qquad (5)$$

with U_B^T being the transpose of U_B. The fact that there are always bases in H_A and H_B providing a biorthogonal Schmidt decomposition of $|\psi\rangle$ is equivalent to stating that there are matrices U_A and U_B such that the resulting matrix M' consists of a diagonal block with only nonnegative entries while the rest of the matrix contains only zeroes. For the case of equal dimensions of Alice's and Bob's spaces, $m = n$, this result is also a well-known theorem of matrix algebra.[31]

Obviously, $|\psi\rangle$ is nonentangled, *i.e.* a simple product state, if and only if its Schmidt rank is one. More generally, the Schmidt rank of a pure state can be viewed as a rough characterization of its entanglement. However, since the Schmidt rank is by construction a discrete quantity, it does not provide a proper quantification of entanglement. Therefore finer entanglement measures are desirable. For the case of two distinguishable parties, a useful measure of entanglement is the von Neumann entropy of partial density matrices constructed from the pure-state density matrix $\rho = |\psi\rangle\langle\psi|$:[32]

$$E\big(|\psi\rangle\big) = -\text{tr}_A\big(\rho_A \log_2 \rho_A\big) = -\text{tr}_B\big(\rho_B \log_2 \rho_B\big) , \qquad (6)$$

where the partial density matrices are obtained by tracing out one of the sub-systems, $\rho_{A/B} = tr_{B/A}\rho$. With the help of the biorthogonal Schmidt decomposition of $|\psi\rangle$ one shows that both partial density matrices have the same spectrum and therefore the same entropy, as stated in Eq. (6). In particular, the Schmidt rank of

$|\psi\rangle$ equals the algebraic rank of the partial density matrices. Thus, $|\psi\rangle$ is nonentangled if and only if the partial density matrices of the pure state $\rho = |\psi\rangle\langle\psi|$ are also pure states, and $|\psi\rangle$ is maximally entangled if its partial density matrices are "maximally mixed", $i.e.$ if they have only one non-zero eigenvalue with a multiplicity of $\min\{m, n\}$.

It is important to observe that the entanglement measure[6] of a given state $|\psi\rangle$ does not depend on the bases used in Alice's and Bob's Hilbert spaces to express this state, because the trace operations in the definition of $E(|\psi\rangle)$ are invariant under an eventual change of bases (performed, in general, independently in both spaces). Therefore, entanglement in bipartite systems is a basis-independent quantity.

Thus, the problem of characterizing and quantifying quantum entanglement for pure states in bipartite systems can been seen as completely solved. Unfortunately, the situation is much less clear for mixed states,[7-10] and for multipartite entanglement. The main obstacle in the latter issue is the fact that the biorthogonal Schmidt decomposition in bipartite systems does not have a true analog in the multipartite case. For details we refer the reader to the research literature; a nonexhaustive collection of recent papers includes Refs. 8 and 11–17.

3.   Quantum computing with electron spins in quantum dots

We will now illustrate the phenomenon of quantum entanglement using the example of a specific (possible) realization of a quantum information processing system.[21] The proposal discussed below deals with qubits realized by the spins of electrons residing on semiconductor quantum dots. As we shall see in this and the following section, the indistinguishable character of the electrons gives rise to quantum correlations that are beyond entanglement between distinguishable parties.

An array of coupled quantum dots, shown in Fig. 1, with each dot containing a top-most spin 1/2, was found to be a promising candidate for a scalable quantum computer[21] where the quantum bit (qubit) is defined by the spin 1/2 on the dot. Quantum algorithms can then be implemented using local single-spin rotations and the exchange coupling between nearby spins, shown confined to dots using electrostatic gating in Fig. 1. These electrons can be moved by electrical gating into the magnetized or high-g layer, producing locally different Zeeman splittings. Alternatively, magnetic field gradients can be applied, as produced by a current wire (shown on the left of the dot array). Then, since every dot-spin is subjected to a different Zeeman splitting, the spins can be addressed individually, $e.g.$ through ESR pulses of an additional in-plane magnetic ac field with the corresponding Larmor frequency $\omega_L = g\mu_B B_\perp/h$. Such mechanisms can be used for single-spin rotations and the initialization step. The exchange coupling between the quantum dots can be controlled by lowering the tunnel barrier between the dots. In this figure, the two rightmost dots are drawn schematically as tunnel-

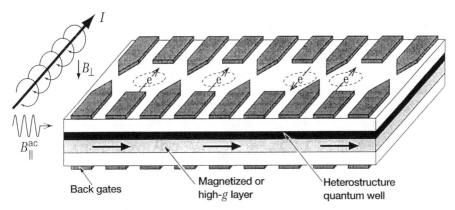

Figure 1. Quantum dot array, controlled by electrical gating. The electrodes (dark gray) define quantum dots (circles) by confining electrons The spin-1/2 ground state (arrow) of the dot represents the qubit. The current-carrying wire on the left can supply a normal magnetic field $B_\perp$.

coupled. Such an exchange mechanism can be used for the *xor* gate operation involving two nearest-neighbor qubits. The *xor* operation between distant qubits is achieved by swapping (via exchange) the qubits first to a nearest-neighbor position. The read-out of the spin state can be achieved via spin-dependent tunneling and SET devices,[21] or via a transport current passing the dot.[44] Note that all single- and two-spin operations, and spin read-out, are controlled electrically via the charge of the electron and not via the magnetic moment of the spin. Thus, no control of local magnetic fields is required, and the spin is used only for storing the information. This spin-to-charge conversion is based on the Pauli principle and Coulomb interaction and allows for very fast switching times (typically picoseconds). A further advantage of this scheme is its scalability to an array of arbitrary size. This proposal is supported by experiments where Coulomb blockade effects,[33] tunneling between neighboring dots,[34,33] and magnetization[35] have been observed, as well as the formation of a delocalized single-particle state in coupled dots.[36] For a detailed review of quantum computing with electron spins in quantum dots see Ref. 23.

The charge of the electron can be used further to transport a spin-qubit along conducting wires.[37] This transport allows one to use spin-entangled electrons as Einstein-Podolsky-Rosen (EPR)[5] pairs, which can be created (*e.g.* in coupled quantum dots or near a superconductor-normal interface), transported, and detected in transport and noise measurements.[37–39] Such EPR pairs represent the fundamental resources for quantum communication.[40]

The electron spin is a natural candidate for a qubit since its spin state in a given direction, $|\!\uparrow\rangle$ or $|\!\downarrow\rangle$, can be identified with the classical bits $|0\rangle$ and $|1\rangle$, while an arbitrary superposition $\alpha|\!\uparrow\rangle + \beta|\!\downarrow\rangle$ defines a qubit. In principle, any quantum two-level system can be used to define a qubit. However, one must be able to control coherent superpositions of the basis states of the quantum computer, *i.e.* no transition from quantum to classical behavior should occur. Thus, the coupling of the environment to the qubit should be small, resulting in a sufficiently large decoherence time T_2 (the time over which the phase of a superposition of $|0\rangle$ and $|1\rangle$ is well-defined). Assuming weak spin-orbit effects, the spin decoherence time T_2 can be completely different from the charge decoherence time (a few nanoseconds), and in fact it is known[41] that T_2 can be orders of magnitude longer than nanoseconds. Time-resolved optical measurements were used to determine T_2^*, the decoherence time of an ensemble of spins, with T_2^* exceeding 100 ns in bulk GaAs.[41] More recently, the single-spin relaxation time T_1 (generally $T_1 \geq T_2$) of a single quantum dot attached to leads was measured via transport to be longer than a few μs,[42] consistent with calculations.[43]

Let us now consider a system of two laterally tunnel-coupled dots having one electron each. Using an appropriate model,[22,24] theoretical calculations have demonstrated the possibility of performing two-qubit quantum gate operations in such a system by varying the tunnel barrier between the dots. An important point to observe here is the fact that the electrons whose spins realize the qubits are indistinguishable particles.[24] In contrast to the usual scenario of distinguishable parties (Alice, Bob, ...) the proper quantum statistics has to be taken into account when a finite tunneling between the dots is inferred.[24,26,27]

In the following section we give an elementary introduction to the theory of "entanglement-like" quantum correlations in systems of indistinguishable particles. We concentrate on the fermionic case and illustrate our findings for the above example of coupled quantum dots.

4. Quantum correlations between indistinguishable particles

For indistinguishable particles a pure quantum state must be formulated in terms of Slater determinants or Slater permanents for fermions and bosons, respectively. Generically, a Slater determinant contains correlations due to the exchange statistics of the indistinguishable fermions. As the simplest possible example consider a wavefunction of two (spinless) fermions,

$$\Psi(r_1, r_2) = (1/\sqrt{2}) \{\phi(r_1)\chi(r_2) - \phi(r_2)\chi(r_1)\} , \qquad (7)$$

with two orthonormalized single-particle wavefunctions $\phi(r)$ and $\chi(r)$. Operator matrix elements between such single Slater determinants contain terms due to the antisymmetrization of coordinates ("exchange contributions" in the language of Hartree-Fock theory). However, if the moduli of $\phi(r)$ and $\chi(r)$ have only vanishingly small overlap, these exchange correlations will also tend to zero for any physically meaningful operator. This situation is generically realized if the supports of the single-particle wavefunctions are essentially centred around locations being sufficiently apart from each other, or the particles are separated by a sufficiently large energy barrier. In this case the antisymmetrization present in Eq. (7) has no physical effect.

Such observations clearly justify the treatment of indistinguishable particles separated by macroscopic distances as effectively distinguishable objects. So far, research in quantum information theory has concentrated on this case, where the exchange statistics of particles forming quantum registers could be neglected, or was not specified at all.

The situation is different when the particles constituting qubits are close together and possibly coupled in some computational process. This is the case for all proposals of quantum information processing based on quantum dots.[21–25] Here qubits are realized by the spins of electrons living in a system of quantum dots. The electrons have the possibility of tunneling eventually from one dot to the other with a probability that can be modified by varying external parameters such as gate voltage or magnetic field. In such a situation the fermionic statistics of electrons and the associated Pauli principle clearly are essential.

Additional correlations in many-fermion systems arise if more than one Slater determinant is involved, that is, when there is no single-particle basis such that a given state of N indistinguishable fermions can be represented as an elementary Slater determinant (*i.e.* a fully antisymmetric combination of N orthogonal single-particle states). These correlations are the analog of quantum entanglement in separated systems and are essential for quantum information processing in non-separated systems.

As an example consider a "swap" process exchanging the spin states of electrons on coupled quantum dots by gating the tunneling amplitude between them.[22,24] Before the gate is turned on, the two electrons in the neighboring quantum dots are in a state represented by a simple Slater determinant, and can be regarded as distinguishable since they are separated by a large energy barrier. When the barrier is lowered, more complex correlations between the electrons due to the dynamics arise. Interestingly, as shown in Refs. 22 and 24, during such a process the system necessarily must enter a highly correlated state that cannot be represented by a single Slater determinant. However, the final state of the gate operation, similarly to the initial one, is given essentially by a single Slater determinant. Moreover, by adjusting the gating time appropriately, one can also perform a "square root of a swap," which turns a single Slater determinant into a "maximally" correlated state in much the same way.[24] Illustrative details of these processes will be given below. At the end of such a process the electrons can again be viewed as effectively distinguishable, but are in a maximally entangled state in the usual sense of distinguishable separated particles. In this sense the highly correlated intermediate state can be viewed as a resource for the production of entangled states.

In the following we give an elementary introduction to recent results in the theory of quantum correlations in systems of indistinguishable particles.[24,26–29] These correlations are analogs of entanglement between distinguishable particles. However, to avoid confusion with the existing literature and in accordance with Refs. 26, 27, and 29 we shall reserve the term "entanglement" for separated systems and characterize the analogous quantum correlation phenomenon in nonseparated systems in terms of the Slater rank and the correlation measure to be defined below.

For the purposes of this article we shall concentrate on elementary results for the case of pure states of two identical fermions. Results for mixed states and more than two fermions can be found elsewhere,[26,27] as can the results for the case of identical bosons.[28,29,27]

We consider the case of two identical fermions sharing an n-dimensional single-particle space H_n resulting in a total Hilbert space $A(H_n \otimes H_n)$ with A denoting the antisymmetrization operator. A general state vector can be written as

$$|w\rangle = \sum_{a,b=1}^{n} w_{ab} f_a^+ f_b^+ |0\rangle , \qquad (8)$$

with fermionic creation operators f_a^+ acting on the vacuum $|0\rangle$. The antisymmetric coefficient matrix w_{ab} fulfills the normalization condition:

$$\mathrm{tr}(\overline{w}w) = -\tfrac{1}{2} , \qquad (9)$$

where the bar stands for complex conjugation. Under a unitary transformation of the single-particle space,

$$f_a^+ \to U f_a^+ U^+ = U_{ba} f_b^+ , \qquad (10)$$

w transforms as

$$w' \to UwU^{\mathrm{T}} , \qquad (11)$$

where U^{T} is the transpose (not the adjoint) of U. For any antisymmetric $n \times n$ complex matrix w there is a unitary transformation U such that $w' = UwU^{\mathrm{T}}$ has nonzero entries only in 2×2 blocks along the diagonal.[26,31] That is,

$$w' = \mathrm{diag}[Z_1, \ldots, Z_r, Z_0] \text{ with } Z_i = \begin{bmatrix} 0 & z_i \\ -z_i & 0 \end{bmatrix} , \qquad (12)$$

$z_i \neq 0$ for $i \in \{1, \ldots, r\}$, and Z_0 being the $(n-2r) \times (n-2r)$ null matrix. Each 2×2 block Z_i corresponds to an elementary Slater determinant in the state $|w'\rangle$. Such elementary Slater determinants are the analogs of product states in systems consisting of distinguishable particles. Thus, when expressed in such a basis, the state $|w\rangle$ is a sum of elementary Slater determinants where each single-particle basis state enters not more than one term. This property is analogous to the biorthogonality of the Schmidt decomposition discussed above. The matrix of Eq. (12) represents an expansion of $|w\rangle$ into a basis of elementary Slater determinants with a minimum number r of non-vanishing terms. This number is analogous to the Schmidt rank for the distinguishable case. Therefore we shall call it the *(fermionic) Slater rank* of $|w\rangle$,[26] and an expansion of the above form a *Slater decomposition* of $|w\rangle$.

We now turn to the case of two fermions in a four-dimensional single-particle space. This case is realized in a system of two coupled quantum dots hosting in total two electrons that are restricted to the lowest orbital state on each dot. In such a system, a simple correlation measure can be defined as follows:[24,26] for a given state described by Eq. (8) with a coefficient matrix w_{ab} one defines a dual state $|w^*_{ab}\rangle$ characterized by the dual matrix

$$\tilde{w}_{ab} = \frac{1}{2} \sum_{c,d=1}^{4} \varepsilon^{abcd} \overline{w}_{cd} , \tag{13}$$

with ε^{abcd} being the usual totally antisymmetric unit tensor. Then the *correlation measure* $\eta(|w\rangle)$ can be defined as

$$\eta(|w\rangle) = |\langle \tilde{w} | w \rangle| = \left| \sum_{a,b,c,d=1}^{4} \varepsilon^{abcd} w_{ab} w_{cd} \right| = |8(w_{12}w_{34} + w_{13}w_{42} + w_{14}w_{23})| . \tag{14}$$

Obviously, $\eta(|w\rangle)$ ranges from zero to one. Importantly, it vanishes if and only if the state $|w\rangle$ has the fermionic Slater rank one, *i.e.* $\eta(|w\rangle)$ is an elementary Slater determinant. This statement was proved first in Ref. 24; an alternative proof can be given using the Slater decomposition of $|w\rangle$ and observing that

$$\det w = \left[\langle \tilde{w} | w \rangle / 8 \right]^2 . \tag{15}$$

The quantity $\eta(|w\rangle)$ measures quantum correlations contained in the two-fermion state $|w\rangle$ that are beyond simple antisymmetrization effects. This correlation measure is in many respects analogous to the entanglement measure "concurrence" used in systems of two distinguishable qubits.[45] These analogies are discussed in detail in Ref. 27, including also the case of indistinguishable bosons. An important difference between just two qubits, *i.e.* two distinguishable two-level systems, and the present case of two electrons in a two-dot system is that in the latter system both electrons eventually can occupy the same dot while the other is empty. Therefore the total Hilbert space is larger than in the two-qubit system, and a generalized correlation measure becomes necessary. Furthermore, as in the two-qubit case, the correlation measure η defined here for pure states of two fermions has a natural extension to mixed fermionic and bosonic states.[26,27]

The expansion of the form of Eq. 12 for a two-fermion system has a recently derived analog in two-boson systems.[28,29] Moreover, the fermionic analog of the biorthogonal Schmidt decomposition of bipartite systems was also used earlier in studies of electron correlations in Rydberg atoms.[46]

We note that the aforementioned double occupancies have temporarily given rise to some controversy about the suitability in principle of such systems as quantum gates. These concerns were eliminated in a recent theoretical study — see Ref. 24 and references therein.

Let us now have a closer look at a specific quantum gate operation, namely the swap process described above. This operation interchanges the contents of the qubits on two dots, for example,

$$|\uparrow\downarrow\rangle \rightarrow |\downarrow\uparrow\rangle , \tag{16}$$

where obvious notation has been used for the spin state on each dot. As we shall see below, the "square root" of such a swap operation provides an efficient way to generate entangled states. Moreover, the "square root of a swap" can be combined with further single-qubit operations to produce an exclusive-*or* (*xor*, or controlled-*not*) gate, which has been shown to be sufficient for the implementation of any quantum algorithm.[47]

Both the initial and the final state in the above example of a quantum gate operation are single Slater determinants. In the beginning and the end of the operation the tunneling amplitude between the dots is exponentially small. The swap process is performed by temporarily gating the tunneling with a pulse-shaped time dependence as shown in Fig. 2. In the presence of a finite tunneling amplitude, *i.e.* during the swap operation, a finite probability for both electrons being on the same dot necessarily occurs. However, this double occupancy probability can be suppressed very efficiently *in the final swapped state* provided that the dynamics of the system is sufficiently close to its adiabatic limit. In fact, this quasi-adiabatic regime is remarkably large.[24] As a result, a clean swap process can be performed even if the tunneling pulse is switched on and off on a time scale close to the natural time scale of the problem given by $\hbar/U_H$ where U_H is an effective repulsion between electrons on the same dot. In the middle of the swap process the system is in a highly correlated quantum state with the correlation measure η being close to its maximum.

Next let us look at the "square root of a swap", which is obtained from the

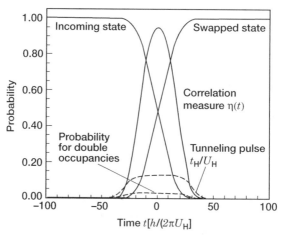

Figure 2. A swap process as a function of time. The tunneling amplitude $t_H(t)$ is plotted in units of the effective repulsion U_H between electrons on the same dot. The square amplitude of the incoming state $|\uparrow\downarrow\rangle$ and the outgoing state $|\downarrow\uparrow\rangle$ are shown as thick lines. The probability to find both electrons on the same dot is necessarily finite *during* the swap process but exponentially suppressed after it. The measure of entanglement $\eta(t)$ is also shown.

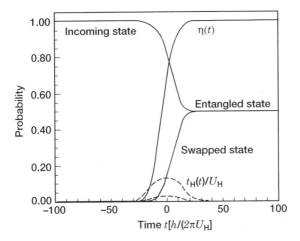

Figure 3. A square root of a swap, which is obtained from the situation of Fig. 2 by halving the pulse duration T. The probability of double occupancies is again strongly suppressed after the tunneling pulse. The resulting state is a fully correlated complex linear combination of the incoming state $|\uparrow\downarrow\rangle$ and the outgoing state $|\downarrow\uparrow\rangle$ of the full swap. The quantum mechanical weights of the latter states are plotted as thick solid lines.

situation of Fig. 2 by halving the pulse duration T. The probability of double occupancies again is strongly suppressed after the tunneling pulse. As shown in Fig. 3, the resulting state is a maximally correlated ($\eta = 1$) complex linear combination of the incoming state $|\uparrow\downarrow\rangle$ and the outgoing state $|\downarrow\uparrow\rangle$ of the full swap with both states having the same weight. The quantum mechanical weights of the latter states are plotted as thick solid lines.

After the tunneling amplitude is switched off again to exponentially small values, both dots carry one electron each. As explained above, due to the high tunneling barrier between the dots, the two electrons in this situation can be considered as effectively distinguishable. In this sense the resulting state in Fig. 3 can be seen as a usual entangled state for distinguishable parties. However, *during* the gate operation such a view is not possible, since there is necessarily a finite amplitude for doubly occupied dots. Since the amplitude of such spin singlet states contributes to the correlation measure $\eta(t)$, the intermediate state during the gate operations shown in Figs. 2 and 3 can, loosely speaking, be interpreted to contain spin as well as orbital entanglement. At the end of the square root of the swap, however, the correlations are due purely to the spin degree of freedom. As a result, the double-dot two-qubit system is also an efficient entangler.

5. Summary

We have given an elementary introduction to the notion of quantum entanglement between distinguishable parties. Entanglement phenomena can be illustrated using the example of the recently proposed spin-qubits in quantum dots.[21] As long as each dots carries one (valence) electron only with high barriers to the neighboring dots, the particles constituting the qubits can be seen as effectively distinguishable, and the usual concepts and theory of quantum entanglement applies. However, two-qubit quantum gate operations in such systems are performed by temporarily lowering the tunneling barriers. In such a situation, the indistinguishable character and proper statistics of the electrons have to be taken into account. Therefore, the question arises how to describe "entanglement" (or, more precisely, quantum correlations analogous to entanglement between distinguishable parties) in systems of indistinguishable particles. In this article we have provided a simple introduction to this kind of question and reported on some elementary results. Interesting questions for further research include experimental manifestations of entanglement-like quantum correlations between fermions using the full antisymmetrized (or, in the case of bosons, symmetrized) Fock space. In particular, possible generalizations of Bell inequalities to the case of indistinguishable particles might complement the approach outlined in this article and suggest experimental studies and applications.

Acknowledgments

We thank D. Bruss, G. Burkard, J. I. Cirac, K. Eckert, M. Kus, M. Lewenstein, A. H. MacDonald, W. K. Wootters, L. You, and P. Zanardi for useful discussions. This work was supported by the Deutsche Forschungsgemeinschaft, the Robert A. Welch Foundation, the U.S. NSF (DMR-0115927), the NCCR Nanoscience, the Swiss NSF, DARPA, and ARO.

References

1. A. Peres, *Quantum Theory: Concepts and Methods*, Dordrecht: Kluver Academic Publishers, 1995.
2. A. Steane, " Quantum computing," *Rep. Prog. Phys.* **61**, 117 (1998).
3. A. Ekert, P. Hayden, and H. Inamori, "Basic concepts in quantum computation," quant-ph/ 0011013.
4. R. F. Werner, "Quantum information theory — an invitation," quant-ph/0101061.
5. A. Einstein, B. Podolsky, and N. Rosen, "Can quantum-mechanical description of physical reality be considered complete?" *Phys. Rev.* **47**, 777 (1935).

als dedicated and committed to the success of the SA. In contrast, the latter are those who are perhaps equally determined to sabotage it. Therefore, the participation of the former individuals should be positively encouraged, while, if possible, the latter individuals should be excluded from any involvement in the child's activities.

Even if it enjoys full parental support and is staffed by able people, the child's success as a venture is obviously dependent upon a cohesive team effort by its members. Staff from both parents have to learn to work together to achieve common goals. Porter, in his study of alternative corporate diversification strategies, concluded that *transferring skills* and *sharing activities* are critically important.[29] Since the LMF is seeking technological diversification opportunities through the notional temporary acquisition of the NBF in the strategic alliance, his approach is directly relevant to the present discussion. A strategic alliance clearly provides a vehicle for transferring skills, or in this context, more often exchanging complementary skills between the two parents. This exchange will primarily occur in their technology development activities. However, some complementary skill transfers will be needed in the reverse direction, both to facilitate the technology transfer process between the two, and to enable the NBF's staff to acquire the development, production, and marketing knowhows they are seeking. Since a successful SA requires mutual trust and an attitude of give-and-take between its partners, it is most important that this climate be established and maintained equitably. One fear of firms participating in SAs, which is a manifestation of the Trojan Horse syndrome, is *technology bleedthrough.* That is, Partner A's acquisition of Partner B's technology which, after the alliance is terminated, A then uses as a competitive weapon against B. Doz *et al.*[25] use the metaphor of the *organizational membrane* to address this issue. They suggest that the differences in strategic intents of the partners provide the differences in pressure between the two sides of the membrane. They imply that, like a membrane that is designed selectively to allow the passage across it of some molecules and not others, the organizational membrane should be designed to allow the passage across it of only those skills and activities that have been contractually agreed between the partners.

Even if information exchanges stay within the spirit of the contract, the relative knowhow that each partner acquires will depend upon two other factors. First, as the above authors put it, this exchange is also dependent upon each organization's receptivity or *porosity.* Some people and organizations are more able to learn from others because they are more receptive or porous to new ideas. Since each partner is seeking skills from the other, each must have the porosity to absorb these skills. These will typically be a mix of explicit and tacit skills. For example, the manufacturing scale-up knowhow sought by an NBF in its alliance with an LMF embraces explicit equipment-embodied skills and tacit experienced-based knowledge of the problems of scaling-up chemical and biological processes. When implementing an SA, it is important that each parent recognizes the attributes of the knowledge or information and skills it is seeking from its partner, and that, insofar as possible, it tries to staff the child with people with the abilities to absorb them. Alster suggests another role requirement for the alliance to succeed.[28] This is enacted by *buffer* or liaison individuals between the child and its parents. Given the differences in cultures and decision-making dynamics that can often exist between the two parents themselves and also their child, this is a crucial role. Without its enactment, the SA may be ossified by the bureaucratic red tape that can develop between the participants. In common with other semiautonomous corporate ventures, the SA faces the twin threats of a *strategic reversal* and the *emergent trap* of Chapter 12.

Terminating or Extending the Alliance

Many SAs are designed to be temporary liaisons in pursuit of specific strategic objectives. Once these objectives have been achieved, they may be terminated under the conditions of the divorce settlement as defined in the original alliance agreement. If the SA has been successful, from both partners' viewpoints, the divorce settlement should be harmonious. If both partners found their marriage to be productive, they may wish to extend it or enter into a new one based upon evolving organizational, technological, and market needs and opportunities. They may even enter into an agreed merger or sellout deal. SAs offer the best option of survival to some firms. McKenna cites examples of computer firms that failed because they did not form alliances; Alster argues that well-designed SAs, of variable duration, often provide better mechanisms for technology commercialization to both parties.[28,30]

13.6 SUCCESSIVE FACTORS FOR STRATEGIC ALLIANCES

Despite such optimistic observations, some researchers take a pessimistic view of SAs, and claim that low chances of success can be expected from them.[26,31,32] Such conflicting observations are consistent with the results of a study by Forrest and Martin.[33] They studied 20 LMFs and 40 NBFs which, between them, had participated in almost 1500 alliances. The LMFs reported a success rate of 47.5% from 346 alliances. This is consistent with W. R. Grace's success rate of 40–50% with mainly R&D-based alliances.[34] In contrast, the NBFs reported a success rate of 83% from 1146 alliances. The lower perception of success for the LMFs appears to be derived, as will be discussed below, from their more rigorous standards of accountability. Respondents were asked to cite the reasons for success and failure in their alliances. The citings were similar for both DBCs and LMFs, and could be summarized as the three Cs of *continuing, compatibility,* and *commitment*. That is, there was agreement between the two parties on the *ends* to be sought, the compatibility to work well together, and the sustained commitment of the required resources to the alliance by both parties jointly to provide the *means* to achieve this end. It is of interest to note the interorganizational cultural compatibility did not figure largely as a success factor. What the results did suggest is that the ability of the two-partner companies to work together is dependent upon the presence of individuals who can bridge any gap between the two corporate cultures involved rather than upon an overall cultural compatibility. Many DBCs now have managers with large company experiences, who may be adept at *gap-bridging* or *boundary-spanning* roles. Such roles are analogous to impedance-matching circuits in electronic systems and synchromeshes in automobile gearboxes in facilitating effective linkages between two disparate subsystems. The absence of such individuals may expose an alliance to cultural disparities between its partners that destroys it. Burrill also reports successful experiences of SAs in the biotechnology industry and expresses similar ideas in stating:

> Something of a marriage between the best aspects of large companies and the best aspects of small companies is occurring in all technology industries, . . .[35]

Kanter expresses similar ideas in her *six Is* for partnership success.[36] The relationship is *important*, there is agreement on long-term *investment*, the marriage partners are *interdependent*, the organizations are *integrated*, each is kept *informed* of the plans and directions of the other, and the partnership is *institutionalized* by a framework of support mechanisms.

Such observations are echoed by Niederkofler in his discussion of the evolution of SAs.[37] He argues that an alliance will succeed if it constitutes a good strategic and operational fit to the goals of both parents. Therefore, the initial negotiations must identify the strategic fits for both parents, and the alliance agreement must lay a sound foundation for its operational implementation, as well as the creation and maintenance of goodwill and trust among all the parties involved. Since these goals may change as it evolves, the original contract should incorporate the flexibility to change both the strategic and operational fits of the alliance to its parents in order to accommodate to their changing needs. Such flexibility should be present if the climate of goodwill and trust has been built up between the parties involved. Obviously, a time may come when the alliance can no longer serve the needs of one or both of its parents. Should that situation arise, the appropriate exit terms of the alliance agreement should be followed, to dissolve the marriage by an "amicable divorce." In this situation, the child might continue to operate as an independent venture severed from its parents or be purchased by a third firm. Alternatively, it may be dissolved, and its component parts divided and shared between parents under mutually acceptable "community property" rules.

FURTHER READING

J. D. Lewis, *Partnerships for Profits: Structuring and Managing Strategic Alliances.* New York: Free Press, 1990.
P. Lorange and J. Roos, *Strategic Alliances.* Cambridge, MA: Blackwell, 1992.

REFERENCES

1. A. Lorenz, "Boeing and Airbus Join Up to Build 'Super' Jumbo Jet." *The Sunday Times* December 20 (1992).
2. "Now For the Really Big One." *London: Economist* January 9, pp. 57–60 (1993).

3. B. Merrifield, "Strategic Alliances in the Global Marketplace." *Research Technology Management* **32**(1), 15–20 (1989).

4. G. Tassey, "Structural Change and Competitiveness: The U.S. Semiconductor Industry." *Technological Forecasting and Social Change* **37**, 85–93 (1990).

5. A. Nanda and C. Bartlett, "Corning Incorporated: A Network of Alliances." Boston: Harvard ICCH 1990.

6. P. A. Abetti, "Technology A Key Strategic Resource." *Management Review* February, pp. 37–41 (1989).

7. W. G. Cutler, "Acquiring Technology from Outside." *Research Technology Management* **34**(3), 11–18 (1991).

8. A. H. Rubenstein, *Managing Technology in the Decentralized Firm.* New York: Wiley, 1989, pp. 274 *et seq.*

9. M. J. C. Martin and D. A. Othen, "Developing University Technology: Some Observations and Comments." *Management of Technology. III. The Key To Global Competitiveness,* Proceedings of the Third International Conference on Management and Technology. Norcross, GA: Industrial Engineering and Management Press, 1992, pp. 48–56.

10. M. J. C. Martin and P. J. Rossen, "R&D Philosophy at ABCO." *Four Cases on the Management of Technological Innovation and Entrepreneurship,* Technological Innovation Studies Program. Ottawa, Ont.: Department of Industry, Trade and Commerce, 1984.

11. H. I. Fusfield and C. S. Haklisch, "Cooperative R&D for competitors." *Harvard Business Review* **63**(6), 60–76 (1985).

12. W. M. Evan and P. Olk, "R&D Consortia: A New U.S. Organizational Form." *Sloan Management Review* **24**(3), 37–46 (1990).

13. H. I. Fusfield, *The Technical Enterprise: Present and Future Patterns.* Cambridge, MA: Ballinger, 1986.

14. D. Dimanescu and J. W. Botkin, *The New Alliance: America's R&D Consortia.* Cambridge, MA: Ballinger, 1986.

15. J. Rhea, "New Directions for Industrial R&D Consortia." *Research Technology Management* **34**(5), 16–26 (1991).

16. W. E. Souder and S. Nassar, "Managing R&D Consortia for Success." *Research Technology Management* **33**(5), 44–50 (1990).

17. R. W. Smilor and D. V. Gibson, "Accelerating Technology Transfer in R&D Consortia." *Research Technology Management* **34**(1), 44–49 (1991).

18. A. Larson, "Partner Networks: Leveraging External Ties to Improve Entrepreneurial Performance." *Journal of Business Venturing* **6**, 173–188 (1991).

19. A. L. Saxenian, "The origin and dynamics of production networks in Silicon Valley." *Research Policy* **20**, 423–437 (1991).

20. Office of Technology Assessment, *Commercial Biotechnology: An International Analysis.* Washington, DC: U.S. Government Printing Office, 1984.

22. U.S. Department of Commerce, International Trade Administration, *High Technology Industries: Profiles and Outlooks—Biotechnology.* Washington, DC: U.S. Government Printing Office, 1984.

22. J. E. Forrest and M. J. C. Martin, "Strategic Alliances as Viable Growth Strategies for

High Technology Firms." *Proceedings of the 1987 IEEE Conference on Management and Technology.* Norcross, GA: IEEE, 1987.

23. J. Freeman and S. R. Barley, "Inter-organizational Relations in Biotechnology." In R. Loveridge and M. Pitt (Eds.), *The Strategic Management of Technological Innovation.* New York: Wiley, 1990, pp. 127–156.

24. K. R. Harrigan, *Strategies for Joint Ventures.* Lexington, MA: Heath, 1985.

25. Y. Doz *et al.*, "Strategic Partnerships: Success or Surrender?" Unpublished presentation. London: AIB-EIBA Joint Annual Meeting, November 20–23, 1986.

26. P. J. Killing, *Strategies for Joint Venture Success.* New York: Praegar, 1983.

27. K. R. Harrigan, *Strategies for Joint Ventures.* Lexington, MA: Heath, 1985, p. 375.

28. N. Alster, "Strategic Partners: Seeking the Right Chemistry" *Electronic Business* **15**(10) (1986).

29. M. E. Porter, "From Competitive Advantage to Corporate Strategy." *Harvard Business Review* **65**(3), 43–59 (1987).

30. R. McKenna, "Market Positioning in High Technology." *California Management Review* **27**(3), 82–108 (1985).

31. Y. Doz, "Technology Partnerships between Large and Small Firms: Issues and Pitfalls," Unpublished presentation. New Brunswick, NJ: Conference on Strategic Alliances, Rutgers University, October 24–26, 1986.

32. N. Alster, "Dealbusters: Why Partnerships Fail." *Electronic Business* **15**(7), 70–75 (1986).

33. J. E. Forrest and M. J. C. Martin, "Strategic Alliances Between Large and Small Research Intensive Organizations: Experiences in the Biotechnology Industry." *R&D Management* **22**(1), 41–53 (1992).

34. M. F. Wolff, "Forging Technology Alliances." *Research Technology Management* **32**, 9–11 (1989).

35. G. S. Burrill, *Biotech 89: Commercialization.* New York: Mary Ann Liebert, 1988, p. 44.

36. R. M. Kanter, *When Giants Learn to Dance.* New York: Simon & Schuster, 1989, Chapter 6.

37. M. Niederkofler, "The Evolution of Strategic Alliances: Opportunities for Managerial Influence." *Journal of Business Venturing* **6**, 237–257 (1991).

____14

EPILOGUE: TOWARDS SUSTAINED INNOVATION AND ENTREPRENEURSHIP IN TECHNOLOGY-BASED FIRMS

Innovative firms tend to have a style of management that is open to new ideas, ways of handling staff that encourage innovation, systems that are customer focussed and which reward innovation, skills at translating ideas into action and so forth . . . One has to really believe an organization cares in order to invest the energy needed to help it change. Such commitment derives from superordinate goals . . . this is probably the most underpublicized "secret weapon" of great companies.

RICHARD TANNER PASCALE AND ANTHONY G. ATHOS,
THE ART OF JAPANESE MANAGEMENT

14.1 INTRODUCTION

In this final chapter we review some overall approaches to the effective management of the innovation and entrepreneurship process in technology-based firms. It would be utopian to expect to end this text by suggesting that there is some ubiquitous guaranteed system for managing the corporate technological innovation and entrepreneurship process. Like the proverbial free lunch, such a system cannot be expected to exist this side of paradise. Nevertheless, after acknowledging this reservation, it is possible to suggest some general guidelines for enhancing the process.

14.2 THE CTO AND THE INNOVATION MANAGEMENT FUNCTION

As was suggested in Chapter 4 (Section 4.2), it is useful to postulate an *Innovation Management Function (IMF)* typically led by the *Chief Technology Officer (CTO)* as the responsible vehicles for the formulation, coordination, and implementation of the innovation and entrepreneurial policies and strategies of the technology-based firm. Although the titles of IMF and IMCS may rarely exist, such roles are certainly enacted in many firms in practice, and the title of CTO is used in some of them.

Frohman, based upon a study of nine companies that have sought to exploit technology, found that successful companies selected projects which supported business goals, reinforced technological leadership, and solved customer's problems. The successful companies had planning/decision systems and organizational structures that reinforced technological strategies and senior management assumed responsibility for technological planning and decision making, which was integrated with business planning. He also suggested that the leadership of the overall technological development should be placed in the hands of such a CTO, who should report to the corporate CEO or someone immediately below that level.[1] Adler and Ferdowes[2] describe the roles and responsibilities of a sample of 25 senior managers holding CTO positions in U.S. Fortune 100 firms. Nineteen of them reported to the Chairman, President, Chief Executive Officer (CEO), or Chief Operating Officer (COO) in their firms, and the remaining six reported to the Executive Vice President or Vice Chairman. All emphasized that their responsibilities embraced product and process technology in both centralized laboratories and SBUs. These results are consistent with Bhalla's recommended CTO job description, which is:[3]

1. To be responsible for the development and execution of a business–technology pan for each SBU in the firm.
2. To tailor the technology organizations (such as corporate and divisional R&D laboratories) in the firm, including their human resources, to provide the most effective support for SBU and corporate objectives.
3. To review, develop, and administer the technology portfolio and budget of the firm, in collaboration with top management.
4. To maintain a technology overview role and forecasting function, in order to react to external developments. In particular, to identify potential technology threats and opportunities and exploit emerging synergies between SBUs.

Although Bhalla does not include them, two other requirements should be added to the above job description:

5. To identify external technology sources for the firm, opportunities for technology development cost sharing, and other potential synergistic alliances between SBUs and external organizations.

And, of increasing importance nowadays:

6. To ensure that future technological innovations are congruent with responsible environmental stewardship and sustainable development.

Needless to say, the preceding job description invests the CTO with considerable power and influence in a firm. This degree of power and influence may be strongly resisted by others there, particularly the CEOs and Vice Presidents of R&D, Engineering, and Technology of the individual SBUs in the decentralized firm. Therefore, although it is becoming increasingly recognized that the CTO role is needed in technology-based firms, even when such appointments are made, not all CTOs may be granted such broad job descriptions. Rubenstein discusses the role and responsibilities of CTOs in some detail, based upon his more than three decades of research and consulting activities in technology management.[4] He, too, is supportive of a strong CTO role in the high-technology firm, and makes the following observation:

> If top management really wants their corporate CTO to look after the entire technology management program of the company, to assure its technical vitality, to have the various labs complement each others' work, and to provide the company with ability to respond to technical threats and to capitalize on technical breakthroughs and opportunities, then it either will have to give him the formal power he needs to resolve major conflicts or must be prepared to pay the continuing role of arbitrator between the CTO and divisional management (Rubenstein,[4] p. 107)

Regardless of whether a CTO position and an IMF is *formally* established in a firm, such a role and function must, at least, be *informally* enacted. The success or failure of this role enactment will be influenced by the appositeness of the firm's organizational structure, and the remainder of the chapter discusses these two issues. Both issues can be usefully discussed within the evolutionary framework introduced in Chapter 2.

14.3 NATURAL AND ARTIFICIAL EVOLUTION

In the Chapter 7 Supplement we briefly discussed Margaret Boden's ideas on the role of that artificial intelligence research might play in improving our understanding of the creative process. Recent work, especially that in neural computing, has focused increasing attention upon the extent of the homology between natural and artificial intelligence. In Chapter 2 we suggested that technological innovation can be viewed in the context of Popper's evolutionary epistemology of science and that the highest status of any innovation (whether it be product, process, or service), like that of a scientific theory, is that of an as *yet unrefuted conjecture*. Popper's approach essentially invokes a parallel homology between biological (or natural) versus scientific and technological (or artificial) evolutions. Just as it is fruitful to compare and contrast natural and artificial intelligence, it is also fruitful to compare and contrast *natural* and *artificial* evolutions, to improve insights in both cases. Comparing natural (or biological) and artificial (or technological) evolutions, we can identify the following similarities and differences:

1. Natural mutations are spontaneous and random and survive or fail through a Darwinian learning process. In contrast, artificial mutations (recall Chapter 2, Section 2.10) succeed or fail through a Lamarckian learning process. Being

Lamarckian rather than Darwinian, artificial evolution is much *faster* than its natural counterpart.

2. In both cases, survival is the reward of the *fitting*, rather than the fittest (again recall Chapter 2, Section 2.10 and Chapter 6), often based upon symbiotic relationships with other organisms or artefacts (recall Chapter 3, Section 3.14).

3. In both cases, these processes encourage and reward creative activities which generate diversity and variety in alternative evolutionary paths to generate a natural ecosphere and artificial *technosphere* with rich variations in adaptations and scopes for nichemanship.

4. As well as being slow, natural terrestrial evolution is a largely incremental process with rare evolutionary discontinuities, perhaps precipitated by extraterrestrial interventions. For example, the revolutionary discontinuity in the ecosphere that led to the extinction of the dinosaurs more than 50 million years ago, may have been precipitated by the impact of an asteroid or comet on the Yucatan Peninsula of Mexico. In contrast, revolutionary discontinuities occur frequently in artificial evolution with major implications for the organizational structures required.

5. *Most important of all,* natural evolution and artificial evolution are *not mutually independent* processes. As indicated in Appendix 4.2, there is growing evidence that the artificial technosphere is degrading the natural ecosphere, with potentially disastrous impacts on the life on earth in future years. Therefore, increasing attention must be paid to the adoption of approaches to technological innovation which will ensure that this potentially catastrophic degradation is reversed and, ideally, eliminated. Approaches to the design of products and processes which seek to reflect and replicate the designs of natural ecosystems are being developed under the rubric of *industrial ecology* and *industrial metabolism*.[5,6] These approaches can be expected to play increasing roles in the designs of innovative products, processes and services in future years.

14.4   ALTERNATIVE ORGANIZATIONAL STRUCTURES

As in the natural ecosphere, which embraces a diverse variety of flora and fauna, there are differing artificial evolutionary radiations or industry groupings, which are shaped by the differences in technologies, applications, and markets. These groupings often also change with discontinuities in technological evolutions, and dictate the organizational requirements of the member firms in each of these groupings. They are primarily dependent upon two sets of factors:

First, three *time–cost* factors:

1. The *length* of Generational Life Cycle (GLC). In the civil airframe industry the overall GLC is long, though GLCs may be shorter in engines, avionics, etc. In contrast, GLCs are short in microelectronics.

2. The *start-up* cost of the GLC.

3. The *unit* cost of the product.

Second, three *diversity* factors:

1. The *diversity* of complementary assets and cotechnologies required to design the product.
2. The *diversity* of complementary assets and cotechnologies required to manufacture, distribute, and market the product.
3. The *diversity* of the product lines.

A *project organization or project design consortium* appears appropriate when generational life cycles are long, both upfront and unit costs are high and the diversity of complementary subtechnologies and assets is low. It is probably most applicable in the aircraft industry (recall Chapter 13, Section 13.1 and the cost of the superjumbo) and in engineering turnkey projects. Note that, although these situations typically require the assembly of complementary subtechnologies and assets, they are not diverse. Such industries typically have well-established supplier networks that can be mobilized to perform specific projects. Again as we saw in Chapter 13, the design consortium is becoming a feature of the semiconductor industry. When life cycles and costs are shorter and a portfolio of projects are perform, the *matrix structure,* as discussed in Chapters 8 and 9, becomes appropriate. As lifecycles shorten and costs fall, it may be appropriate implement innovations in a more traditional functional organization. However, as we saw in the operations of Texas Instruments (Chapter 12, Section 12.4), they are best realized in participative frameworks which have some resemblances to good university faculties. We might therefore label them *collegial functional organizations;* a term that will be discussed further later. As diversity increases, in product lines and the manufacturing, marketing, and distribution resources required, an *intrapreneurial organization* is sought, as exemplified by 3M. Finally, when innovation also requires a diversity of subtechnologies and complementary assets, an *interpreneurial organization,* able to cooperate in a network or spider's web of alliances, as exemplified by Corning, the biotechnology industry, and to a lesser extent more recently, the computer industry.

14.5 SOME OBSERVATIONS FROM RECENT STUDIES

The conceptual generalizations of Sections 14.3 and 14.4 are of little practical value unless can be supported by specific observations on current practices in successful technology-based firms. Therefore we now link them to some recent studies of successful firms, to conclude the chapter with suggested prescriptions for the *successful* management of innovation and entrepreneurship in technology-based firms.

From their study of the management practices of a sample of Japanese and U.S. firms, Pascale and Athos[7] drew conclusions broadly congruent with those of Frohman cited earlier. They expressed these conclusions in their 7-S framework of management. They suggested that successful Japanese and U.S. firms (including TI and 3M) have the style, staff, skills, systems, structure, and strategy to stimulate and nurture innovation. These six Ss are embedded in the seventh—the superordinate goals or the values enshrined in the organizational culture which elicit a commitment to innovation from the firm as a whole and each of its individual members. There are alternative approaches to enshrining this culture, as reflected in the contrasting approaches of TI and 3M, but they suggest that this is (no longer) the *secret weapon* of the successfully innovative Japanese and Western companies. Another analysis of corporate practices for stimulating innovation and entrepreneurship, based upon the 7-S framework, was *In Search of Excellence*.[8] Received with critical acclaim when first published, it lost some favor after the appearance of an article reporting the declining performances of its *excellent* companies.[9] Nevertheless, many of its observations remain valid and are reflected in those of later studies. Excellent firms, like excellent baseball bats, can be expected to experience losses of form and experience less than excellent seasons. It is their hitting lifelong RBIs that ensures their ultimate entries into baseball's "Hall of Fame." High technology could also claim to have its "Hall of Fame" firms that sometimes stumble in their pursuits of excellence. Two later studies are of particular interest since, when considered together with the above and other writings, they suggest a useful set of guidelines for managing innovation and entrepreneurship in technology-based firms. First, Jelinek and Schoonhoven[10] conducted in-depth studies of five leading U.S. based high-technology firms in the electronics industry, Hewlett-Packard (HP), Intel, Motorola, National Semiconductor, and Texas Instruments (TI). Second, Majone also conducted an in-depth study of three U.S. based firms, two in the electronics industry, GE Medical Systems (GEMS) and Motorola Communications (COMM), together with the diversified Corning Incorporated.[11]

Learning-by-Striving to be Best in Class

Several writers have echoed Popper's error-elimination epistemology in emphasizing the role of organizational learning.[12,13] For example, Majone, in his in-depth study of three U.S. technology-based firms (GE Medical Systems, Motorola Communications and Corning), emphasizes *learning-by-trying* as the key to success.[11] Some of these writers mainly concentrate on *continuous improvement* learning, specifically associated with learning or experience curves. However, Botkin *et al.*[14] describe two levels of learning:

1. *Maintenance* learning associated continuous improvement and an advance down the experience curve, which can be viewed as an exercise in the Japanese *kaizen*.

2. *Innovative* learning which is required to deal with the frequent episodes of discontinuous and sometimes turbulent changes that characterize typical contemporary technological evolutions. It can be identified with generational innovations and also with more radical or revolutionary changes. Innovative learning is characterized by the capacity to anticipate and participate in change. This requires imaginative insight and foresight into technology and market coevolutions, to identify both the nature and *timing* of discontinuous

change. Witness the debates over the last decade or so concerning the convergencies of computer, telecommunications, and media technologies and the emergence of the $3.5 trillion information superhighway-based industry. It will also be required in pursuing the, by no means, utopian ideals of industrial ecology and metabolism for responsible environmental management.

Firms that Majone describe as *Best in Class* practice both maintenance and innovative learning. The first through incremental improvements in their established lines; the second by anticipating and participating in technological discontinuities, through pioneering the introduction of new products and product lines. That is, rather than being satisfied with developing products, organizational structures, and strategies that fit the evolving technosphere, they actively and creatively *seek and strive* to lead and shape this coevolution by conserving, stretching, and levering their technology and management competencies. Hamel and Prahalad[15,16] express similar ideas in their concepts of *expeditionary marketing* and strategy as *stretch and leverage*. It is also manifested physically in Rothwell and Gardiner's *robust design* concept in, for example, the Boeing 747 Series that have evolved over a 35 year life span.[17]

Successful firms possess the capacity and will to create continuously radical, generational, and incremental innovations, typically following an offensive first-to-marketing strategy. All five firms in the Jelinek-Schoonhoven and Majone studies can claim to have pioneered radical and generational innovations. Although GEMS entered the emergent CAT scanner market as a defensive innovator, following its introduction of the CT 8800 and purchase of the EMI-Thorn operation, it moved to a first-to-market strategy. After a late entry, it also followed an offensive strategy in the magnetic resonance (MR) medical electronics market segment. Intel's pioneering efforts in designing, manufacturing, and marketing successive generations of microprocessors are well known, but most readers are probably less familiar with similar achievements by Corning in optical fibers. Majone points out that Corning not only pioneered radical optical fiber innovations, but introduced seven generational innovations in processing technology over 15 years! *Kaizen* or continuous improvements through incremental innovations is particularly associated with Japanese firms, but was again manifested in these sample firms. TI and National Semiconductor are especially identified with learning or experience curve cost/price reduction strategies in the semiconductor industry, but the other firms showed that they were able to protect their leaderships in pioneering radical or generational innovations with successions of incremental improvements in products and processes. The tight interplay between product and process design innovations and improvements is a notable feature of the semiconductor industry where, in Jelinek's and Schoonhoven's words, success depends upon winning the *end game of manufacturing*. Also, as Corning's above-cited achievements attest, the skilled management of manufacturing plays an important role in all firms. The sustained continuous creation of innovation must be based upon the management of the interfaces among all three functions of R&D, manufacturing, and marketing that form an eternal triangle or *ménage à trois* (Figure 14.1). Note that firms do not necessarily

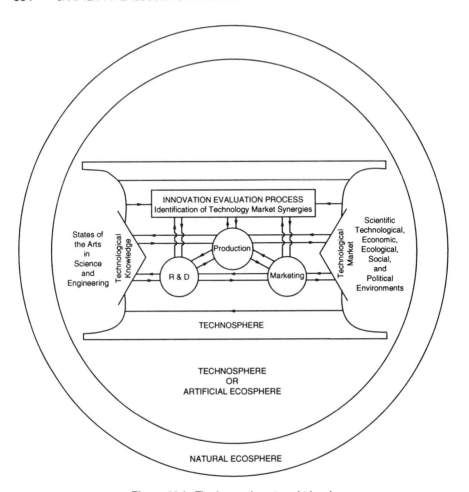

Figure 14.1. The innovation eternal triangle.

have to grow large to strive in this way. A small company pursuing a technological nichemanship strategy can strive to the *Best in Niche* firm.

Continuous Creation Based Upon Local Knowledge

Both maintenance and innovative learning must be based upon creative thinking, and Wagner[18] uses a metaphor from cosmology to illustrate this requirement. The theory of continuous creation of matter was conjectured in the 1950s to explain observations in our expanding universe. Most cosmologists claim that subsequent observations have refuted it in favor of the conjecture that our universe began with the explosion of an immensely dense concentration of matter in a big bang, more than 10 billion years ago, and that none has been created since then. However, some, including Sir Fred Hoyle, still support the continuous creation theory, arguing that matter may be created in certain localities in the universe rather than diffusely throughout space–time. Clearly, the diversity of outputs and approaches, needed to ensure sustained success for technology-based firms, must be based upon the *continuous creation* of innovation, analogous to continuous creation in cosmology. This too is typically based upon local pockets of knowledge within the organization.

In their study, Jelinek and Schoonhoven argue that, given the rapid rate of technological change, most senior managers are unlikely to have an up-to-date grasp of the state-of-the-art, but must rely on *local knowledge*. That is, engineers, scientists, and others who may be quite junior in the organizational hierarchy are likely to be most aware of evolving technological possibilities, manufacturing methods and market needs. This knowledge is typically context specific, so it exists in local pockets of expertise. Moreover, since individuals participating throughout the innovation process should be encouraged to make creative contributions, it must be both broadly and deeply rooted with successful contributions, from whatever sources, given commensurate recognition and rewards. It also implies that new employees should be quickly integrated into the corporate culture so that they feel able to make real contributions while their prior experiences are still up to date.

Jelinek and Schoonhoven also argue that such firms must induce a strong culture that strikes a careful balance between challenge and support, encouragement and demanding standards. This can be stressful for all concerned, and can sometimes lead to domestic disruptions and "hi-tech burnout." Since learning-by-striving generates *failures* as well as successes, the culture must also practice tolerance of failure. From his study of 150 new products, Gulliver[19] concluded that failure is the ultimate teacher; Thomas Edison, one of history's leading inventor-entrepreneurs, claimed that he failed his way to success. Although Jelinek and Schoonhoven emphasize the importance of a strong culture, it should be noted that not all firms entirely share this view. Cannon and Honda, for example, encourage the creation of *multiple* cultures by hiring significant numbers of midcareer managers from other firms to provide countercultures, to avoid the dangers of excessive inbreeding that is a risk of a single dominant organizational culture.[20]

Garvin suggests the following as important organizational learning attributes:

> A learning organization is an organization skilled at creating, acquiring, and transferring knowledge, and at modifying its behavior to reflect new knowledge and insights. . . . Organizations that do past the definitional test . . . become adept at translating new knowledge into new ways of behaving. These companies actively manage the learning process to ensure that it occurs by design rather than by chance.[21]

That is, as well as creating and acquiring knowledge, the organization must be skilled at transferring and disseminating knowledge, particularly if it is locally based. Jelinek and Schoonhoven also emphasize this point. In their sample of firms, the formal organizational structures are supported by numerous quasiformal and informal structures to facilitate this dissemination. The former are temporary teams or task forces, usually with dotted-line linkages to the formal organizational structure, formed to undertake specific missions in support of overall operations. They are essentially informal *collegial* groupings that come together to attack specific problems that lie outside normal operating procedures. Since participations in these groupings are likely to based upon the members' problem-solving abilities, they can often prove the foundation of local knowledge needed for sustained innovation.

Focused Leadership and Stratocracy: Domes and Jazz Orchestras

Although continuous creation may be fostered by the organizational culture which stretches individuals and accesses local knowledge, that alone is insufficient to ensure sustained innovation and corporate success. Innovations unquestionably must be deeply rooted in the organization if they are to thrive, but the adage to *think locally and act globally,* must also be honored. Innovations must also simultaneously be effectively filtered by higher-level managers to ensure that such proposals are congruent with a corporate focus, and often a global perspective coupled with responsible environmental stewardship. Establishing and maintaining this focus, as well as the organizational culture, is a leadership imperative for senior management. A consistent persistent strategic focus is a characteristic of successful firms. Nevertheless, the requirements of local knowledge-based innovation dictate a departure from the leadership and decision-making structures of the past. The joint requirements of strategic focus and local knowledge dictate that senior managers should communicate with individuals well down the corporate hierarchy if both parties are to play effective roles. It is also implies that senior managers should be open and willing to describe the rationales of their decisions, to encourage further new ideas, and to some extent, to give others genuine power to help steer and influence the organization. Bahrami and Evans[22] describe this hierarchical spanning requirement as *stratocracy.* Tomasko[23] suggests that this spanning requirement is best achieved by integrating organizations in a *dome-like* rather than *pyramid* organizational structures. Clearly the collegial linkages cited above are likely to be more effective in a dome-like structure, with strong lateral as well as vertical struts, rather than a pyramid-like organizational structure. Matrix structures, interpreneurial and networking alliance approaches can also be viewed as methods for structuring the organization as a dome rather than a pyramid.

The need for this organizational flexibility or fleet-footedness is also exemplified in Kanter's *When Giants Learn to Dance,*[24] and Drucker uses a symphony orchestra metaphor for organizational structure to convey a similar idea.[25] Rather than work through a rigid hierarchy, like orchestra conductors coordinating individual instrumentalists, senior managers must interact with individual pockets of local knowledge within the framework of an overall score. Although this metaphor is compelling, it is also inexact. Given that innovation is based upon local knowledge combined with strategic direction, the score of a classical music symphony, which imposes quite rigid requirements on both instrumentalists and conductor, does not apply. Rather, the metaphor of a large orchestra playing jazz is more apposite. The conductor may choose an established orchestral work, say *Porgy and Bess* or *Rhapsody in Blue,* but allow individual jazz instrumentalists to extemporize around the score in light of their preferences, playing skills, and current musical fashions. In other words, perhaps successful corporate innovation is closer to Basie than Beethoven!

Frequent Organizational Mutations

Although their conductors interact directly with individual orchestra members, whether they are playing classical or jazz music, one should recall that orchestras have hierarchical structures based upon the conductor, orchestra leader, first violin, and so on. Similarly, despite the importance of local knowledge and stratocracy, a formal hierarchical organizational structure is required both to implement new innovations and to maintain ongoing present operations in a strategic framework. Given the rapid rate and unpredictable, sometimes turbulent, nature of technological changes, this formal structure is itself subject to frequent changes. Such changes do not reflect dissatisfaction with the past performance of the current structure, but the recognition that changes in the firm's technological, social, and market environments require a changed organizational fit. They can be viewed as deliberately engendered *organizational mutations* designed to maintain their adaptiveness to their technosphere. The technology-based firm in some industries is perhaps homologous to a *potent virus* in that it often has to undergo frequent *organizational mutations* to maintain and consolidate its competitive advantage in its technosphere. This adaptiveness is particularly important given the often unpredictable *chaotic features* of both the innovation process and technological change that favor strategic incrementalism over rigid forecasting and planning; Peters goes so far as to recommend that firms should be in a permanently revolutionary state.[20,26,27] Reflecting such considerations, most of the Jelinek–Schoonhoven and Majone firms in the electronics industry indulge in frequent internal structural reorganizations for this reason, while Corning is more likely to seek such changes through changes in its spider's web alliance network.

Based upon these observations and studies, we can conclude with some comments on the effective management of innovation and entrepreneurship in the technology-based firm.

14.6 THE EFFECTIVE MANAGEMENT OF INNOVATION AND ENTREPRENEURSHIP IN TECHNOLOGY-BASED FIRMS

Whether or not a technology-based firm has a CTO formally appointed and an IMF formally instituted, all must have a cadre of top managers who implicitly enact these roles. The function of this cadre is to oversee and, where appropriate, participate in the activities considered throughout this book. It should ensure that the firm practices the following *seven vital virtues* to stimulate and encourage innovation and entrepreneurship.

1. **Ensure** that continuous innovation creation, locally conceived but universally nurtured, is respected and accepted by all as a necessary condition for sus-

tained corporate success. Based upon the innovation chain and a critical functions venture team, it will require entrepreneurial, intrapreneurial, or interpreneurial linkages to ensure success. Therefore, the CTO and IMF must ensure that these linkages are present to be available for use as and when needed.

2. **Ensure** that the corporate culture or cultures be consciously engendered and sustained by top management to encourage this continuous creation. It should simultaneously provide individuals with support and encouragement, together with challenge and demanding standards to stretch their capabilities.

3. **Recognize** that a "Best in Class" performance requires more than fast and sustained maintenance learning. It is achieved by striving to stretch the technological and managerial capabilities of the organization above and beyond their expectations, by making maximum use of its competences and resources, and by shaping the changing requirements of technology–market coevolutions more than matching them.

4. **Ensure** that top management provides a focus and strategic direction for the above, not through a rigid hierarchical pyramid, but rather through a dome-like organizational structure that encourages multiple lateral formal and informal connections. The structure should itself be flexible, so allowing for the possibly frequent structural changes needed to match the changing requirements of technology–market coevolutions.

5. **Recognize** that successful participation in technology–market coevolution is dependent upon fast and sustained organizational learning through technological and managerial learning. This is based upon persistent learning-by-trying with failures as well as successes. Learning-by-failing should be tolerated in the context of the overall learning-by-trying process.

6. **Recognize** that fast organizational learning is also based upon fast innovation. The organizational structure and culture should engender fast innovation.

7. Given that underlying chaotic mechanisms may be inherent in technology–market coevolutionary processes, it is imprudent to place total confidence in the even best laid future visions and strategic plans.[20] As the Prussian General von Block reputedly put it: *"No plan survives first contact with the enemy."* Therefore, *ensure* that the signals of change are continuously monitored to detect technological discontinuities and paradigm shifts early enough for the firm to participate in and shape them profitably. Above all, *ensure* that the firm and its members are open minded towards the future, and are *prepared to be prepared* for the unexpected. In a world of rapid technological change, the unexpected is the only certainty!

14.7 AFTERWORD

The Preface of this book began by commenting that public and business concern for technological innovation had been aroused by the problems of declining growth and increasing unemployment. Many writers, like Freeman and Soete,[28] take the view that the stimulation and implementation of technological change offers the best means of ameliorating, if not eradicating, these problems. Writing this book will have been worthwhile if it helps present and future technological innovators and entrepreneurs create wealth for themselves, work for others, and worth for society. To any such readers, whether they are planning or have already launched their own innovative ventures as entrepreneurs, intrapreneurs, or interpreneurs, go this author's good wishes for success in their endeavors.

REFERENCES

1. A. L. Frohman, "Technology as a Competitive Weapon." *Harvard Business Review* **60**(1), 97–104 (1982).

2. P. S. Adler and K. Ferdowes, The Chief Technology Officer. *California Management Review* **32**(3), 55–62 (1990).

3. Bhalla, *The Effective Management of Technology.* Reading, MA: Addison-Wesley, 1987.

4. A. H. Rubenstein, *Managing Technology in the Decentralized Firm.* New York: Wiley, 1989, Chapter 3.

5. H. Tibbs, *Industrial Ecology: An Environmental Agenda for Industry.* Cambridge, MA: Arthur D. Little, 1991.

6. R. U. Ayres and U. Simonis, *Industrial Metabolism.* New York: U.N. University Press, 1992.

7. R. T. Pascale and A. G. Athos, *The Art of Japanese Management.* New York: Simon & Schuster, 1981.

8. T. J. Peters and R. H. Waterman, *In Search of Excellence.* New York: Harper & Row, 1982.

9. "Who's Excellent Now?" *Business Week* November 5, pp. 76–78 (1984).

10. M. Jelinek and C. B. Schoonhoven, *The Innovation Marathon Lessons from High Technology Firms.* Cambridge, MA: Basil/Blackwell, 1990.

11. J. G. Morone, *Winning in High-Tech Markets: The Role of General Management.* Boston: Harvard Business School Press, 1993.

12. M. Jelinek, *Institutionalizing Innovation: A Study of Organizational Learning Systems.* New York: Prager, 1979, Chapter 5.

13. P. M. Senge, *The Fifth Discipline.* New York: Doubleday, 1990.

14. J. W. Botkin, M. Elmandjra, and M. Malitza, *No Limits to Learning: Bridging the Human Gap.* Elmsford, NY: Pergamon, 1979.

15. G. Hamel and C. K. Prahalad, "Corporate Imagination and Expeditionary Marketing." *Harvard Business Review* **69**(4), 81–92 (1991).

16. G. Hamel and C. K. Prahalad, "Strategy as Stretch and Leverage." *Harvard Business Review* **71**(2), 75–84 (1993).

17. R. Rothwell and P. Gardiner, "Re-Innovation and Robust Designs: Producer and User Benefits." *Journal of Marketing Management* **3**(3), 372–386 (1988).

18. H. E. Wagner, "The Open Corporation." *California Management Review* **33**(4), 46–60 (1991).

19. F. R. Gulliver, "Post-Project Appraisals Pay." *Harvard Business Review* **65**(2) (1987).

20. R. Stacey, "Strategy as Order Emerging from Chaos." *Long Range Planning* **26**(1), 10–17 (1993).

21. D. A. Garvin, "Building a Learning Organization." *Harvard Business Review* **71**(4), 78–91 (1993).

22. H. Bahrami and S. Evans, "Stratocracy in High-Technology Firms." *California Management Review* **30**(1), 51–66 (1987).

23. R. Tomasko, *Rethinking the Corporation.*   New York: AMACOM Books, 1993.

24. R. M. Kanter, *When Giants Learn to Dance.*   New York: Simon & Schuster, 1989.

25. P. F. Drucker, "The Coming of the New Organization." *Harvard Business Review* **66**(1), 45–53 (1988).

26. J. B. Quinn, "Innovation and Corporate Strategy: Managed Chaos" *Technology in Society* **7**(3), 167–183 (1986).

27. T. Peters, "Crazy Ways for Crazy Days." *BBC Television Broadcast* December 6 (1993).

28. C. Freeman and L. Soete (Eds.), *Technical Change and Full Employment.* Oxford: Basil-Blackwell, 1987.

INDEX

6. E. Schrödinger, "Die gegenwärtige Situation in der Quantenmechanik," *Naturwiss.* **23**, 807 (1935).
7. For a recent review see M. Lewenstein, D. Bruss, J. I. Cirac, M. Kus, J. Samsonowicz, A. Sanpera, and R. Tarrach, "Separability and distillability in composite quantum systems — a primer," *J. Modern Opt.* **77**, 2481 (2000).
8. For a recent review see B. M. Terhal, "Detecting quantum entanglement," quant-ph/0101032.
9. A. Peres, "Causality, randomness, and the microwave background," *Phys. Rev. Lett.* **76**, 1413 (1996).
10. M. Horodecki, P. Horodecki, and R. Horodecki, "Separability of mixed states: Necessary and sufficient conditions," *Phys. Lett. A* **223**, 1, (1996).
11. V. Coffman, J. Kundu, and W. K. Wootters, "Distributed entanglement," *Phys. Rev. A* **61** 052306 (2000).
12. A. Acin, A. Adrianov, L. Costa, E. Jané, J. I. Latorre, and R. Tarrach, "Generalized Schmidt decomposition and classification of three-quantum-bit states," *Phys. Rev. Lett.* **85**, 1560 (2000); A. Acin, E.Jane, W. Dür, and G. Vidal, "Optimal distillation of a Greenberger-Horne-Zeilinger state," *Phys. Rev. Lett.* **85**, 4811 (2000); A. Acin, D. Bruss, M. Lewenstein, and A. Sanpera, "Classification of mixed three-qubit states," *Phys. Rev. Lett.* **87**, 040401(2001).
13. H. A. Carteret, A. Higuchi, and A. Sudbery, "Multipartite generalization of the Schmidt decomposition," *J. Phys. A: Math. Gen.* **41**, 7932 (2000).
14. W. Dür, G. Vidal, and J. I. Cirac, "Three qubits can be entangled in two inequivalent ways," *Phys. Rev. A* **62**,062314 (2000).
15. W. K. Wootters, "Entangled chains," *Phys. Rev. A* **63**, 052032 (2001).
16. K. M. O'Connor and W. K. Wootters, "Entangled rings," *Phys. Rev. A* **63**, 052302 (2001).
17. A. V. Thapliyal, "Multipartite pure-state entanglement," *Phys. Rev. A* **59**, 3336 (1998).
18. J. S. Bell, "On the Einstein-Podolsky-Rosen paradox," *Physics* **1**, 196 (1964).
19. A. Aspect, J. Dalibard, and C. Roger, "Experimental test of Bell's inequalities using time-varying analyzers," *Phys. Rev. Lett.* **49**, 1804 (1982); A. Aspect, "Experimental tests of Bell's inequalities," *Europhys. News* **22**, 73 (1991).
20. P. W. Shor, *Proc. 35th Annual Symp. Foundations Computer Science*, IEEE Computer Society Press, Santa Fe, 1994, pp. 124-134; A. Ekert and R. Josza, "Quantum computation and Shor's factoring algorithm," *Rev. Mod. Phys.* **68**, 733 (1996).
21. D. Loss and D. P. DiVincenzo, "Quantum computation with quantum dots," *Phys. Rev. A* **57**,120 (1998).
22. G. Burkard, D. Loss, and D. P. DiVincenzo, "Coupled quantum dots as quantum gates," *Phys. Rev. B* **59**, 2070 (1999).
23. G. Burkard, H.-A. Engel, and D. Loss, "Spintronics and quantum dots for quantum computing and quantum communication," *Fortschr. Phys.* **48**, 965 (2000).

24. J. Schliemann, D. Loss, and A. H. MacDonald, "Double-occupancy errors, adiabaticity, and entanglement of spin qubits in quantum dots," *Phys.Rev. B* **63**, 085311 (2001).

25. X. Hu, R. de Sousa, and S. Das Sarma, "Decoherence and dephasing in spin-based solid state quantum computers," cond-mat/0108339.

26. J. Schliemann, J. I. Cirac, M. Kus, M. Lewenstein, and D. Loss, "Quantum correlations in two-fermion systems," *Phys. Rev. A* **64**, 022303 (2001).

27. K. Eckert, J. Schliemann, D. Bruss, M. Lewenstein, and D. Loss, "Quantum correlations in systems of indistinguishable particles," quant-ph/0203060.

28. Y. S. Li, B. Zeng, X. S. Liu, and G. L.Long, "Entanglement in a two-identical-particle system," *Phys. Rev. A* **64**, 054302 (2001).

29. R. Paskauskas and L. You, "Quantum correlations in two-boson wavefunctions," *Phys. Rev. A* **64**, 042310 (2001).

30. P. Zanardi, "Entangled fermions," *Phys. Rev. A* **65**, 042101 (2002).

31. M. L. Mehta, *Elements of Matrix Theory*, Delhi: Hindustan Publishing Corporation, 1977.

32. C. H. Bennett, H. J. Bernstein, S. Popescu, and B. Schumacher, "Concentrating partial entanglement by local operations," *Phys. Rev. A* **53**, 2046 (1996).

33. F. R. Waugh, M. J. Berry, D. J. Mar, R. M. Westervelt, K. L. Campman, and A. C. Gossard, "Single-electron charging in double and triple quantum dots with tunable coupling ," *Phys. Rev. Lett.* **75**, 705 (1995); C. Livermore, C. H. Crouch, R. M.Westervelt, K. L. Campman, and A. C. Gossard, "The Coulomb blockade in coupled quantum dots ," *Science* **274**,1332 (1996).

34. L. P. Kouwenhoven, G. Schön, and L. L. Sohn, eds., *Mesoscopic Electron Transport*, NATO ASI Series E Vol. 345, Dordrecht: Kluwer Academic Publishers, 1997.

35. T. H. Oosterkamp, S. F. Godijn, M. J.Uilenreef, Yu. V. Nazarov, N. C. van der Vaart, and L. P. Kouwenhoven, "Changes in the magnetization of a double quantum dot," *Phys. Rev. Lett.* **80**, 4951 (1998).

36. R. H. Blick, D. Pfannkuche, R. J. Haug, K. von Klitzing, and K. Eberl, "Formation of a coherent Mode in a double quantum dot," *Phys. Rev. Lett.* **80**, 4032 (1998); R. H.Blick, D. W. van der Weide, R. J. Haug, and K. Eberl, "Complex broadband millimeter wave response of a double quantum dot: Rabi oscillations in an artificial molecule," *Phys. Rev.Lett.* **81**, 689 (1998); T. H. Oosterkamp, T. Fujisawa, W. G. van der Wiel, K. Ishibashi, R. V. Hijman, S. Tarucha, and L. P. Kouwenhoven, "Microwave spectroscopy of a quantum-dot molecule," *Nature* **395**, 873 (1998); I. J. Maasilta and V. J. Goldman, "Tunneling through a coherent 'quantum antidot molecule'," *Phys. Rev. Lett.* **84**, 1776 (2000).

37. G. Burkard, D. Loss, and E. V. Sukhorukov, "Noise of entangled electrons: Bunching and antibunching," *Phys. Rev. B* **61**,R16303 (2000).

38. P. Recher, E. V. Sukhorukov, and D. Loss, "Andreev tunneling, Coulomb blockade, and resonant transport of nonlocal spin-entangled electrons," *Phys. Rev. B* **63**,165314 (2001).

39. D. Loss and E. V. Sukhorukov, "Probing entanglement and nonlocality of electrons in a double-dot via transport and noise," *Phys. Rev. Lett.* **84**, 1035 (2000).

40. C. H. Bennett and D. P. DiVincenzo, "Quantum information and computation," *Nature* **404**, 247(2000).

41. J. M. Kikkawa, I. P. Smorchkova, N. Samarth, and D. D. Awschalom, "Room-temperature spin memory in two-dimensional electron gases," *Science* **277**, 1284 (1997); J. M. Kikkawa and D. D. Awschalom, "Resonant spin amplification in *n*-type GaAs," *Phys. Rev. Lett.* **80**, 4313 (1998); D. D. Awschalom and J. M. Kikkawa, "Electron spin and optical coherence in semiconductors," *Physics Today* **52**(6), 33 (1999).

42. T. Fujisawa, Y. Tokura, and Y. Hirayama, "Transient current spectroscopy of a quantum dot in the Coulomb blockade regime," *Phys. Rev. B* **63** 081304 (2001).

43. A. V. Khaetskii and Yu. V. Nazarov, "Spin relaxation in semiconductor quantum dots," *Phys. Rev. B* **61**, 12639 (2000).

44. H.-A. Engel and D. Loss, "Detection of single spin decoherence in a quantum dot via charge currents," *Phys. Rev. Lett.* **86**, 4648 (2001); H.-A. Engel and D. Loss, "Single spin dynamics and decoherence in a quantum dot via charge transport," cond-mat/0109470.

45. S. Hill and W. K. Wootters, "Entanglement of a pair of quantum bits," *Phys. Rev. Lett.* **78**, 5022 (1997); W. K. Wootters, "Entanglement of formation of an arbitrary state of two qubits," *Phys. Rev. Lett.* **80**, 2245 (1998).

46. R. Grobe, K. Rzazwski, and J. H. Eberly, "Measure of electron-electron correlation in atomic physics," *J. Phys. B* **27**, L503 (1994); M. Y. Ivanov, D. Bitouk, K. Rzazwski, and S. Kotochigova, "Classical chaos and its quantum measures in Rydberg states of multielectron atoms," *Phys. Rev. A* **52**, 149 (1995).

47. D. P. DiVincenzo, "Two-bit gates are universal for quantum computation," *Phys. Rev. A* **51**, 1015 (1995).

Quantum Computation with Quasiparticles of the Fractional Quantum Hall Effect

D. V. Averin and V. J. Goldman
Dept. of Physics, State University of New York at Stony Brook
Stony Brook, NY 11794-3800, U.S.A.

1. Introduction

"Topological" quantum computation with anyons has been suggested as a way of implementing intrinsically fault-tolerant quantum computation.[1-4] The intertwining of anyons, quasiparticles of a two-dimensional electron system (2DES) with nontrivial exchange statistics, induces unitary transformations of the system wavefunction that depend only on the topological order of the underlying 2DES. These transformations can be used to perform quantum logic, the topological nature of which is expected to make it more robust against environmental decoherence. The aim of this work is to propose a specific and experimentally feasible approach for implementation of the basic elements of the anyonic quantum computation using adiabatic transport of the fractional quantum Hall effect (FQHE) quasiparticles in systems of quantum antidots.[5]

An antidot is a small hole in the 2DES produced by electron depletion, which localizes FQHE quasiparticles at its boundary due to the combined action of the magnetic field and the electric field created in the depleted region. If the antidot is sufficiently small, the energy spectrum of the antidot-bound quasiparticle states is discrete, with finite excitation energy Δ. When Δ is larger than the temperature kT, modulation of the external gate voltage can be used to attract quasiparticles one by one to the antidot.[5,6] In this regime multi-antidot systems can be used to perform quantum logic based on the adiabatic manipulation of individual quasiparticles.

2. FQHE qubits

In analogy to Cooper-pair qubits,[7-9] information in FQHE qubits of this type can be encoded by the position of a quasiparticle in the system of two antidots. The FQHE qubit, illustrated in Fig. 1, is then the double-antidot system gate-voltage tuned near the resonance, where the energy difference ε between the quasiparticle states localized at the two antidots is small, $\varepsilon \ll \Delta$. At energies smaller than Δ, the dynamics of such a double-antidot system is equivalent to the dynamics of a common two-state system (qubit). The quasiparticle states localized at the two antidots are the $|0\rangle$ and $|1\rangle$ states of the computational basis of this qubit. The gate electrodes of the structure can be designed to control separately the energy difference ε and the tunnel coupling Ω of the resonant quasiparticle states.

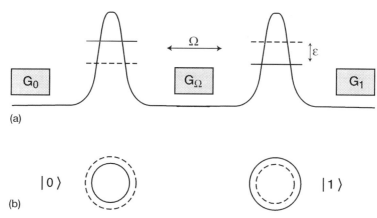

Figure 1. Schematic energy profile (a) and structure (b) of the double antidot FQHE qubit. Solid (dashed) horizontal lines in (a) indicate the edges of the incompressible electron liquid when the quasiparticle is localized at the left (right) antidot. Displacement of the electron liquid is quantized due to quantization of the single-particle states circling the antidots. Shaded rectangles in (a) are the gate electrodes controlling the energies of the antidot-bound states ($G_{0,1}$) and their tunnel coupling (G_{Ω}).

3. FQHE logic gates

The most natural approach to the construction of two-qubit gates with the FQHE qubits is to use fractional statistics[10,11] of the FQHE quasiparticles. Due to fractional statistics, intertwining of the two quasiparticle trajectories in the course of time evolution of the two qubits realizes controlled-phase transformations with nontrivial values of the phase. The precise result of this operation depends on the nature of the FQHE state. In this work, we discuss the most basic and robust Laughlin state with the filling factor $v = 1/m = 1/3$, where the quasiparticles have abelian statistics and the intertwining of trajectories leads to multiplication of the state wavefunction by the phase factor $e^{\pm 2\pi i/3}$. The sign of the phase depends on the direction of the magnetic field and the direction of rotation of one quasiparticle trajectory around the other.

A possible structure of the controlled-phase gate is shown in Fig. 2. Each of the columns of four antidots contains two qubits, and the arrows denote trajectories of quasiparticle transfer through the system. The transfer leads to the transformation of the quantum state of the two qubits and its shift from the gate input (left column in Fig. 2) to the output (right column). The quasiparticle transfer can be achieved by standard adiabatic level-crossing dynamics. If a pair of antidots is coupled by the tunnel amplitude Ω, a gate-voltage–induced variation of the energy difference ε through the value $\varepsilon = 0$ (slow on the time scale Ω^{-1}) leads to the transfer of a quasiparticle between these antidots. Correct operation of the

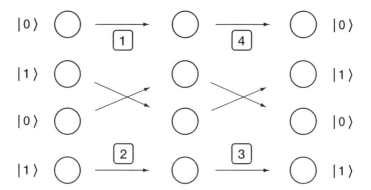

Figure 2. Antidot implementation of the two-qubit controlled-phase gate. The states |0⟩ and |1⟩ are the computational basis states of the two qubits. The arrows show the quasiparticle transfer processes for each basis state during the gate operation. The arrow numbering denotes the time sequence of these processes.

controlled-phase gate in Fig. 2 requires that the gate voltage pulses applied to the antidots are timed so that the state of the upper qubit is propagated at first halfway through the gate, then the state of the lower qubit is propagated through the whole gate, and finally the state of the upper qubit is transferred to the output. In this case, if the quasiparticle of the upper qubit is in the state |1⟩, trajectories of the quasiparticle propagation in the lower qubit encircle this quasiparticle, and the two states of the lower qubit acquire an additional phase difference $\pm 2\pi/3$, conditioned on the state of the upper qubit. We take the direction of magnetic field to be such that the state |1⟩ of the lower qubit acquires a positive extra phase $2\pi/3$. Assuming that the parameters of the driving pulses are adjusted in such a way that the dynamic phases accumulated by the qubit states are integer multiples of 2π, the transformation matrix P of the gate can be written as

$$P = \text{diag} [\ 1,\ 1,\ 1,\ e^{2\pi i/3}\ ] \tag{1}$$

in the basis of the four gate states |00⟩, |01⟩, |10⟩, |11⟩.

The controlled-phase gate P, given by Eq. (1), combined with the possibility of performing arbitrary singe-qubit transformations, is sufficient for universal quantum computation. To demonstrate this explicitly, we construct a combination of the P gate with single-qubit gates that reproduces the usual controlled-*not* (*c-not*) gate C. The *c-not* gate is known to be sufficient for universal quantum computation.[12] Since the gates P and C are not equivalent with respect to single-qubit transformations, two applications of P are required to reproduce C.[13] To find appropriate single-qubit transformations that should complement the two P gates, it is convenient first to reduce P to the conditional z-rotation R of the second qubit through the angle $-\pi/3$: $R = \text{diag} [\ 1,\ 1,\ e^{-\pi i/3},\ e^{\pi i/3}\ ]$. We notice that $R = S(-\pi/3)P$, where $S(\alpha)$ is an unconditional shift of the phase of the state |1⟩ of the first qubit by α. After this reduction, it is straightforward to find the necessary single-qubit transformations from the requirements that the state of the first qubit is unchanged by C, while the conditional action of C on the second qubit is given by the Pauli matrix σ_x. These two requirements do not specify the necessary transformations uniquely. One possible choice is to use the transformations that correspond physically to modulation of the tunnel coupling between the states of the second qubit (*i.e.*, involve only matrices σ_x, σ_y). In this case, we obtain

$$C = S(\pi/2)\,(U_-)^\dagger\, S(-\pi/3)\, P\, U_-\, U_+ S(-\pi/3)\, P\,(U_+)^\dagger\,, \tag{2}$$

where $U_\pm = [\mathbf{1}]_1 \otimes [\exp\{-i\varphi(\sigma_x \pm \sigma_y)/\sqrt{2}\}]_2$. Here the subscripts 1, 2 denote the part of the transformation acting on the first and the second qubit, respectively, and the rotation angle φ is given by the condition $\cos(2\varphi) = 1/\sqrt{3}$, $0 \le \varphi \le \pi/2$. Physically, the transformations S can be implemented as pulses of the gate voltage applied to the antidot $|1\rangle$ of the first qubit, while the U's represent pulsed modulation of the amplitude of the tunnel coupling between the two antidots of the second qubit that keeps the phase of this coupling fixed.

4. Decoherence mechanisms

As the next step, we analyze decoherence mechanisms in the antidot qubits. At low temperatures, the energy gap in the FQHE liquid exponentially suppresses quasiparticle excitations in the bulk of the sample. Due to this suppression only sample edges and external metallic gate electrodes support low energy excitations that can give rise to dissipation and decoherence. A qubit is coupled to both the gate electrodes and the edges by the Coulomb interaction. The charge q of the qubit quasiparticle ($q = e/m$, where m is an odd integer for the primary Laughlin FQHE liquids) induces a polarization charge on the gate electrodes that fluctuates in the course of qubit time evolution. The current thus induced in the electrodes with finite resistance R leads to energy dissipation and decoherence. This decoherence mechanism associated with the "electromagnetic environment" of the structure (see, *e.g.*, Ref. 14) is generic for most of the solid-state qubits. In the FQHE qubits its strength should be lower than in other charge-based qubits, due to the smaller charge of the FQHE quasiparticles. Indeed, if the gate electrode is close to an antidot (on the scale of the distance d between the two qubit antidots), the amplitude of the variations of the induced charge is roughly equal to the quasiparticle charge q. In this "worst case" scenario, the limitation on the quality factor of qubit dynamics introduced by the gate electrode is equal to $e^2 R/\hbar m^2$, and is on the order of 10^{-3} for realistic values of the resistance R and for the $m = 3$ qubits considered in this work. Optimization of the gate structure of the qubit should further reduce the strength of this type of decoherence by reducing electrostatic gate-qubit coupling.

   The Coulomb interaction also couples qubit dynamics to edge excitations of the FQHE liquid. The edge supports one-dimensional (1D) chiral plasmon modes[15] propagating with velocity v. In the situation of interest here, when the qubit-edge distance L is much larger than the qubit size d, the coupling operator V can be expressed directly in terms of the 1D density of charge $\rho(x)$ carried by plasmon modes: $V = \sigma_z \int dx U(x) \rho(x)$. In this expression, σ_z represents the position of the quasiparticle on one or the other antidot of the qubit, and $U(x)$ is the variation (with the quasiparticle position) of the electrostatic potential created by the qubit at point x along the edge. A representative estimate of the dissipation/decoherence rate introduced by this coupling is given by the decay rate Γ of the excited antisymmetric superposition of the antidot states. Assuming that the qubit dipole is perpendicular to a straight edge, and that the electric field is not screened between the edge and the qubit, we can evaluate Γ directly:

$$\Gamma = (d/L)^2 \, (e^2/4\pi\kappa\kappa_0\hbar v)^2 \, (\Omega/4\pi\hbar m^3) \, \exp[-L\Omega/\hbar v] \,, \qquad (3)$$

where κ is the material dielectric constant. This equation shows that the edge-related limitation $\hbar\Gamma/\Omega$ on the qubit quality factor can vary widely, depending on the system geometry and qubit energy parameters. For a realistic set of numbers — $\kappa \approx 10$, $v \approx 10^5$ m/s, $\Omega \approx 0.1$ K, $d \approx 100$ nm (see discussion below) — we obtain $\hbar\Gamma/\Omega \approx 10^{-3}$ for the edge that is $L \approx 3$ μm away from the qubit.

To evaluate qubit parameters we summarize the basic set of requirements necessary for correct operation of the FQHE qubits and gates described above as $kT \ll \varepsilon$ and $\Omega \ll \Delta$. The antidot excitation energy Δ is estimated as $\Delta \approx \hbar u/r$, where r is the antidot radius and $u \approx 10^4$–10^5 m/s is the velocity of quasiparticle motion around the antidot.[16] This estimate means that at a temperature $T = 0.05$ K the radius r should be smaller than 100 nm. Since the tunnel coupling Ω decreases rapidly with the tunneling distance s between the antidots, $\Omega \propto \exp[-eBs^2/12\hbar]$,[17] the fact that Ω should be at least larger than T means that s should not exceed a few magnetic lengths $l_B = (\hbar/eB)^{1/2} \sim 10$ nm for typical values of the magnetic field B. Although these requirements on the radius r and antidot spacing s can be satisfied with present-day fabrication technology, they present a formidable challenge. It should be noted that these requirements are not specific to our FQHE scheme, but characterize all semiconductor solid-state qubits based directly on the quantum dynamics of individual quasiparticles, and not collective degrees of freedom (as used, e.g., in the case of superconductors).

5.   Conclusions

We believe that the fabrication challenges facing FQHE qubits are well compensated by the advantages of the FQHE approach. First among these advantages is the energy gap of the FQHE liquid that suppresses quasiparticle excitations and the associated decoherence in the bulk of the 2DES, and allows control of the remaining sources of decoherence through the system layout — see the discussion above. The second advantage is the topological nature of statistical phase that makes it possible to entangle qubits without their direct dynamic interaction. This possibility should lead to simpler design of the FQHE quantum logic circuits in comparison to other solid-state qubits, where control of the qubit-qubit interaction typically presents a difficult problem.

Acknowledgments

This work was supported in part by the NSA and ARDA under an ARO contract.

References

1. A. Yu. Kitaev, "Fault-tolerant quantum computation by anyons," quant-ph/9707021.
2. J. Preskill, "Fault-tolerant quantum computation", in: H.-K. Lo, S. Papesku, and T. Spiller, eds., *Introduction to Quantum Computation and Information*, Singapore: World Scientific, 1998, pp. 213-269.
3. S. Lloyd, "Quantum computation with abelian anyons," quant-ph/0004010.
4. M. H. Freedman, A. Yu. Kitaev, M. J. Larsen, and Z. Wang, "Topological quantum computation," quant-ph/0101025.
5. V. J. Goldman and B. Su, "Resonant tunneling in the quantum Hall regime: measurement of fractional charge," *Science* **267**, 1010 (1995).
6. I. J. Maasilta and V. J. Goldman, "Tunneling through a coherent quantum antidot molecule," *Phys. Rev. Lett.* **84**, 1776 (2000).
7. D. V. Averin, "Adiabatic quantum computation with Cooper pairs," *Solid State Commun.* **105**, 659 (1998).
8. Yu. Makhlin, G. Schon, and A. Shnirman, "Josephson-junction qubits with controlled couplings," *Nature* **398**, 305 (1999).
9. Y. Nakamura, Yu. A Pashkin, and J. S. Tsai, "Coherent control of macroscopic quantum states in a single-Cooper-pair box," *Nature* **398**, 786 (1999).
10. B. I. Halperin, "Statistics of quasiparticles and the hierarchy of fractional quantized Hall states," *Phys. Rev. Lett.* **52**, 1583 (1984).
11. D. Arovas, J. R. Schrieffer, and F. Wilczek, "Fractional statistics and the quantum Hall effect," *Phys. Rev. Lett.* **53**, 722 (1984).
12 A. Barenco, C. H. Bennett, R. Cleve, D. P. DiVincenzo, N. Margolus, P. Shor, T. Sleator, J. A. Smolin, and H. Weinfurter, "Elementary gates for quantum computation," *Phys. Rev. A* **52**, 3457 (1995).
13. Yu. Makhlin, "Nonlocal properties of two-qubit gates and mixed states and optimization of quantum computation," quant-ph/0002045.
14. G. L. Ingold and Yu. V. Nazarov, "Charge tunneling rates in ultrasmall junctions," in: H. Grabert and M. Devoret, eds., *Single Charge Tunneling*, New York: Plenum, 1992, pp. 21-107.
15. X. G. Wen, "Theory of the edge states in fractional quantum Hall effects," *Int. J. Mod. Phys. B* **6**, 1711 (1992).
16. I. J. Maasilta and V. J. Goldman, "Energetics of quantum antidot states in the quantum Hall regime," *Phys. Rev. B* **57**, R4273 (1998).
17. A. Auerbach, "Comparison of tunneling rates of fractional charges and electrons across a quantum Hall strip," *Phys. Rev. Lett.* **80**, 817 (1998).

Photonics with Chips

Arto V. Nurmikko

Div. of Engineering, Brown University, Providence, RI 02912, U.S.A.

1. Introduction and overview

Advances in optical telecommunication within the past decade have fundamentally changed the way information can be exchanged and distributed globally. Although subject to forces of the commercial marketplace, it is likely that "transparent optical networks" will continue to lead the communication revolution in the foreseeable future. Among the contemporary challenges of such networks is the delivery by light waves of high information content/speed to an individual residential or other user, the so-called "last mile" bottleneck. Current access strategies include approaches such as "fiber-to-curb," "fiber-to-home," and so on.

Underlying the current push for expansion of optical networks is a broader issue, which concerns the "moving interface" between microelectronics and photonics. Research is well underway, for example, in the development of high-speed optoelectronic modems that will bring this interface within a personal computer. Likewise, optical cables are being implemented and are under aggressive development for applications to ultrahigh-speed backplane inter-connections for large- and medium-frame computers, featuring virtues such as a high degree of parallelism and low interference and power dissipation. It appears probable that chip-to-chip optical interconnects will replace or at least augment electrical wiring in the near future, particularly for system-on-a-chip (SOC) architectures.

Gazing into the crystal ball gets more intriguing, albeit less clear, if we consider the microelectronics/photonics interface in terms of longer term information technologies on the chip scale. The subject is open-ended, but as a concrete example, we note the challenges for future digital microelectronic chips imposed by the "wiring problem." We raise in this article the question of the prospects of photonics on the chip scale. The phrase is meant to focus the question on the possible future role of photonics as a complement or as an outright alternative to any number of tasks and functions that are presently performed by microelectronic circuits. As evidenced by other chapters in this book, not only has the "conventional" behemoth of the microelectronics industry continued its astounding pace of progress unabated, but suggestions about a shifting of gears into the world of nanoelectronics are beginning to show promise at the basic research level. In particular, there have been many intriguing developments at the intersection of nanoscale materials science and nanotechnology. Hence, any predictions of chip-scale photonics need to be weighed against the possibilities of

340

intersecting, indeed complementary, opportunities to possible future nanoelectronics.

The concept of chip-scale photonics is not meant to imply a single-minded definition in this article. Rather, we will consider some specific building blocks as possible cornerstones that are being laid today for photonic circuits and their functionality. Since subwavelength components are likely to play a major role in the ultimate implementation of dense wavelength division multiplexed (DWDM) chip-scale photonic systems, some of these might be referred to as "nanophotonic" approaches.

The organization of this article is as follows. We will first examine some current concepts and their possible implementation for free-space optical interconnects in providing for low-power interference-free microelectronic chip-to-chip and intrachip data transfer. We will then raise the question of all-photonic chip-scale components, which might be better able to perform logic or other functional operations at ultrahigh speeds not attainable by micro/nanoelectronic means. Key elements in such a nanophotonic chip include compact high-efficiency optical switches and high-density optical waveguides, as well as imbedded active elements, operating at extremely high data rates (up to THz) and exploiting the DWDM degree of freedom unique to lightwave technology.

2.   Free-space optical interconnects for electronic chips

Problems with the scaling of off-chip electrical interconnections with increasing microelectronic chip complexity, density, and performance have been appreciated for some time. Interconnect scaling is becoming an issue with on-chip wiring as well, especially for global interconnects, due to issues of power dissipation and speed. The scaling for both RC- and LC-type transmission line regimes roughly states that miniaturization of the entire interconnect wiring system does not change the RC or LC time constant, for example, whereas an "optical wire" would have no such restrictions. Such scale invariance implies little savings in reducing the wire size (diameter and length) in terms of speed and power dissipation. Motivated by this recognition, there has been active research in the past few years into the possibilities of using optics for interconnects for all but the shortest wiring distances on a chip (say <1 mm). In particular, optical interconnects appear especially well-suited for high-bandwidth input/output (I/O) wiring. The potential for optics to solve electrical interconnect problems has been advanced by the remarkable progress in photonics component technologies for long-distance optical telecommunication. While the component requirements for on-chip or off-chip (modular ICs) can be quite different from those of the telecom applications, the present technology allows flexible adaptation of the latter to the former.

The obvious and fundamental virtues of optics (at the level of pure photonics) include small propagation delays, near absence of signal distortion (group velocity some fraction of the speed of light c), absence of electromagnetic interference, and virtually no latency. Likewise, there is an intrinsically perfect voltage isolation between the input and output points.

The technical issues that relate to implementation of optical interconnects, on or off a microelectronic chip, come in at different technical layers. At the base hardware level, the question relates to the type of optoelectronic device technology that might be suitable for the purpose at hand, including the key issue of integration with the microelectronic chip backplane. At the systems level, the fundamental challenge concerns the architecture of the optical interconnect subsystem and its compatibility with the electronic chip architecture for a given processor type.

At the component level, one might expect some hierarchy of optical interconnect solutions. On-chip and off-chip optical wires can be envisioned as dielectric waveguides (rigid ridge guides that can be fabricated by planar processing techniques, or flexible optical fibers). Such guides would be subject to time- and wavelength-division multiplexing for enhanced data transfer, especially in conjunction with ultrashort laser pulses (<1 ps). An alternative is free-space (or "imaging") optical interconnects that utilize optical beam fan-outs by using refractive or diffractive optical elements with sub-wavelength features, such as arrays of microlenses.

Two types of optoelectronic devices have been demonstrated so far as interconnects with silicon CMOS chips. The quantum well III-V semiconductor electroabsorptive modulator approach (sometimes denoted as CMOS-SEED) grew out of the use of arrays of such planar electro-optical switches as relatively dense two-dimensional arrays for spatial light modulators, also referred to as smart pixels.[1] With an external light beam and the use of low-power photodetectors, a compact multielement optical I/O package can be constructed. In one demonstration, 4000 such elements acting effectively as optical I/O "pins" were flip-chip bonded to a CMOS chip.[2] Bonding optical interconnect devices directly over active silicon gates has also been demonstrated by several groups. Although the number of pins in a microelectronic chip may be larger (including the need for multiple pins for ground and power), it is interesting to note that the pin density is unlikely to exceed the optical case. At present, the power dissipation in the modulator and detector (photodiode) circuits is still somewhat too large but is projected to reach a point (on the order of 1 mW or less), where thousands of such "optical pins" could be imbedded on a CMOS chip without need for special heat-sinking.[3]

Vertical-cavity surface-emitting diode lasers (VCSELs) represent a rapidly developing class of active optical devices that is making significant entries into many optical telecommunication applications. Small-area devices (the diameter of the active region being a few microns) with laser threshold currents corresponding to tens of microwatts have been demonstrated, with data transfer (modulation) rates up to about 10 Gb/s per device. Monolithic integration of VCSELs and photodetectors by flip-chip bonding methods has been shown to be feasible. Such integration creates hybrid microelectronic/optoelectronic I/O circuits at power

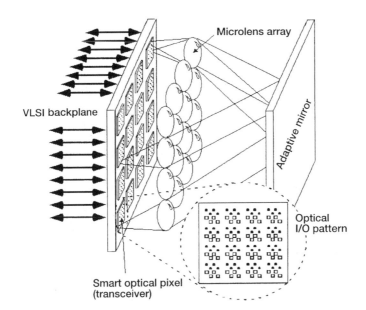

Figure 1. Schematic of FSOC interconnect arrangement utilizing arrays of smart pixels and micro-optical refractive/reflective optics.[5]

consumption levels that suggest applications to interconnect technologies.[4] Several efforts are currently underway to fabricate 2D arrays of VCSEL-driven optical interconnects, where the ultimate power dissipation issues appear to be limited mainly by the photoreceiver (thermal noise in a photodiode and its preamplifier) in the overall electrical–optical–electrical wiring.

Planar SEED and VCSEL optical interconnect drivers lend themselves to exploitation of the basic geometrical virtue of optics in the third spatial dimension perpendicular to the chip. Free-space optical communication (FSOC) is particularly attractive in this instance. Figure 1 shows a schematic illustration of such an approach at the design/concept level.[5] With "smart pixel" optoelectronic device arrays, arbitrary interconnect patterns (fabric) are in principle possible. The use of microlens arrays and related nanostructured optical elements to fan out and image many beams from a single VCSEL for space multiplexing is a field of activity well under way, exploiting high-resolution lithographies (nanoimprint and electron beam lithography). The left-hand panel of Fig. 2 shows an illustration of a portion of such a fabricated planar microlens array (by grayscale lithography) and its far-field fan-out (from 1 to 64 beams).[6] These and related diffractive optical elements, with sub-wavelength index of refraction texture, are subject to further miniaturization for implementing free-space optical interconnect schemes. Estimates have been made for an interconnect technology for I/O application on an electronic chip with data transfer rate density beyond 1 Tb/sec/cm^2, at a density of

~1000 optical I/O "wires" per cm^2. It is not obvious how electrical interconnects (*e.g.* a high-performance bus) can be scaled to this regime of aggregate I/O bandwidth. Several authors have made comparisons between electrical interconnects (on-chip, off-chip) and free-space optical interconnects in terms of delay and energy (power dissipation). While such estimates clearly include many assumptions (and would depend on the architecture employed), it has been argued that SEED- or VCSEL-based optical interconnects can offer a power-delay product a factor of 30 better (lower) than future all-electrical equivalents.[7]

The design of a future optical interconnect system, based either on the free-space approach outlined above or on the use of guided-wave optics, must also adapt to the overall architectural constraints of the optoelectronic-VLSI (OE-VLSI) system-on-chip. Basically, one looks for a balance between the specific computational tasks of the microelectronic chip and the communication tasks and capabilities of the optical subsystem. Similar to the case of optical telecom, it is not difficult to envision an optical "bus" with a huge available bandwidth (> Tb/s) that carries the aggregate data between chips (in a modular system) or specific functional blocks/locations on a single chip. In simplest terms, the problem is one of optimizing the distribution system of the data traffic at the input and output to the optical bus. This optimization is sometimes referred to as the "firehose problem,"[1] whose architectural solution is largely influenced by the space-time utilization of the optical I/O bandwidth of each OE-VLSI chip interface, given the significantly lower electrical I/O bandwidth to the same chip.[5,8] There are current efforts to use information theoretic analyses for the design of such architecture, as well as approaches that take a computer architect's view of today's digital signal processors for the design of optical interconnects. A great deal of potentially applicable prior science exists, *e.g.* in the area of application mapping approaches

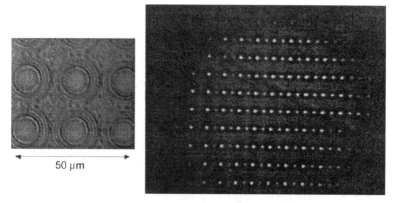

50 µm

Figure 2. Portion of an array of planar diffractive optical microlenses, with sub-wavelength features (left); experimental demonstration of multiple-beam fan-out by the array.[6]

to embedded multiprocessors, where techniques are available for performance estimates and design-space exploration for selecting the architectural template and optimizing inter-processor communication and synchronization.

3. Chip-scale photonics

We now speculate on the prospects of an *all-photonic* chip, based on the general idea of eventual ultra-miniaturization of the wavelength-multiplexed transparent optical networks down to the chip scale. The main benefit of such a chip would be based on the prospect of very high speed operation (>>100 GHz) and the ability to deal with several optical wavelengths simultaneously, as in dense wavelength division multiplexing (DWDM). Figure 3 shows an artist's sketch of a possible future chip-scale all-photonic processor, with some endowed functionality such as signal-processing logic. (A discussion of optical memory is beyond the scope of this article). While the implementation of the concept is likely to take on any number of different looks, with architecture defined by the design and integration of specific elements, some generic key device elements can be identified. In the following sections we will discuss two such elements: the seminal problem of implementing a low-energy ultrafast optical switch and the imbedding of optical wiring on such a chip. Additional important issues include the employment of chip-scale diffractive optics to capture and fully implement the DWDM opportunities, I/O coupling to the chip, and the need for a stable optical on-board clock. Any or all of these devices may require features on the subwavelength scale, both for compactness and exploitation of near-field optical effects.

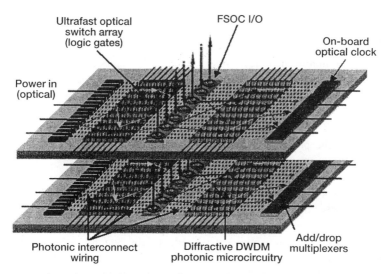

Figure 3. Artist's sketch of an all-photonic chip, listing some of the key device components and circuit elements.

4.   The ultrafast optical switch

The concept of an ultrafast light-by-light switch, based on the nonlinear optical response of a material, is an old subject. Numerous demonstrations of optical gates that operate at picosecond or subpicosecond speeds can be found in the literature over the past three decades. For any chip-scale application, however, where a possibly large density of such switches is envisioned, *e.g.* for interconnected photonic logic gates, the speed of the switch must be matched with low energy dissipation (including insertion losses). When compared with microelectronic switches (*e.g.* in silicon CMOS), switching energies in the fJ regime will be required in the optical analog. Only recently have high-speed optical switches begun to appear on the horizon, in terms of approaching such a low-energy requirement. We note that when optical computing was under rather intense research efforts about a decade ago, the lack of high-performance optical switches was probably as much of a reason as anything that ultimately impeded the actual physical implementation of practical optical computing architectures in early 1990's. The search for optical switches proved largely unsuccessful, basically because the simultaneous requirements of large nonlinearity, low loss, and short lifetime were physically incompatible.

The physics of actuation of the simplest nonlinear optical gate is usefully categorized according to resonant or nonresonant interaction of the medium with the optical light wave. Nonresonant interaction (featuring third-order susceptibility $\chi^{(3)}$ such as in the nonlinear Kerr effect) is attractive because one operates in the largely transparent regime. However, the required optical intensities I and optical path lengths tend to be very large ($I > 1$ MW/cm^2, where typically power-length products of ~1 W·km are needed if silica fiber is used as the medium). Moreover, there are arguments from a general sum rule point of view[9] that make it unlikely that new "magic" materials may be found that possess values of $\chi^{(3)}$ that significantly exceed those of a very large number of solids, liquids, and gases already studied during the past three decades.

Resonant nonlinear optical effects in semiconductors imply optically induced population changes at the band edges in either absorbing or gain media, leading to saturation of absorption or optical gain coefficients as well as associated changes in the index of refraction. Here we examine two specific candidates for a resonant optical switch: the switching in a semiconductor optical amplifier (via index change) and the prospect of using a saturable absorber medium in an optical microcavity.

- *Semiconductor optical amplifier in an interferometric switch*

An effective optical switch can be constructed from a semiconductor optical amplifier (SOA), imbedded in a phase-sensitive optical interferometer.[10] With pulses of photoexcitation above the bandgap (the "pump"), the gain is driven towards saturation by induced emission, and an index of refraction change ensues. The dynamics of the process have been studied in detail[11] to show that rather small electron-hole population changes (on the order of 10^{17} cm^{-3}, or about 10% of the equilibrium inversion density) result in an index change of $\Delta n \sim 10^{-3}$. As one consequence of the small-signal modulation regime, the recovery of the induced change can be quite fast, well below the ~1-ns recovery time of the full population inversion. When the SOA is inserted within an optical-fiber interferometer as a "nonlinear mirror in the loop," fast optical switching can be achieved. In principle, switching windows below 10 ps and repetition rates up to 100 GHz are feasible, as shown in the switching data of Fig. 4.[12] To date, the SOA fiber-loop mirror has been demonstrated in many laboratories to achieve a multitude of switching functions including an add-drop multiplexer, wavelength conversion, all-optical regenerative memory, and an approach to optical logic. In their studies of the SOA dynamics, Ueno *et al.* have reported record low-power implementation of an SOA-based optical switch in InGaAsP at 1548/1560 nm (pump/probe) at pump pulse energies of 6 fJ and a repetition rate of 42 GHz.[13] Elsewhere, these optical gates have been operated at data rates in excess of 100 Gb/s in wavelength-division multiplexing applications.[14]

An example of the use of SOA-interferometric fiber-loop switches to optical logic is the recent demonstration of reversible optical logic.[15] The scheme is based on the concept of the Fredkin-Toffoli gate for conservative logic,[16] which has been considered as a universal gate to realize all standard Boolean functions. Figure 5 shows the schematic of this arrangement, where three input ports and three output ports complete the reversible logic whose outputs can be used to determine uniquely the input state to the logic gate. As usual, the SOA is temporally displaced from the fiber loop center to create a switching window in time in which a control pulse via port C can reduce the gain of the SOA. This gain change can induce a π radians differential phase shift between the two counter-propagating signal pulses in the loop and switch the signal from the input port A or B to the output port A' or B'. Note the use of three different wavelengths for A, B, and C, in the spirit of "wavelength division multiplexing" which is a fundamental asset of optoelectronics *vs.* microelectronics. Figure 5 also includes the truth table for this universal gate, which so far has been operated at (pump) pulse energies of 100 fJ.

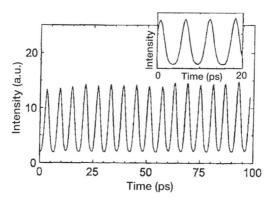

Figure 4. Example of high-speed optical switching by optical phase modulation in a SOA-based InGaAsP gate.[12]

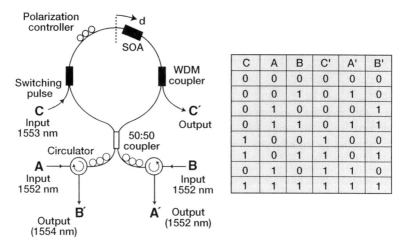

C	A	B	C'	A'	B'
0	0	0	0	0	0
0	0	1	0	1	0
0	1	0	0	0	1
0	1	1	0	1	1
1	0	0	1	0	0
1	0	1	1	0	1
0	1	0	1	1	0
1	1	1	1	1	1

Figure 5. Schematic of a specialized all-photonic logic gate based on an SOA interferometer (left) and the truth table for the three-wavelength I/O device (right).[15]

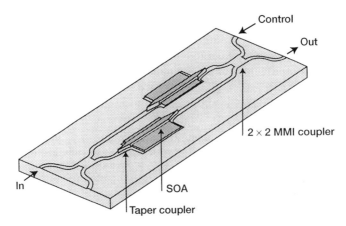

Figure 6. Layout of a monolithic SOA-based Mach-Zehnder interferometer high-speed optical switch, featuring twin-waveguide tapered mode couplers to the active InGaAsP QW region.[18]

For a chip-scale compact low-energy version of such a logic gate or other comparable arrangements for achieving specialized logic operations, *e.g.* for quantum computing,[17] a monolithic version is required. With advances in planar optical waveguide techniques, the implementation of the basic interferometric switch has already been demonstrated.[18] Figure 6 shows one such monolithic arrangement, using two SOAs within a ridge-waveguide Mach-Zehnder inteferometer in a co-propagating mode for both the pump and probe pulses (traveling to the left and the right in the drawing, respectively). The design takes advantage of a "twin-waveguide" arrangement, where the SOA section is reached by light within the guides though a taper mode converter, which on entry pushes the waveguide mode up the chip to the active region for optical amplification. The interferometer includes a pair of 2×2 MMI (multimode interference) power splitters. In the first demonstrations, the device has been operated at the somewhat large switching energies of ~100 fJ in the 1550-nm range. However, significant energy reduction can be expected through design optimization and the choice of wavelengths and materials. The overall footprint of the device is approximately 200×1000 μm^2, the length dictated largely by that of the SOA sections.

Research efforts are currently underway to develop monolithic optical logic elements, for example, the Fredkin gate, as a basis for specialized optical computation/logic analysis at very high speeds (potentially up to the THz range). Apart from its functionality, this multiwavelength gate is useful in the context of potentially incorporating a finite number of such devices on a single chip. The footprint of the SOA-based interferometric switch, as described above, is large by the standards of microelectronic components. The Fredkin gate is an example of a multifunctional logic element that utilizes the chip real estate more effectively than a collection of simple binary logic gates. An important space-saving issue concerns the implementation of optical waveguides with small radii of curvatures. In addition to conventional designs based on high-index-contrast ridge waveguides, "wiring strategies" such as those based on photonic crystals look attractive. At the same time, it seems clear that other geometries and perhaps other gain media for the SOA are needed as an alternative to the present edge-emitting configuration, for possible future ultrahigh-speed reasonably compact chip-scale "designer" optical logic circuits.

- *Exciton-polaritons and optical switching*

Passive saturable semiconductor absorbers that operate on the basis of simple electron-hole band filling effects are of limited utility where the performance (speed, saturation intensity) is determined by a combination of interband and intraband processes. The former generally restrict the response time. Techniques such as proton implantation of low-temperature epitaxy have been used to shorten the electron-hole pair lifetime in GaAs-based quantum well absorbers and to incorporate them as passive mode-locking elements in solid-state lasers.[19] The bleaching of the absorption is magnified by quasi-2D exciton effects in quantum wells (phase-space filling and exchange effects) but the associated figure of merit $\Delta\chi/\chi \sim \chi^{(3)}$ is limited fundamentally by an oscillator sum-rule argument for Wannier excitons. For typical III-V semiconductors, this argument makes it unlikely that a simple excitonic saturable absorber will make a viable optical switch in any high speed optical logic or chip-scale processor functions.

Recently, several theoretical ideas have been proposed for enhancing the performance of a resonant excitonic saturable absorber for switching applications.

The basic idea is to increase the effective oscillator strength of the (exciton) dipole, either by imbedding the absorber in an optical microcavity for enhanced light-matter coupling, or by creating a hybrid electronic layered material composed of inorganic and organic semiconductors,[20,21] or both. Here we briefly summarize the ideas by Agnonovich et al.[21] that, as of this writing, remain to be experimentally realized.

Optical resonances in organic semiconductors are dominated by Frenkel excitons, whose effective Bohr radii are of atomic dimensions (in contrast, a Wannier exciton in GaAs has a radius of ~10 nm). These resonance possess very large oscillator strengths but require very high excitation levels for bleaching of absorption. The hybridization of a Frenkel and a Wannier exciton can be calculated to possess both a high oscillator strength (which implies a large resonant $\chi^{(3)}$ as well as a short radiative lifetime, among other things) and a relative ease of saturation (similar to that of a Wannier exciton). This combination makes them interesting candidates for versatile high-performance optical switches for chip-scale photonics. The approach to implementing the hybrid exciton absorber offers some choices. The most immediate approach involves precision epitaxy where organic and inorganic semiconductor quantum wells are proximate so that their electronic wavefunctions overlap due to tunneling. This proximity can be difficult in practice as most inorganic semiconductors (e.g. GaAs) display a high density of charged surface states, which is at variance with the required interface perfection for the hybrid exciton concept.

On the other hand, enclosing the two types of semiconductor quantum wells within an optical microcavity (e.g. a one-dimensional structure such as a Fabry-Perot cavity), considerably relaxes the material-synthesis problems. In this instance, the optical field within the cavity is tuned to the resonance of the two types of excitons (QW layers separated by an optical wavelength at antinodes of the field), so as to provide a photon-mediated coupling for the hybridization.[21] In the "strong coupling regime" of the microcavity (frequency splitting of the polariton modes exceeding the unhybridized exciton linewidths), we may expect the benefits of both types of excitons for facilitating a high-performance optical switch. With anticipated short radiative lifetimes, (on the order of 10 ps) a large fraction of the excitation energy deposited into the system by the switching pump pulses should be released as radiant electromagnetic energy and thus lead to significantly reduced heat generation in the switch. We note that the effects of a microcavity have been considered also in the context of modifying the dipole-dipole interaction,[22] discussed in the context of an optical "wire" in the next section, but suggesting yet another physical mechanism for a high-performance excitonic optical switch.

5.   Optical wiring : Photonic crystals and near-field guides

Versatile dielectric waveguides based on planar processing techniques are commonly used in modern optoelectronic devices for circulators, isolators, power splitters, and many other passive and active devices. Layered inorganic glasses, polymers, crystalline semiconductors, and other material classes are being used, *e.g.* in ridge waveguides, with vertical confinement defined by index contrasts between layers and lateral confinement defined by lithography and etching. Typically, the lateral waveguide dimensions are on the order of a wavelength for single-mode guides, while minimum bending radii tend to be on the order of many wavelengths for acceptably small radiation losses. This situation is unlikely to change as it is fundamentally limited by the finite index of refraction difference between the guide and cladding layers for practical optoelectronic materials. Hence photonic circuits of even modest functionality will be difficult to realize on a scale, say, smaller than about 1 cm^2.

For high-density optical wiring on a photonic chip, it is likely that some combination of photonic crystal guides and those based on near-field effects (subwavelength dimensions) will be an integral part of such a futuristic scenario. The development of photonic crystals as "optical wires" is under active research worldwide and will not be reviewed here. Suffice it to say that within the past year or so, relatively high-performance guides based on two-dimensional photonic crystals have been demonstrated.[23] Typically, periodic arrays of airholes etched into a planar semiconductor waveguide with (unpatterned) lanes of "defects" define the guiding structures, which have been demonstrated with abrupt bends beyond 60° redirection of the wire. One example of such a fabricated arrangement is shown in Fig. 7, involving a GaAs guide clad by Al_xO_y and SiO_2

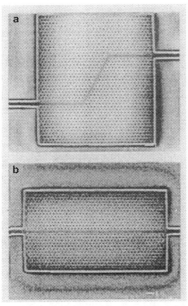

Figure 7. Plan view of two types of photonic crystal waveguides (60° bend and straight) composed of "defect rows" in an air-hole type photonic crystal in GaAs planar waveguide (from Ref. 24).

layers.[24] In a more sophisticated scheme of bending a photonic wire, Fabry-Perot-like resonant cavities have been embedded at the bend location for emphasis on the wavelength selectivity of such structures.[25] In general, one can imagine more complex photonic wiring schemes using these types of wavelength-size guides, including the extension to directions perpendicular to the plane of the chip. Waveguide couplers, ring resonators, routers, add/drop filters and other building blocks can be expected in future applications of photonic crystal guides. Likewise, active media can be embedded into such structures[26] for implementing ultralow threshold lasers or local "repeater" stages (e.g. with a semiconductor amplifying medium) to combat insertion losses.

The photonic crystal guides have their own "I/O problem" that stems from the large optical insertion losses when these wires are connected to conventional index guides (e.g. ridge geometry). The problem is fundamentally that of a mode mismatch between the very tightly confined photonic crystal mode and the loosely bound index-guided mode, which typically restricts the coupling efficiency to a few percent at best, i.e. an unacceptably low value. The solution to the problem requires the design and fabrication of suitable "mode transformers," among which are generic tapered geometries for the conventional guides inserted within an "adiabatic" transition region into the photonic crystal guide.[27]

Photonic crystal guides continue to present a practical fabrication challenge: most test demonstrations involve electron beam lithography with relatively deep submicron features etched into the layered material of choice. Such process strategies are unlikely to be commercially viable and thus ongoing efforts are also focusing on other types of methods in the fabrication of the periodic structures that underlie the photonic crystal concept. Holographic lithography, nanoimprint techniques, and self-assembled nanostructures are examples of such contemporary research thrusts.

A very different strategy for developing high-density photonic wiring for future on-chip applications is based on the use of mixed electromagnetic and electronic modes (polaritons) in defining the subwavelength diameter guide. Recently, surface and volume plasma modes have been chosen as the electronic excitation in demonstrations of this general approach, but opportunities with other material resonances possessing large cross-sections for optical excitation exist as well. We cite here one class of proposed approaches for "near field" guiding by plasmon excitations in simple metals that takes advantage of advances made in the synthesis of various metallic nanoparticles by chemical and physical means, e.g. enhanced Raman scattering from spherical Ag (or silvered dielectric) particles. This approach suggests the possibility of a subwavelength-sized waveguide composed of a linear chain of metal nanoparticles (spherical, cylindrical, etc.) where light propagates by electrodynamic interparticle coupling.[28] The particle spacing d is considered to be small enough so that dipole-dipole coupling dominates the energy transfer between nearest neighbor particles, in close analogy to Forster resonant energy transfer in molecular dyes (donor-acceptor pairs). The interaction, with coupling strength scaling as d^{-3}, is nonradiative and by definition occurs in the near-field regime.

As with any metallic particles at optical frequencies, attenuation by the decay of the plasmon to single electron excitations is fundamentally large at room temperature. Thus, for an array of 25-nm-radius Ag particles spaced by 75 nm in vacuum, attenuation losses on the order of 3 dB / 0.5 μm have been estimated at the surface plasmon energy in the visible (green) range. Given the coherent propagation of electromagnetic energy in such a guide around sharp corners (bending radius much below the wavelength of light) even such high losses might be acceptable for submicron optical interconnects (including a 3D interconnect architecture). In designing the supporting medium for such a nanoparticle array guide, one might also take advantage of the recently observed "squeezing" of the electromagnetic energy in the immediate vicinity of the nanoparticles.[29] Elsewhere, such anomalous near-field spatial distributions of plasmon-polariton waves near sharp boundaries have been studied recently in enhanced optical transmission through subwavelength apertures,[30] a physical process that may have applications to high-spatial-resolution optical imaging or beam steering on the chip scale. We emphasize that other types of electronic resonances in nanostructured media may provide alternative candidates for the dipolarly-coupled near-field photonic wires, with considerably smaller losses than those that are intrinsic to plasmon excitations in metals.

6. Conclusions

We have taken a rough cut across many efforts aimed at pushing the microelectronic-photonic interface towards the chip scale. A significant portion of the discussion has been devoted to ideas whose implementation is still at a rudimentary stage compared with the high degree of sophistication found in microelectronics. We have also omitted key components for a future ultrahigh-speed chip-scale photonic processor, such as adaptive diffractive spectral elements and ultrashort-pulse semiconductor lasers. In the intersection of nanomaterials and nanoscience with basic optical physics and photonic technologies, a wide spectrum of exciting research opportunities exists for translating the artist's sketch of Fig. 3 into some form of eventual reality. This reality would offer technical solutions for signal processing and specialized computing which, due to fundamental limitations, may not be accessible by future nano- and microelectronic systems.

References

1. D. A. B. Miller, "Rationale and challenges for optical interconnects to electronic chips," *Proc. IEEE* **88**, 728 (2000).
2. A. L. Lentine, K. W. Goossen, J. A. Walker, L. M. F. Chirovsky, L. A. D'Asaro, *et al.*, "Arrays of optoelectronic switching nodes comprised of flip-chip-bonded MQW modulators and detectors on silicon CMOS circuitry," *IEEE Photon. Technol. Lett.* **8**, 221 (1996).

3. A. V. Krishnamoorthy and D. A. B. Miller, "Scaling optoelectronic–VLSI circuits into the 21st century: A technology roadmap," *IEEE J. Selected Topics Quantum Electronics* **2**, 55 (1996).

4. O. Sjolund, D. A. Louderback, E. R. Hegblom, J. Ko, and L. A. Coldren, "Monolithic integration of substrate input output resonant photodetectors and vertical-cavity lasers," *IEEE J. Quantum Electronics* **35**, 1015 (1999).

5. M. W Haney, M. P Christensen, P. Milojkovic, G. J. Fokken, M. Vickberg, *et al.*, "Description and evaluation of the FAST-Net smart pixel-based optical interconnection prototype," *Proc. IEEE* **88**, 819 (2000).

6. D. Prather, (private communication); see also J. N. Mait, A. Scherer, O. Dial, D. Prather, and X. Gao, "Diffractive lens fabricated with binary features less than 60 nm," *Opt. Lett.* **25**, 381 (2000).

7. G. I. Yayla, P. J. Marchand, and S. C. Esener, "Speed and energy analysis of digital interconnections: comparison of on-chip, off-chip, and free-space technologies," *Appl. Optics* **37**, 205-227 (1998).

8. Yue Liu, E. M. Strzelecka, J. Nohava, M. K. Hibb-Brenner, and E. Towe, "Smart-pixel array technology for free-space optical interconnects," *Proc. IEEE* **88**, 764 (2000).

9. M. G. Kuzyk, "Physical limits on electronic nonlinear molecular susceptibilities," *Phys. Rev. Lett.* **85**, 1218 (2000).

10. A. Ehrhardt, M. Eiselt, G. Grosskopf, L. Kuller, R. Ludwig, *et al.*, "Semiconductor laser amplifier as optical switching gate," *J. Lightwave Technol.* **11**, 1287 (1993); M. G. Kane, I. Glesk, J. P. Sokoloff, and P. R. Prucnal, "Asymmetric optical loop mirror: analysis of an all-optical switch," *Appl. Opt.* **33**, 6833 (1994); R. J. Manning and D. A. O. Davies, "Three-wavelength device for all-optical signal processing," *Opt. Lett.* **19**, 889 (1994).

11. R. J. Manning, A. D. Ellis, A. J. Poustie, and K. J. Blow, "Semiconductor laser amplifiers for ultrafast all-optical signal processing," *J. Opt. Soc. Am. B* **14**, 3204 (1997).

12. S. Nakamura, Y. Ueno, and K. Tajima, "168-Gb/s all-optical wavelength conversion with a symmetric-Mach–Zehnder-type switch," *IEEE Photon. Technol. Lett.* **13**, 1091 (2001).

13. Y. Ueno, S. Nakamura, and K. Tajima, "Record low-power all-optical semiconductor switch operation at ultrafast repetition rates above the carrier cutoff frequency," *Opt. Lett.* **23**, 1846 (1998).

14. K. L. Hall, K. A. Rauschenbach, "100-Gbit/s bitwise logic," *Opt. Lett.* **23**, 1271 (1998); C. Schubert, S. Diez, J. Berger, R. Ludwig, U. Feiste, *et al.*, "160-Gb/s all-optical demultiplexing using a gain-transparent ultrafast-nonlinear interferometer," *IEEE Photon. Technol. Lett.* **13**, 475 (2001).

15. A. J. Poustie and K. J. Blow, "Demonstration of an all-optical Fredkin gate," *Opt. Commun.* **174**, 317 (2000).

16. E. Fredkin and T. Toffoli, "Conservative logic," *Int. J. Theor. Phys.* **21**, 219 (1982).

17. A. Ekert and R. Jozsa, "Quantum computation and Shor's factoring algorithm," *Rev. Mod. Phys.* **68**, 733 (1996).
18. Fengnian Xia, J. K Thomson, M. R Gokhale, P. V Studenkov,.Jian Wei, W Lin, and S. R Forrest, "An asymmetric twin-waveguide high-bandwidth photodiode using a lateral taper coupler," *IEEE Photon. Technol. Lett.* **13**, 845 (2001).
19. See, for example, D. A. Yanson, M. W. Street, S. D. McDougall, I. G. Thayne, and J. H. Marsh, "Terahertz repetition frequencies from harmonic mode-locked monolithic compound-cavity laser diodes," *Appl. Phys. Lett.* **78**, 3571 (2001).
20. V. M. Agranovich, H. Benisty, and C. Weisbuch, "Organic and inorganic quantum wells in a microcavity: Frenkel-Wannier-Mott excitons hybridization and energy transformation," *Solid State Commun.* **102** , 631 (1997).
21. V. M. Agranovich , D. M. Basko, G. C. LaRocca, and F. Bassani, "Excitons and optical nonlinearities in hybrid organic–inorganic nanostructures," *J. Phys. Cond. Matter* **10**, 9369 (1998).
22. G. S. Agarwal and S. Dutta Gupta, "Microcavity-induced modification of the dipole-dipole interaction," *Phys. Rev. A* **57**, 667 (1998).
23. See, for example, N. Fukaya, D. Ohsaki, and T. Baba, "Two-dimensional photonic crystal waveguides with 60° bends in a thin slab structure," *Jpn. J. Appl. Phys.* **39**, 2619 (2000).
24. E. Chow, S. Y. Lin, J. R. Wendt, S. G. Johnson, and J. D. Joannopoulos, "Quantitative analysis of bending efficiency in photonic-crystal waveguide bends at 1.55 μm wavelengths," *Opt. Lett.* **26**, 286 (2001).
25. S. Olivier, H. Benisty, M. Rattier, C. Weisbuch, M. Qiu, *et al.*, "Resonant and nonresonant transmission through waveguide bends in a planar photonic crystal," *Appl. Phys. Lett.* **79**, 2514 (2001).
26. M. Meier, A. Mekis, A. Dodabalapur, A. Timko, R. E. Slusher, J. D. Joannopoulos, and O. Nalamasu, "Laser action from two-dimensional distributed feedback in photonic crystals," *Appl. Phys. Lett.* **74**, 7 (1999).
27. M. Palamaru and Ph. Lalanne, "Photonic crystal waveguides: Out-of-plane losses and adiabatic modal conversion," *Appl. Phys. Lett.* **78**, 1466 (2001).
28. J. Takahara, S. Yamagishi, H. Taki, A. Morimoto, and T. Kobayashi, "Guiding of a one-dimensional optical beam with nanometer diameter," *Opt. Lett.* **22**, 475 (1997); M. Quinten, A. Leitner, J. R. Krenn, and F. Aussenegg, "Electromagnetic energy transport via linear chains of silver nanoparticles," *Opt. Lett.* **23**, 1331 (1998); M. L. Brongersma, J. W. Hartman, and H. A. Atwater, "Electromagnetic energy transfer and switching in nanoparticle chain arrays below the diffraction limit," *Phys. Rev B* **62**, R16356 (2000).
29. J. R. Krenn, A. Dereux, J. C. Weeber, E. Bourillot, Y. Lacroute, *et al.*, "Squeezing the optical near-field zone by plasmon coupling of metallic nanoparticles," *Phys. Rev. Lett.* **82**, 2590 (1999).
30. A. Nahata, T. Thio, R. Linke, T. Ishi, and K. Ohashi, to be published in *Digest Conf. Lasers Electro-Optics CLEO-2002*, p. 93.

Metacrystals: Three Dimensional Systems of Interacting Quantum Dots

D. Johnstone

Air Force Office of Scientific Research
801 N. Randolph Street, Arlington, VA 22203, U.S.A.

1. Introduction

Extensive work has put the nanoscopic quantum dot squarely in the center of future developments of a new class of engineered semiconductors. The quantum dot, as a starting tool for engineered semiconductors, is unique in that it has discrete energy levels similar to atoms. As with semiconductors synthesized from atoms, the wavefunction overlap between closely spaced quantum dots can cause energy bands to form. Research up until this point focused on techniques to obtain quantum dots of uniform size and uniform spacing. The challenge now is to determine how properties determined by the size, composition, spacing, and interlayer material can be manipulated to obtain a desired band structure. Furthermore, it is conceivable that properties such as the effective mass and band gap not only can be tailored through proper control of the dot composition, but also can be controlled dynamically through externally applied signals.

To some extent, the parallel between interacting quantum dots and bulk semiconductors is appropriate. However, interacting quantum dots can give access to potentially easier control over the aggregate properties, with the added aspect that there is now a medium separating what were the atoms, and are now the quantum dots. The trend that started with quantum wells continues with quantum dots, where the transition energies can shift by adjusting various perturbations applied via the interlayer material.

The freedom to tailor the material properties comes from the ability to select quantum dot size, spacing, and composition. Once fabrication methods mature to the point where they can select and control sizes or composition of alternating sublattices of dots, we will be able to synthesize the systems of quantum dots analogous to compound semiconductors. Further control can be exercised by the appropriate selection of the spacing and orientation of the quantum dots, as the crystal system controls the properties of bulk semiconductors.

Several other areas of investigation show considerable promise. It would be worthwhile to explore the range of structural properties that do not have apparent analogs in the standard semiconductor realm, such as structures with periodicity over several quantum dot "atoms." In addition, various polaritons are expected to interact in the same frequency range as a consequence of reduced size, resulting in unique properties. The Kondo effect has been observed in quantum dots, where the spin on an odd number of electrons in a quantum dot interacts with the spin of the surrounding electron gas, causing conductance peaks.[1] Another consideration is access to greater nonlinear properties achieved from synthesis of quantum dot systems designed with higher degrees of non-centrosymmetry.

2. Perspectives and prospects in superlattices

Preliminary groundwork clearly demonstrates the potential of systems of interacting quantum dots. Results from quantum well superlattice (SL) structures have laid a foundation for results expected for systems of quantum dots. Accomplishments directly with quantum dot superlattices have been sparse but exciting. There has been an intense flurry of research into Bloch oscillations as a means of accessing terahertz radiation frequencies. We can also get an idea of the magnitude of shifts in the band properties of quantum dot SLs from studies of quantum well SLs under magnetic or electric fields. This section gives a brief cross-section of the current literature.

Quantum well superlattice structures made possible observation of Bloch oscillations, which were predicted in periodic structures in 1928, but not observed until a decade ago because of material constraints. Now, with the advent of molecular beam epitaxy, low-defect superlattices have been grown and used for radiation detection above 10 THz, at frequencies inversely proportional to the superlattice period, in a device measuring reduced dc current under THz irradiation.[2] These authors found that increasing the current density improves the responsivity of the superlattice, which drops rapidly when the frequency of the radiation reaches the frequency of the transverse optical (TO) phonons. Emission of multiple electromagnetic harmonics at THz frequencies has also been generated where the cutoff frequency was found to be limited by high applied field strength.[3] Similarly, bias applied to a GaAs/AlAs superlattice results in second and third harmonic generation.[4] Without bias, the structure generates only third harmonics of the input 0.7 THz beam. Other studies have shown that collisions in the electron distribution at discrete values of amplitude of an applied oscillating field result in a self-induced transparency.[5] At high carrier densities, the Bloch oscillations interact with background plasma oscillations, causing the motion of the wave packet to become anharmonic, and breaking up the Bloch oscillations.[6] Oscillating carriers in a coherent plasmon change the static bias field where the Bloch oscillations occur, resulting in the anharmonicity. In 2D and 3D quantum dot superlattices, the width of the miniband depends exponentially on the crystallographic index.[7] Coupling between Bloch oscillating electrons and longitudinal phonons can generate resonant phonon excitation.[8]

Screening of an applied electric field results in a transient frequency shift, or chirp, of Bloch oscillations observed as the electrons transition from coherent oscillatory motion to incoherent drift transport.[9] Miniband structure calculations for a linear array of quantum dots find that, for island spacing greater than 4 nm, the miniband is narrower than the optical phonon energy.[10] Furthermore, the gap between the first and second minibands is greater than the optical phonon energy.

The decreased optical phonon scattering that results leads to structures that are well-suited for Bloch oscillations.

Many of the studies of stacked quantum dots employ GaAs with InGaAs dots. In III-Vs, the strain field minimum occurs above the previously grown island. However, for elastically anisotropic materials such as IV-VIs this is not the case. For IV-VI PbSe islands in $PbSe_{1-x}Eu_xTe$ spacer layers, the material above each dot has three strain minima laterally displaced on the surface.[11] This effect can control the lateral spacing of dots. Changing the spacer layer from 40 nm to 57 nm shifted the lateral dot spacing from 55 nm to 80 nm, respectively. The hexagonal ordering of dots also improved with increasing number of periods.

Optical studies of the decay lifetime of quantum dot systems have shown that the excitation density can affect the exciton lifetime in quantum dot systems that are dense enough to interact. For sufficiently low excitation density, the exciton coherence length can exceed the quantum dot size, preserving the coherent coupling between dots and shortening the recombination lifetime.[12]

Phonon effects can be more pronounced in quantum dot systems than other systems with lower dimensional confinement. Bias applied to InP dots in InGaP significantly reduces photoluminescence from hot electrons if the carriers relax to the ground states via emission of acoustic phonons, when compared to relaxation via emission of longitudinal optical phonons.[13] Phonon confinement in the quantum dot is another consideration. Discontinuity in the phonon dispersion relations of the barrier and dot materials results in the observed confinement. In order to account for the observed behavior, it is also necessary to include optical vibrations that are hybridized modes between the barrier and dot modes.[14]

Acoustic phonons can determine the miniband transport in 2D quantum dot superlattices.[15] As the electric field increases, the conduction in an InGaAs QD superlattice makes a transition from miniband conduction to hopping. Observation of negative differential photocurrent signifies the metal-insulator transition.

3.   Examples of QD achievements

Investigations at the University of New Mexico, University of Southern California, and University of California, Santa Barbara point to specific examples of where the science of quantum dots needs to proceed for the next steps.

Quantum dots in a well demonstrated at the University of New Mexico enhanced the capture of carriers. The approach resulted in threshold current for lasing lower than performance values with quantum wells alone in devices for communications.[16] With the addition of high-reflection coatings on the facets, the threshold current was as low as 16 A/cm^2.

"Punctuated island growth" (PIG) is an effective way to control the size distribution of quantum dots, pioneered at the University of Southern California.[17] PIG deposits the quantum dot material in two steps separated by a period of time in which the initially deposited material has time to equilibrate. This method has resulted in very narrow size distributions, evident from photoluminescence line widths of ~23 meV. Investigations sponsored under the National Nanotechnology Initiative will extend this work to integrate semiconductors and metals in quantum dots for functional devices. A spectrum of fabrication methods will be pursued, including those mentioned below.

Researchers at the University of California, Santa Barbara, have employed holographic lithography to form mesas on which a thin lattice mismatched layer is deposited.[18] The mismatched layer presents a controllable strain field for subsequent quantum dot nucleation. The alignment of the patterned mesas with the substrate orientation can control alignment of the quantum dot arrays (Fig 1). Follow-on work transitions the stressor layer investigations from InAs to nitrides. Nitrides are of interest due to their wide band gap, and large piezoelectric and spontaneous polarization properties. These unique properties are being investigated as an approach to dynamic control of interacting quantum dot band properties.

The demonstration at UNM of low threshold lasers has resulted in a commercially viable product. The work at the other two universities is more fundamental, representing important steps toward controllable three-dimensional quantum-dot metacrystals.

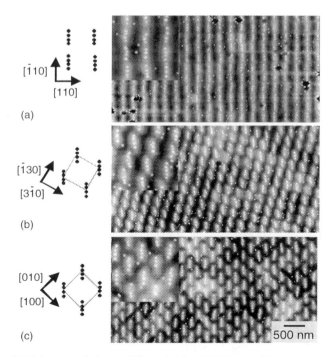

Figure 1. AFM image of three different island lattices with square unit cell and basis of three or four quantum dots aligned along a <110> direction. The unit cell vector directions indicate the orientation of the three different lattices.

4.   Fabrication of metacrystals

The potential for novel new devices indicated from structures that may seem impossible at this point should guide the search for fabrication methods. The potential of metacrystals should be predicted first using modeling and simulation methods as a guide to narrow the investigation of fabrication methods. There is currently an array of possible fabrication methods in development, each very promising. However, each may lend itself to only a limited subset of the range of conceivable interacting quantum dot systems.

For example, if modeling shows that significant control can be achieved over the band structure in a metacrystal by forming an analog of a bulk compound semiconductor, or by forming a lattice of quantum dots with a period extending over several dots, current fabrication methods would prove to be inadequate. A compromise path would be to use one of the versatile methods, with little regard for economy, in order to confirm the potential indicated by modeling, then focus on lowering cost and increasing throughput.

Modeling has been accomplished using a linear combination of strained bulk bands method within an empirical pseudopotential Hamiltonian.[19] This method expands the wave functions of a nanostructure as a linear combination of bulk states from the constituent materials. The envelope-function equation of a superlattice is also described using wave functions and corresponding effective-mass equations formulated from a linear combination of Bloch states only of the material with the smaller band gap.[20]

Similar band structure calculations on quantum dots predict that an indirect gap material makes a transition to direct gap at very small dot size.[21] For silicon quantum dots, the pseudopotential method was used to calculate the single particle states and evaluate the screened electron-hole Coulomb interaction.[22] An appropriate size-dependent indirect gap symmetry forms optically forbidden excitons. It was found that the direct electron-hole Coulomb interaction lowers the energy of otherwise spin-forbidden excitons relative to the optically active excitons, and the Coulomb correlation becomes more important for small dots.

A recent issue of *Materials Research Bulletin* has a very good review of representative nanofabrication and microfabrication methods.[23] A few of the methods are summarized here to point out the types of issues involved.

Embossing can replicate patterns formed on a mold by pressing the mold onto a deformable resist followed by etching of the thin spots in the resist. This method of nanoimprint lithography (NIL) has been used to form 10-nm dots spaced 40 nm apart. One form of guided self-assembly uses a similar mask placed in close proximity to a viscous film. The mask draws pillars from the film, balancing gravitational force and surface tension. The volume of film limits the size of the pillars. Although several pillars can form under a single template protrusion, the diameter and spacing limitation predictions are on the order of 100 nm. Atomic force microscopes have evolved for use in dip-pen nanolithography (DPN). DPN uses a probe tip to transport ink to a sample, making use of the meniscus formed at the tip to reduce size fluctuations that would otherwise take place with pressure variations. Lines and dots have been drawn with 12-nm width, spaced apart with 5-nm resolution. This is an exciting approach to demonstrate some of the structures based on modeling. Developing parallel probe tip arrays addresses the obvious limitation of speed.

5. Conclusions

The perspective afforded by current work taking place in quantum dots is useful to point the way toward future payoffs, both low-hanging fruit ready to pick, and longer term, requiring plenty of nurture.

The most intriguing area to investigate is the range of band properties that can be achieved with a metacrystal. Tunable band properties have been demonstrated before, but quantum dot systems may offer greater sensitivity to external perturbations, especially considering unique features such as discrete energy levels, delta function density of states, and greater freedom from lattice mismatch problems. Although the interface states are a prime concern for quantum dot systems, positive control of the strain may allow introduction of a single misfit dislocation in a quantum dot, accessing deeper energy levels. If a fabrication method can be found to synthesize structures with different materials on alternating lattice sites, the properties of compound metacrystals can be characterized, with the potential to further control the band properties.

Studies of quantum well superlattices have revealed other possibilities to extend the realm of semiconductors using quantum dots. The frequency of Bloch oscillations goes up as the superlattice period decreases. Quantum dot systems have already demonstrated higher frequencies. Although not well understood, phonons will likely play a bigger part in quantum dot systems. For example, the frequency of TO phonons limits the frequency of Bloch oscillations. However, reducing the size of quantum dots reduces the highest frequencies of phonons due to the quantum confinement.[24]

Current research reports are still full of surprises. One of the points of contention is the existence or importance of the 'phonon bottleneck' where the rate of relaxation of carriers to lower energy states is limited by participation of phonons. If the energy difference between two quantum dot states is equal to the energy of an LO phonon, relaxation takes place quickly with emission of an LO phonon or LO and LA phonon. Otherwise, relaxation is much slower than for quantum wells. The scattering rate for LA phonons depends greatly on the sensitivity of quantum dots to phonon-induced changes in effective mass. With phonon-induced changes in effective mass taken into account, the phonon emission rate can be two orders of magnitude higher than the emission rate based predominantly on deformation potential.[25] Other differences predicted between QD and other low-dimensional systems are the non-existence of interface LO phonons in spherical QDs,[14] and the enhancement of exciton energy by localization in a quantum dot superlattice.[26]

As the name implies, metacrystals are the next logical successor to heterostructures, quantum wells, and individual quantum dots. Metacrystals represent a transformation of simple quantum dots to a more highly organized form, into a new but related discipline dealing critically with the nature, structure, and behavior of the original quantum dots.

References

1. M. Pustilnik, Y. Avishai, and K. Kikoin, "Quantum dots with even number of electrons: Kondo effect in a finite magnetic field," *Phys. Rev. Lett.* **84**, 1756 (2000).

2. S. Winnerl, "GaAs/AlAs superlattices for detection of terahertz radiation," *Microelectronics J.* **31**, 398 (2000).

3   M. W. Feise and D. S. Citrin, "Semiclassical theory of terahertz multiple-harmonic generation in semiconductor superlattices," *Appl. Phys. Lett.* **75**, 3536 (1999).

4. S. Winnerl, E. Schomburg, S. Brandl, O. Kus, K. F. Renk, M. Wanke, S. Allen, A. Ignatov, V. Ustinov, A. Zhukov, and P. Kop'ev, "Frequency doubling and tripling of terahertz radiation in a GaAs/AlAs superlattice due to frequency modulation of Bloch oscillations," *Appl. Phys. Lett.* **77**, 1259 (2000).

5. Y. Romanov and Y. Romanova, "Dynamic electron localization and self-induced transparency in semiconductor superlattices," *Phys. Solid State* **43**, 539 (2001).

6. F. Löser, B. Rosam, K. Leo, Y. Kosevich, and K. Kohler, "Bloch oscillations under influence of coherent plasmon coupling," *Tech. Digest Quantum Electronics Laser Science Conf.* (2000).

7. I. Dmitriev and R. Suris, "Electron localization and Bloch oscillations in quantum-dot superlattices under a constant electric field," *Semicond.* **35**, 212 (2001).

8. A. Ghosh, L. Jonsson, and J. Wilkins, "Bloch oscillations in the presence of plasmons and phonons," *Phys. Rev. Lett.* **85**, 1084 (2000).

9. M. Forst, G. C. Cho, T. Dekorsy, and H. Kurz, "Chirped Bloch oscillations in strain-balanced InGaAs/InGaAs superlattices," *Superlatt. Microstruct.* **26**, 83 (1999).

10. C. Pryor, "Quantum wires formed from coupled InAs/GaAs strained quantum dots," *Phys. Rev. Lett.* **80**, 3579 (1998).

11. G. Springholz, M. Pinczolits, G. Bauer, H. H. Kang, and L. Saslamanca-Riba, "Phase diagram of lateral and vertical ordering in self-organized PbSe quantum dot superlattice grown MBE," *J. Crystal Growth* **227-228**, 1126 (2001).

12. T. Nishimura, S. Lan, K. Akahane, M. Kawabe, and O. Wada, "Coherent and incoherent carrier dynamics of InGaAs quantum dots analyzed by transient photoluminescence," *J. Luminescence* **87-89**, 494 (2000).

13. I. Kozin, I. Ignatiev, S. Nair, H. Ren, S. Sugou, and Y. Masumoto, "LO phonon resonances in photoluminescence spectra of InP self-assembled quantum dots in electric field," *J. Luminescence* **87-89**, 441 (2000).

14. X. Li and Y. Arakawa, "Confined optical phonons in semiconductor quantum dots," *Solid State Commun.* **109**, 351 (1999).

15. H. Song, Y. Okada, K. Akahane, S. Lan, H. Xu, and M. Kawabe, "Metal-insulator transition in an $In_{0.4}Ga_{0.6}As/GaAs(311)B$ quantum dot superlattice," *Phys. Lett. A* **284**, 130 (2001).

16. A. Stintz, G. Liu, H. Li, L. Lester, and K. Malloy, "Low-threshold current density 1.3-μm InAs quantum-dot lasers with the dots-in-a-well (DWELL) structure," *IEEE Photonics Technol. Lett.* **12**, 591 (2000).

17. I. Mukhametzhanov, Z. Wei, R. Heitz, and A. Madhukar, "Punctuated island growth: An approach to examination and control of quantum dot density, size, and shape evolution," *Appl. Phys. Lett.* **75**, 85 (1999).

18. H. Lee, J. Johnson, M. He, J. Speck, and P. Petroff, "Strain-engineered self-assembled semiconductor quantum dot lattices," *Appl. Phys. Lett.* **78**, 105 (2001).

19. L. Wang and A. Zunger, "Linear combination of bulk bands method for large-scale electronic structure calculations on strained nanostructures," *Phys. Rev. B* **59**, 15806 (1999).

20. B. W. Kim, Y. I. Jun, and B. Jung, "Envelope-function equation and motion of wave packet in a semiconductor superlattice structure," *ETRI J.* **21**, 1 (1999).

21. V. A. Singh, V. Ranjan, and M. Kapoor, "Semiconductor quantum dots: Theory and phenomenology," *MRS Bulletin* **22**, 563 (1999).

22. F. A. Reboredo, A. Franceschetti, and A. Zunger, "Dark excitons due to direct Coulomb interactions in silicon quantum dots," *Phys. Rev. B* **61**, 13073 (2000).

23. C. Mirkin and J. Rogers, "Emerging methods for micro- and nanofabrication," *MRS Bulletin* **26**, 506 (2001).

24. S. Ren, Z. Gu, and D. Lu, "Quantum confinement of phonon modes in GaAs quantum dots," *Solid State Commun.* **113**, 273 (2000).

25. V. Pipa, V. Mitin, and M. Stroscio, " Acoustic phonon bottleneck in quantum dots: Role of deformation variation of electron effective mass," *Solid State Commun.* **117**, 713 (2001).

26. S. Lan, K. Akahane, K. Jang, T. Kawamura, Y. Okada, and M. Kawabe, "Realization and characterization of two-dimensional quantum dot superlattices forming 'artificial crystals'," *Microelectronics Eng.* **47**, 131 (1999).

InGaAs/GaAs Quantum Well Microcavities with Spatially Controlled Carrier Injection

S. N. M. Mestanza, A. A. Von Zuben, and N. C. Frateschi

Universidade Estadual de Campinas — Unicamp, Campinas, SP, Brazil

1. Introduction

Micro-optical-cavity lasers are desired for optoelectronic integration. In particular, high quality factor (Q) cavities are obtained with circular microdisks and/or micro-cylinders. These devices are suitable for processing light emission parallel to the substrate in a fashion similar to electronic transport between transistors in integrated microelectronic circuits. A limitation in these devices is the lack of light directionality, poor control over its spectral behavior, and the small external differential quantum efficiency. Recently, there has been interest in optical cavities with boundaries based on chaotic billiards, particularly stadium structures, where enhancement in light directionality is proposed.[1]

In the semiclassical limit, it has been shown by Heller that the eigenmodes of the Helmholtz equation for the stationary solution of the Schrödinger equation have higher amplitude along classical periodic orbits, called scars.[2] Bogomolny also showed that the averaged wave-functions for very high quantum numbers lead to scars, which he completely mapped for the stadium boundary.[3] In our case, the semiclassical limit is the ray-tracing limit, achieved for cavity dimensions much greater than the wavelength of light in the semiconductor medium. It is interesting to investigate the role of scars in laser performance, for one may be able to control power directionality utilizing selected periodic orbits. Plus, one may expect to see consequences of this control on other laser properties, such as the light emission spectrum.

InGaAs/GaAs strained quantum well structures embedded in InGaP optical confining layers are suitable for the fabrication of microcavity structures. Very high gain is available in these structures plus very low surface recombination velocity is observed at the InGaP/GaAs interfaces. Light ion irradiation, such as in He$^+$ implantation, has been shown to provide highly resistive layers without optical damage to the InGaP material, making it suitable for changing the carrier injection geometry in these devices.[4]

In this work, we present our recent development of microcavities based on InGaAs/GaAs/InGaP quantum well structures. These structures have circular, elliptical, and stadium cross-sections. We first present the results of uniformly pumped devices. The dependence of the light power and spectra on injected current are presented. An increase in power directionality is observed for both the elliptical and the stadium cylinders. Finally, carrier injection along the diamond scar is forced following light He$^+$ implantation. Higher directional power as well as a great enhancement in side-mode suppression are observed. Apparently, beating between two subsequent interacting allowed scar modes is responsible for this effect.

2. Fabrication

Figure 1 shows a cross-section micrograph of the epitaxial structure used in this work. The epitaxial structure is as follows, starting from the substrate: n^+-GaAs buffer; n-InGaP (1.05 µm, 1×10^{18} cm^{-3}); n-InGaP (0.15 µm, 5×10^{17} cm^{-3}); undoped GaAs (0.10 µm); undoped In$_{0.21}$Ga$_{0.79}$As quantum well (8 nm) for emission at $\lambda \sim 0.98$ µm and $\lambda \sim 0.90$ µm at 300 K for the transition between first and second levels, respectively; undoped GaAs (0.10 µm); p-InGaP (0.10 µm, 2×10^{17} cm^{-3}); p-GaAs etch-stop layer (6 nm, 2×10^{17} cm^{-3}); p-InGaP (50 nm, 2×10^{17} cm^{-3}); p-InGaP (0.91 µm, 6×10^{17} cm^{-3}); p-InGaP (0.19 µm, 1×10^{18} cm^{-3}); p^+-GaAs (0.1 µm, 3×10^{19} cm^{-3}) and contact p^{++}-GaAs layer (0.1 µm, $>5\times10^{19}$ cm^{-3}). The first processing step involves the p-side metallization by Ti/Pt/Au lift-off. Subsequently, the entire structure is RIE dry etched in a SiCl$_4$/Ar plasma. Finally, the n-side is metallized by Au/Ge-Ni-Au. The cylinder cross-sections are disks, ellipses, and stadia as shown in the inset of Fig. 1. Reference radii of $R = 10$ µm, 15 µm, and 20 µm were utilized.

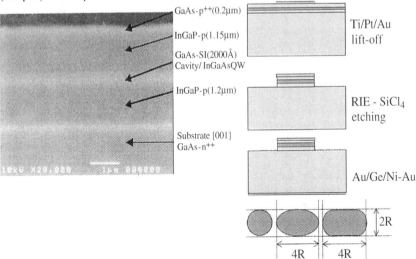

Figure 1. Cross-section scanning electron microscope micrograph of the InGaAs/GaAs/InGaP laser structure. The inset shows schematically the processing and the cylindrical cavity cross-section geometry. Devices with reference radii R = 10, 15, and 20 µm were fabricated.

3. Disk, ellipse, and stadium results

Figure 2 shows the measurement of light output power (L) versus injected current (I) for the structures with $R = 20$ μm. Light is captured by a multi-mode fiber, 90-μm core, placed over the substrate 1 mm away from the structures. The measurements for the ellipse and the stadium are done along the major axis. All measurements were done at 10 °C in pulsed mode (1 μs, 1% duty cycle). A typical threshold current around 15 mA is observed for both the ellipse and the disk. The stadium structure presents a small increase in d^2L/dI^2 at around 24 mA. The threshold current value for the stadium is better confirmed by the spectrum measurement. It is interesting to observe the obvious increase in output power with respect to the disk for the non-circular cylinders. The power along the minor axis shows systematically 10× less power for both non-circular cylinders, indicating an increase in directional power. Thus, the increase in directional power in the non-circular structures cannot be related to classically chaotic borders.

Figure 3 shows the spectra obtained for the lasers above driven under different current injection values. The resolution bandwidth for all spectra in this work is 0.1 nm. Emission centered at $\lambda = 1$ μm corresponding to the transition between the first quantum well levels is observed. Both the ellipse and disk operate under multimode conditions, with typical mode separations of 0.1–0.2 nm,

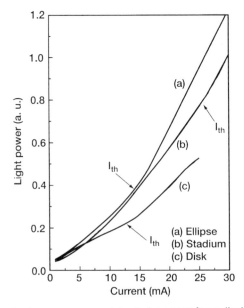

Figure 2. Light output power versus injected current for cylindrical structures with reference radius $R = 20$ μm.

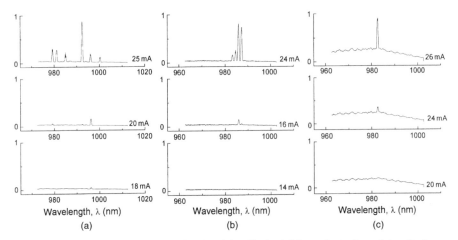

Figure 3. Spectra obtained for circular (a), elliptical (b), and stadium (c) cylinders under different injection currents. All structures have a radius of R = 20 µm.

corresponding to wavelengths that are sub-multiples of the photon round trip length. Some stadium structures showed a higher tendency to operate single-mode.

Devices with smaller radii exhibited laser emission near $\lambda = 0.9$ µm, corresponding to a transition between the first excited quantum well levels. Multi-mode behavior is always observed for all structures. Figure 4 shows a typical spectrum for a stadium structure with $R = 15$ µm, operated at 70 mA. These smaller structures presented typical threshold currents between 50–100 mA.

4. Ion implantation effects

In order to improve injection along a semiclassical scar, He$^+$ ion implantation was employed as shown in Fig. 5. First, a 1.5-µm-thick photoresist layer protects a region along a diamond scar. Subsequently, He$^+$ implantation is performed to isolate the unprotected regions. The choice of He$^+$ ion energy was based on calculations of radiation defect distributions by TRIM code. For 100 keV He$^+$ ions, the maximum defect density is located at a depth of 0.54 µm, and the radiation damage formation occurs only within the InGaP layer. An incidence angle of 15° was used to avoid channeling effects. A sheet resistance in excess of 1 MΩ is obtained after a dose of ~10^{13} cm^{-2}. Subsequently, the laser is processed as described above. Structures with and without implantation were fabricated side by side for comparison. Reference radii of $R = 10$ µm, 15 µm, and 20 µm were utilized.

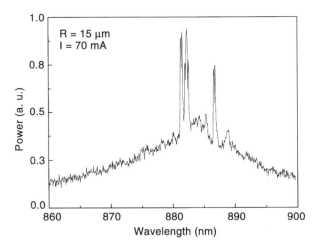

Figure 4. Spectrum for a stadium structure with R =15 µm, operated at a current of I = 70 mA.

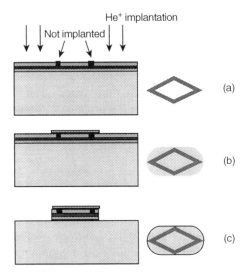

Figure 5. Scheme of the processing for injection along a diamond scar using He⁺ irradiation. Photoresist protection along the diamond scar followed by a 10^{13} cm^{-2} dose of 100 keV He⁺ ions (a); p-side metalization (b); and SiCl$_4$/Ar RIE etching of the entire structure followed by n-side metalization (c). Structures with and without implantation were fabricated with reference radii of R = 10, 15, and 20 µm.

The spectra for these stadium structures were investigated for different radii with drive currents about 20% above threshold. All measurements were done as described above. Figure 6(a) shows the results for the devices without implantation. As expected, multi-mode behavior is observed in all cases. For comparison, Fig. 6(b) shows the spectra obtained for the corresponding neighbor devices with ion implantation forcing the injection along the diamond scar. A great enhancement in side-mode suppression is observed for the second case. Also, it is clear that the suppression is further augmented for smaller disks.

In order to try to correlate the spectral behavior with the scars in these structures, we performed far-field measurements. Again, a 90-μm-core multimode fiber was placed over the substrate, 1 mm away from the devices. A micro-translation stage was used to move the fiber and the peak power from an optical spectrum analyzer was recorded as a function of the angle. The lasers were operated 20% above threshold for this measurement.

Figure 7(a) shows the normalized far-field plot for a device of radius $R = 15$ μm, with implantation, for the light emission along the major and minor axes. Along both directions, a maximum at the center is observed. This behavior is observed for all devices with $R < 20$ μm. For $R = 20$ μm, a minimum was also observed at the center along the minor axis. Both far-field plots are expected for the four Bogomolny scar orbits shown in Fig. 8. Figure 7(b) shows the far-field plot along the minor axis for devices with and without implantation. The data are normalized to the maximum value measured for the emission at $\theta = 0°$, which occurs for the unimplanted device. We observe that a higher suppression of the lateral emission is obtained.

One explanation for the increase in side-mode suppression is suggested by a simple analysis of the scar orbits shown in Fig. 8. For a mode to survive, the photon lifetime in the cavity has to be longer than the round-trip time. Photon

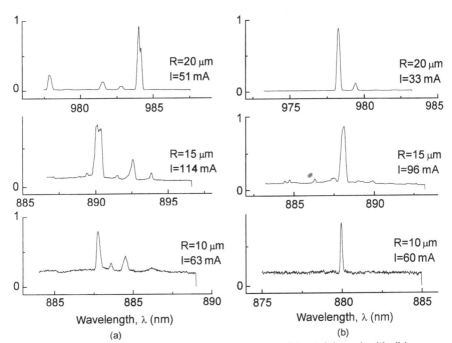

Figure 6. Spectra obtained for stadium structures without (a) and with (b) scar injection. All lasers are driven 20% above threshold. The resolution bandwidth is 0.1 nm.

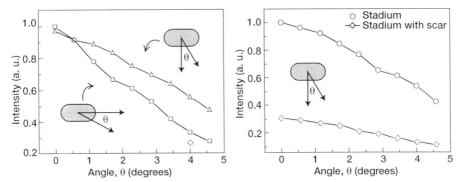

Figure 7. Normalized far-field distributions (R = 15 μm and driving current 20% above threshold). Stadium with scar injection measured along the minor and major axes (a). Far-field plot normalized to maximum output power for a stadium with scar and without scar (b).

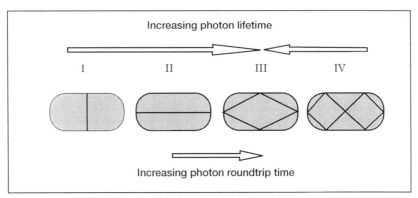

Figure 8. Periodic scar orbits compatible with the far-field measurements. The orbits' photon roundtrip times increase from left to right. Photon lifetime is largest for scar III.

lifetime can be calculated easily knowing the typical reflectivity obtained with the dry-etched structures. This calculation is accomplished by summing the contribution to the electric field at a point of the orbit after many roundtrips. Expanding the electromagnetic power near quasi-resonances, one obtains the photon lifetime at this particular quasi-resonance by a simple Fourier transform. Orbits I and II with short round-trip times and very short photon lifetimes have very small normal incidence reflectivity. In fact, the modal separation below threshold shows no difference for the emission along the minor and major axes, indicating that these orbits are very lossy. Therefore, scars III and IV are the best candidates for stimulated emission. Since the photon round-trip time for orbit IV is longer and the lifetime can be demonstrated to be shorter than for orbit III, we expect that the last one is the most favorable for lasing. The strength of mode III is further increased by the selective injection scheme. Therefore, multimode behavior involving modes along orbit III that are separated by ~0.1 nm is expected. However, orbit IV is also a high-Q orbit that interacts as a perturbation with the lasing mode in orbit III. The coupling between these two orbits results in beating with coincident modes separated by 5–13 nm for $R = 20$ μm to 10 μm, respectively. These coincident modes have higher photon-semiconductor interaction, hence higher gain. Also, since these modes are spaced further from the maximum gain mode, an enhancement of side-mode suppression is expected. For the smaller structures, the suppression is yet more evident because the gain-bandwidth spectrum for transitions involving the second quantum well level is half that for transitions involving the fundamental mode.

5. Conclusion

Micro-cylindrical-cavity lasers based on InGaAs/GaAs/InGaP quantum well structures were fabricated. Elliptical and stadium cylinders show higher directional power than circular structures. Stadium devices with carrier injection along scars of periodic orbits were fabricated using light He$^+$ implantation. Higher directional power as well as a great enhancement in side-mode suppression are observed for these devices. We propose that the beating between two subsequent scar interacting modes is increasing the mode separation within the gain-bandwidth spectrum. Suppression is more evident for smaller devices with transitions involving the second level where the gain bandwidth is much smaller.

Acknowledgments

This work was financed by the Fundação de Amparo à Pesquisa do Estado de São Paulo (FAPESP). The lithography masks were fabricated at the Instituto de Tecnologia de Informática (ITI) Campinas – Brazil.

References

1. C. Gmachl, F. Capasso, E. E. Narimanov, J. U. Nockel, A. D. Stone, J. Faist, D. L. Sivco, and A. Cho, "High-power directional emission from microlasers with chaotic resonators," *Science* **280**, 1556 (1998).
2. E. J. Heller, "Bound-state eigenfunctions of classically chaotic Hamiltonian systems: Scars of periodic orbits," *Phys. Rev. Lett.* **53**, 1515 (1984).
3. E. B. Bogomolny, "Smoothed wavefunctions of chaotic quantum systems," *Physica D* **31**, 169 (1988).
4. I. Danilov, L. L. Pataro, M. P. P. de Castro, and N. C. Frateschi, "Electrical isolation and transparency in ion-irradiated p-InGaP/GaAs/InGaAs structures," *J. Appl. Phys.* **88**, 7354 (2000).

Contributors

Abramo, A.
DIEGM, Univ. di Udine, Udine, Italy

Akama, S.
Tokyo Institute of Technology, Yokohama, Japan

Allibert, F.
Institut de Microélectronique, Electromagnétisme et Photonique, ENSERG Grenoble, France

Asryan, L. V.
A. F. Ioffe Physico-Technical Institute, St Petersburg, Russia

Averin, D. V.
Dept. of Physics and Astronomy, SUNY at Stony Brook, Stony Brook, NY, U.S.A.

Aydin, C.
Dept. of Physics, Brown University, Providence, RI, U.S.A.

Belenky, G.
Dept. of Electrical Engineering, SUNY at Stony Brook, Stony Brook, NY, U.S.A.

Benschop, J.
ASML Inc., Veldhoven, The Netherlands

Boolchand, P.
Dept. of Electrical and Computer Engineering and Computer Science University of Cincinnati, Cincinnati, OH, U.S.A.

Brown, G.
Air Force Research Laboratory Materials Directorate, Wright Patterson AFB, OH, U.S.A.

Cahay, M.
Dept. of Electrical and Computer Engineering and Computer Science Univ. of Cincinnati, Cincinnati, OH, U.S.A.

Calle, F.
Dpto. Ingeniería Electrónica, Univ. Politécnica de Madrid, Madrid, Spain

Calleja, E.
Dpto. Ingeniería Electrónica, Univ. Politécnica de Madrid, Madrid, Spain

Chudnovskiy, F.
NY State Center for Advanced Sensor Technology, Stony Brook, NY, U.S.A.

Cristoloveanu, S.
*Institut de Microélectronique, Electromagnétisme et Photonique, ENSERG
Grenoble, France*

Dalla Serra, A.
DIEGM, Univ. di Udine, Udine, Italy

de la Houssaye, P. R.
SPAWAR Systems Center San Diego, San Diego, CA, U.S.A.

Dorojevets, M.
*Dept. of Electrical and Computer Engineering, SUNY at Stony Brook
Stony Brook, NY, U.S.A.*

Dyakonov, M. I.
*Laboratoire de Physique Mathématique
Université de Montpellier II, Montpellier, France*

Esseni, D.
DIEGM, Univ. di Udine, Udine, Italy

S. J. Farrell
Cable & Wireless, Covent Garden, London, U.K.

Fernández, S.
Dpto. Ingeniería Electrónica, Univ. Politécnica de Madrid, Madrid, Spain

Fiegna, C.
Dept. of Engineering, Univ. di Ferrara, Ferrara, Italy

Frateschi, N. C.
Universidade Estadual de Campinas — Unicamp, Campinas, Brazil

Gaska, R.
Sensor Electronic Technology, Inc., Latham, NY, U.S.A.

Goldman, V. J.
Dept. of Physics and Astronomy, SUNY at Stony Brook, Stony Brook, NY, U.S.A.

Goldstein, J.
Air Force Research Laboratory, Wright-Patterson AFB, OH, U.S.A.

Grasser, T.
Institute for Microelectronics, TU Wien, Vienna, Austria

Green, M. A.
Special Research Centre for Third Generation Photovoltaics
University of New South Wales, Sydney, Australia

Hu, Q.
Dept. of Electrical Engineering and Computer Science
Massachusetts Institute of Technology, Cambridge, MA, U.S.A.

Hybertsen, M. S.
Agere Systems, Murray Hill, NJ, U.S.A.

Ishiwara, H.
Tokyo Institute of Technology, Yokohama, Japan

Iwai, H.
Tokyo Institute of Technology, Yokohama, Japan

Jiménez, A.
Dpto. Ingeniería Electrónica, Univ. Politécnica de Madrid, Madrid, Spain

Johnstone, D.
Air Force Office of Scientific Research, Arlington, VA, U.S.A.

Kashiwagi, I.
Tokyo Institute of Technology, Yokohama, Japan

Kelly, M. J.
School of Electronics, Computing and Mathematics, University of Surrey, Guildford, U.K.

Kikuchi, A.
Tokyo Institute of Technology, Yokohama, Japan

Kobayashi, C.
Tokyo Institute of Technology, Yokohama, Japan

Lagnado, I.
SPAWAR Systems Center San Diego, San Diego, CA, U.S.A.

Ledentsov, N. N.
A. F. Ioffe Physical-Technical Institute, St. Petersburg, Russia

Likharev, K. K.
Dept. of Physics and Astronomy, SUNY at Stony Brook, Stony Brook, NY, U.S.A.

Liu, J. W.
Cable & Wireless, Vienna, VA, U.S.A.

Loss, D.
Dept. of Physics and Astronomy, University of Basel, Basel, Switzerland

Luryi, S.
Dept. of Electrical Engineering, SUNY at Stony Brook, Stony Brook, NY, U.S.A.

Malhotra, A.
Dept. of Electrical and Computer Engineering and Computer Science
University of Cincinnati, Cincinnati, OH, U.S.A

Mastrapasqua, M.
Agere Systems, Murray Hill, NJ, U.S.A.

Maudar, H. A.
Cable & Wireless, Vienna, VA, U.S.A.

Melloch, M. R.
School of Electrical and Computer Engineering, Purdue University
West Lafayette, IN, U.S.A.

Menna, R.
Princeton Lightwave Inc., Cranbury, NJ, U.S.A.

Mestanza, S. N. M.
Universidade Estadual de Campinas — Unicamp, Campinas, Brasil

Modukuru, Y.
Dept. of Electrical and Computer Engineering and Computer Science
University of Cincinnati, Cincinnati, OH, U.S.A

Muñoz, E.
Dpto. Ingeniería Electrónica, Univ. Politécnica de Madrid, Madrid, Spain

Naranjo, F. B.
Dpto. Ingeniería Electrónica, Univ. Politécnica de Madrid, Madrid, Spain

Nurmikko, A. V.
Div. of Engineering, Brown University, Providence, RI, U.S.A.

Ohmi, S.
Tokyo Institute of Technology, Yokohama, Japan

Ooshima, C.
Tokyo Institute of Technology, Yokohama, Japan

Oshima, K.
Tokyo Institute of Technology, Yokohama, Japan

Palestri, P.
DIEGM, Univ. of Udine, Via delle Scienze 208, 33100 Udine, Italy

Pau, J. L.
Dpto. Ingeniería Electrónica, Univ. Politécnica de Madrid, Madrid, Spain

Pavesi, M.
Dept. of Physics, Univ. di Parma, Parma, Italy

Pretet, J.
Institut de Microélectronique, Electromagnétisme et Photonique, ENSERG Grenoble, France

Rakitin, A.
Div. of Engineering, Brown University, Providence, RI, U.S.A.

Razeghi, M.
Dept. of Electrical and Computer Engineering, Northwestern University Evanston, IL, U.S.A.

Reno, J. L.
Sandia National Labs, Albuquerque, NM, U.S.A.

Reynolds, Jr., C. L.
Agere Systems, Breinigsville, PA, U.S.A.

Rigolli, P.
Dept. of Physics, Univ. di Parma, Parma, Italy

Ristic, J.
Dpto. Ingeniería Electrónica, Univ. Politécnica de Madrid, Madrid, Spain

Sánchez-Garcia, M. A.
Dpto. Ingeniería Electrónica, Univ. Politécnica de Madrid, Madrid, Spain

Sangiorgi, E.
DIEGM, Univ. di Udine, Udine, Italy

Sato, K.
Tokyo Institute of Technology, Yokohama, Japan

Schliemann, J.
Dept. of Physics, University of Texas at Austin, Austin, TX, U.S.A.

Selberherr, S.
Institute for Microelectronics, TU Wien, Vienna, Austria

Selmi, L.
DIEGM, Univ. di Udine, Udine, Italy

Shur, M. S.
Dept. of Electrical, Computer and Systems Engineering
Rensselaer Polytechnic Institute, Troy, NY, U.S.A.

Shterengas, L.
Dept. of Electrical Engineering, SUNY at Stony Brook, Stony Brook, NY, U.S.A.

Smith, III, T. P.
Cable & Wireless, Vienna, VA, U.S.A.

Solomon, P. M.
IBM T. J. Watson Research Center, Yorktown Heights, NY 10598, U.S.A.

Sonek, G. J.
Optical Switch Corporation, Bedford, MA, U.S.A.

Spivak, B.
Dept. of Physics, University of Washington, Seattle, WA, U.S.A.

Sze, S. M.
National Nano Device Laboratories
National Chiao Tung University, Hsinchu, Taiwan, R.O.C.

Taguchi, J.
Tokyo Institute of Technology, Yokohama, Japan

Takeda, M.
Tokyo Institute of Technology, Yokohama, Japan

Tang, H.
Dept. of Electrical and Computer Engineering and Computer Science
University of Cincinnati, Cincinnati, OH, U.S.A.

Thachery, J.
Dept. of Electrical and Computer Engineering and Computer Science
University of Cincinnati, Cincinnati, OH, U.S.A.

Trussell, C. W.
Night Vision & Electronic Sensors Directorate, Ft. Belvoir, VA, U.S.A

Von Zuben, A. A.
Universidade Estadual de Campinas — Unicamp, Campinas, Brasil

Wei, Y.
Dept. of Electrical and Computer Engineering
Northwestern University, Evanston, IL, U.S.A.

Widdershoven, F.
Philips Research Leuven, Leuven, Belgium

Williams, B. S.
Dept. of Electrical Engineering and Computer Science
Massachusetts Institute of Technology, Cambridge, MA, U.S.A

Xu, J. M.
Div. of Engineering, Brown University, Providence, RI, U.S.A.

Yamamoto, H.
Tokyo Institute of Technology, Yokohama, Japan

Zaslavsky, A.
Div. of Engineering, Brown University, Providence, RI, U.S.A.

Zia, R.
Div. of Engineering, Brown University, Providence, RI, U.S.A.

Zukauskas, A.
Institute of Materials Science and Applied Research
Vilnius University, Vilnius, Lithuania

Index